NONLINEAR INTEGRAL EQUATIONS AND INCLUSIONS

NONLINEAR INTEGRAL EQUATIONS AND INCLUSIONS

RAVI P. AGARWAL
MARIA MEEHAN
DONAL O'REGAN

Nova Science Publishers, Inc.
Huntington, New York

Senior Editors: Susan Boriotti and Donna Dennis
Coordinating Editor: Tatiana Shohov
Office Manager: Annette Hellinger
Graphics: Wanda Serrano
Book Production: Matthew Kozlowski, Jonathan Rose and Jennifer Vogt
Circulation: Cathy DeGregory and Ave Maria Gonzalez
Communications and Acquisitions: Serge P. Shohov

Library of Congress Cataloging-in-Publication Data
Available Upon Request

ISBN 1-59033-094-3.

227 Main Street, Suite 100
Huntington, New York 11743
Tele. 631-424-NOVA (6682) Fax 631-424-4666
e-mail: Novascience@earthlink.net
Web Site: http://www.nexusworld.com/nova

Printed in the United States of America

CONTENTS

PREFACE

In this book we discuss current topics of interest in the theory of nonlinear integral equations and inclusions. In particular we are interested in establishing existence principles and results, using various fixed point methods, for a range of nonlinear integral equations and integral inclusions.

In the first half of Chapter 1, we remind the reader of some well-known existence principles for functional and functional-differential equations, and consider some applications of these results to integral and integrodifferential equations. The second half of Chapter 1 is devoted to presenting an existence theory for singular functional-differential equations.

We are concerned with the existence of multiple nonnegative and/or positive solutions of various classes of nonlinear integral equations in Chapter 2 and Chapter 3. Krasnoselskii's fixed point theorem and a nonlinear alternative of Leray-Schauder type are the main fixed point theorems used to obtain our results in both of these chapters. In Chapter 2, the existence of multiple, nonnegative, continuous and L^p solutions, of nonlinear Fredholm integral equations – defined on both compact intervals and the half-line – is discussed. In Chapter 3, the results of the previous chapter are extended and additional results presented, which guarantee the existence of positive solutions of singular Fredholm integral equations. Singular Volterra integral equations are also considered. The existence theory presented for these equations, relies on Krasnoselskii's fixed point theorem and the ability to obtain solutions of certain integral inequalities.

Periodic and almost periodic solutions of nonlinear integral equations is the topic considered in Chapter 4. The chapter opens with some admissibilty results and existence principles for a general Fredholm integral operator and equation respectively. A variety of results which establish the existence of one or more periodic or almost periodic solutions of certain classes of nonlinear integral equations, are then discussed. The nonlinear alternative of Leray-Schauder type, Krasnoselskii's fixed point theorem and the Leggett Williams fixed point theorem are the fixed points methods used to obtain the desired results. We see a different application of the Leggett Williams fixed point theorem in Chapter 5, where it is used to establish the existence of three nonnegative solutions of a nonlinear integral equation.

Chapter 6 contains quite a comprehensive theory on the topological structure of the solution set of certain functional and functional-differential equations. Fixed points for operators on Banach spaces and Fréchet spaces, and continuums of solutions in Fréchet spaces with applications, are just some of the topics examined.

A general continuation theory for weakly condensing operators between Banach spaces is considered in Chapter 7. The continuation theorem established at the start of the chapter is used to present some very general existence principles for nonlinear operator equations. We also show that in the case of second order boundary value problems, the theory developed in this chapter for weakly continuous operators, can lead to the same conclusions as the theory of compact (strong) operators.

Chapter 8 studies Volterra and Urysohn integral equations in a ball of a Banach space E when the nonlinear kernel satisfies a Carathéodory condition. The existence principles are based upon the continuation theorem of Mönch [112] and a result by Heinz [79]. More applicable existence results are then derived from the general existence principles by means of differential and integral inequalities.

In Chapter 9 we present fixed point theorems for weakly sequentially continuous maps between Banach spaces, or more generally, metrizable locally convex spaces. An analogue of Mönch's fixed point theorem in the weak topology setting for weakly sequentially continuous maps is also presented, and an existence result is established for operator equations in Banach spaces relative to the weak topology.

The remainder of the book is concerned with integral inclusions. An extension of the results found in Chapter 8, to Hammerstein and Volterra integral inclusions, can be found in Chapter 10. In the first half of Chapter 11, existence principles for Cauchy integrodifferential inclusions are presented. These results, together with the Poincaré operator, are then used to obtain an existence principle for a periodic integrodifferential inclusion.

A multivalued analogue of Krasnoselskii's fixed point theorem in a cone, is established in Chapter 12. This fixed point result is then applied to provide results which establish the existence of positive, continuous and L^p, solutions of a Fredholm integral inclusion. In Chapter 13, a generalisation of the Petryshyn Legget Williams fixed point theorem is presented and an application to integral inclusions to ensure the existence of three nonnegative solutions is considered.

Applicable fixed point theorems which can be used to examine general equations on a noncompact interval, are established in Chapter 14. We then describe how these results may be applied to to prove the existence of one (or more) continuous solution of a multivalued nonlinear equation, defined on the half line. Remaining with integral inclusions on the half line, in the final chapter of the book – Chapter 15 – we discuss the topological structure of the solution set of a Volterra integral inclusion.

CHAPTER 1

FUNCTIONAL AND FUNCTIONAL-DIFFERENTIAL EQUATIONS WITH APPLICATIONS

1.1. Introduction.

It is our intention in Section 1.2 and Section 1.3 of this chapter, to remind the reader of some well-known existence principles for the functional equation

$$y(t) = Ny(t), \tag{1.1.1}$$

and the functional-differential equation

$$\begin{cases} y'(t) = Vy(t) \\ y(0) = y_0, \end{cases} \tag{1.1.2}$$

where both are defined on an interval $I \subseteq \mathbf{R}$. Section 1.2 is dedicated to (1.1.1) and we discuss (1.1.2) in Section 1.3.

In Section 1.2 a general existence principle for (1.1.1) is established using a nonlinear alternative of Leray-Schauder type, where we assume that the operator $N : E(I) \to E(I)$, with $E(I)$ a normed linear function space. In particular we examine the case where N is an integral operator of the form

$$Ny(t) := g(t) + \int_0^T k(t,s)g(s,y(s))\,ds.$$

An existence theory is established for (1.1.1) with this choice of N, and solutions which lie in the spaces $C[0,T]$, $L^p[0,T]$, $L^p[0,\infty)$ and $C_l[0,\infty)$ are discussed. Returning then to the operator equation (1.1.1), we present some additional existence principles for this equation, except now we further assume that the operator $N : E(I) \to E(I)$ is an abstract Volterra operator. The existence of solutions of (1.1.1) in $C[0,T)$ and $L^p[0,T)$, $0 \leq T \leq \infty$, are considered here.

The functional-differential equation (1.1.2) is examined in Section 1.2. We first assume that the operator $V : C[0,T] \to L^p[0,T]$ and the existence of a solution $y \in AC[0,T]$ of (1.1.2) is guaranteed using the nonlinear alternative. The case when the operator $V : C[0,T) \to L^p_{loc}[0,T)$, and a solution $y \in AC_{loc}[0,T)$, $0 \leq T \leq \infty$, of (1.1.2) is sought, is then discussed. One such result follows from the Schauder-Tychonoff theorem, and if it is assumed that V is an abstract Volterra operator, one obtains an additional existence principle.

The results in both Section 1.2 and Section 1.3 are well-known. For further details on both of these equations and their applications to integral and integrodifferential equations, we refer the reader to [46-48,51,59,60,75,76,94, 102,104,126,132,133,137,141] and the references therein.

The material in Section 1.4 is more recent (see [86,87,109]). In this section we examine a variation of the functional-differential equation (1.1.2), namely

$$\text{(1.1.3)} \qquad \begin{cases} y'(t) = Vy(t), \text{ a.e. } t \in I \\ \lim_{t\to 0^+} h(t)y(t) = a, \end{cases}$$

where $I = (0,T]$ or $I = (0,T)$, and V is a singular or singular Volterra operator. In the first half of the section we present an existence principle which guarantees the existence of a solution $y \in C(0,T]$ with $hy \in AC[0,T]$, of (1.1.3) with $I = (0,T]$. However, en route to this result, we examine the somewhat simpler functional-differential equation

$$\text{(1.1.4)} \qquad \begin{cases} (h(t)y(t))' = Ny(t), \text{ a.e. } t \in [0,T] \\ \lim_{t\to 0^+} h(t)y(t) = a, \end{cases}$$

where $N : E_h[0,T] \to L[0,T]$ is a regular Carathéodory operator. (Here $E_h[0,T]$ is the space of functions $y \in C(0,T]$ for which $hy \in C[0,T]$, and $L[0,T] := L^1[0,T]$.) We prove the existence of a solution $y \in C(0,T]$, such that $hy \in AC[0,T]$, of (1.1.4) and then use this result to obtain the desired result for (1.1.3) with $I = (0,T]$. In addition, we consider the functional equation

$$\text{(1.1.5)} \qquad h(t)y(t) = a + N^\star y(t), \quad t \in [0,T],$$

where $N^\star : E_h[0,T] \to C[0,T]$, and present conditions which ensure that (1.1.5) has at least one solution $y \in C(0,T]$, such that $hy \in C[0,T]$. With this result established, further results for certain cases of (1.1.3) follow easily.

The second half of Section 1.4, is dedicated to the more general version of (1.1.3) defined on the open interval $(0,T)$, and given by

$$\text{(1.1.6)} \qquad \begin{cases} y'(t) = Vy(t), \text{ a.e. } t \in (0,T) \\ \lim_{t\to 0^+} h(t)y(t) = a. \end{cases}$$

We show that (1.1.6) has a solution $y \in C(0,T)$ with $hy \in AC_{loc}[0,T)$, by first presenting two existence principles for (1.1.4), where $[0,T]$ has been replaced by $[0,T)$, and then using these results to obtain the result for (1.1.6).

1.2. Functional equations with applications to integral equations.

The aim of this section is to remind the reader of some well known existence principles for nonlinear integral equations. However before we delve into the theory of integral equations, it may be useful to stand back and first take a look at the bigger picture. In this way the reader can become more familiar with the fixed point method used throughout the chapter before becoming acquainted with details of particular integral equations.

Consider the abstract functional equation

$$\text{(1.2.1)} \qquad y(t) = Ny(t)$$

on an interval $I \subseteq \mathbf{R}$, where it is assumed that the operator $N : E \to E$, with $E := E(I)$ a normed linear function space. We would like to prove the existence of a solution $y \in E$ of (1.2.1). Finding a solution of (1.2.1) in E is of course equivalent to showing that $N : E \to E$ has a fixed point. The fixed point approach that we use is a nonlinear alternative of Leray-Schauder type which we now recall.

Theorem 1.2.1. Nonlinear Alternative.
Let C be a convex subset of a normed linear space E, and let U be an open subset of C, with $p^\star \in U$. Then every compact, continuous map $N : \overline{U} \to C$ has at least one of the following two properties:

(*i*) *N has a fixed point in $\overline{U}$;*
(*ii*) *there is an $x \in \partial U$ with $x = (1-\lambda)p^\star + \lambda Nx$, for some $0 < \lambda < 1$.*

Remark 1.2.1. We say that a map $N : E \to E$ is *compact* if $N(E)$ is relatively compact in E, while a map $N : E \to E$ is *completely continuous* if $N(\Omega)$ is relatively compact in E for each bounded set Ω in E.

We want to apply the nonlinear alternative to show that $N : E \to E$ has a fixed point. Let the norm on $E = E(I)$ be denoted by $|.|_E$. Assume that

(1.2.2) $N : E \to E$ is a continuous and completely continuous operator

is true. In addition let the convex set C, the open set U and $p^\star$ be given by $C = E$,

$$U := \{y \in E : |y|_E < M\}, \text{ some } M > 0 \tag{1.2.3}$$

and $p^\star = 0$ respectively. We will describe presently how the constant $M > 0$ is chosen. Now from (1.2.2) and Remark 1.2.1 we see that

$$N : \overline{U} \to E \text{ is a continuous, compact operator.}$$

Thus the hypotheses of the nonlinear alternative are satisfied and therefore at least one of (i) or (ii) hold. To guarantee a fixed point of $N : \overline{U} \to E$ we must ensure that (ii) isn't true.

Consider the functional equation

$$y(t) = \lambda N y(t) \text{ on } I, \text{ with } 0 < \lambda < 1. \tag{1.2.4$_\lambda$}$$

If we choose $M > 0$ in (1.2.3) such that any solution $y \in E$ of $(1.2.4)_\lambda$ satisfies

$$|y|_E \neq M,$$

then condition (ii) cannot hold. Consequently $N : \overline{U} \to E$ has a fixed point $y \in \overline{U}$, or equivalently, (1.2.1) has a solution $y \in E$ such that $|y|_E \leq M$.

How does one find such a constant $M > 0$? One approach is to place a growth condition on the operator N and use this to determine a suitable M. For example, suppose that

$$\begin{cases} \text{there exists } \psi : [0,\infty) \to [0,\infty), \text{ with } \psi \in C[0,\infty) \\ \text{such that } |Ny|_E \leq \psi(|y|_E) \text{ for all } y \in E \end{cases} \tag{1.2.5}$$

holds. Then any solution $y \in E$ of $(1.2.4)_\lambda$ is such that

$$|y|_E \leq |Ny|_E \leq \psi(|y|_E).$$

In particular if $y \in \partial U$ then $|y|_E = M$ and M must satisfy

$$M \leq \psi(M). \tag{1.2.6}$$

However if M is chosen to satisfy

$$\text{there exists } M > 0 \text{ such that } \psi(M) < M, \tag{1.2.7}$$

then (1.2.6) cannot hold, which in turn implies that (*ii*) is not true. Thus N has a fixed point in $\overline{U}$.

We have therefore proved the following theorem.

Theorem 1.2.2. *Let I be an interval of* $\mathbf{R}$ *and $E = E(I)$ a normed linear function space with norm defined by $|.|_E$. Suppose that* (1.2.2), (1.2.5) *and* (1.2.7) *hold. Then* (1.2.1) *has at least one solution $y \in E$ such that $|y|_E < M$.*

Remark 1.2.2. In the above analysis we have shown that any solution $y \in E$ of $(1.2.4)_\lambda$ satisfies $|y|_E \neq M$. We do *not* claim that any solution of $(1.2.4)_\lambda$ satisfies $|y|_E \leq M$.

We also remark at this point that if one was to insist that $\psi \in C[0, \infty)$ as defined in (1.2.5) is also nondecreasing then Theorem 1.2.2 may also be proved using Schauder's fixed point theorem. To see this first recall Schauder's fixed point theorem.

Theorem 1.2.3. Schauder's Fixed Point Theorem.
Let D be a closed, convex subset of a normed linear space E. Then every compact, continuous map $N : D \to D$ has at least one fixed point.

Letting $D := \{y \in E : |y|_E \leq M\}$, we see from (1.2.5), (1.2.7) and the additional assumption that ψ is nondecreasing, that for $y \in D$,

$$|Ny|_E \leq \psi(|y|_E) \leq \psi(M) < M.$$

This inequality along with (1.2.2) implies that

$$N : D \to D \text{ is continuous and compact.}$$

The result is now immediate from Schauder's fixed point theorem.

Now consider (1.2.1) when N is a nonlinear integral operator, that is,

$$y(t) = Ny(t) := g(t) + \int_0^1 k(t,s) f(s, y(s))\, ds \tag{1.2.8}$$

defined on the *compact* interval $I := [0, 1]$. We will use the nonlinear alternative to present a variety of existence principles for this equation. Firstly we are interested in proving the existence of a continuous solution $y \in C[0,1]$ of (1.2.8), and secondly we place conditions on g, k and f such that the integral equation has a solution $y \in L^p[0,1]$, $1 \leq p < \infty$.

Notation. We use $|.|_0$ to denote the norm on $C[0,1]$, that is, for $u \in C[0,1]$, $|u|_0 := \max_{t \in [0,1]} |u(t)|$. For $1 \leq p < \infty$, let $L^p[0,1]$ denote the Banach space of pth–power integrable functions with $||u||_p := \left(\int_0^1 |u(t)|^p\,dt\right)^{\frac{1}{p}}$. Also $L^\infty[0,1]$ denotes the Banach space of essentially bounded measurable functions together with the essential supremum norm denoted by $||.||_\infty$.

Let us first concentrate on proving the existence of solution $y \in C[0,1]$ of (2.1.8). Since we intend to use the nonlinear alternative, the operator N must be such that

$$N : C[0,1] \to C[0,1] \text{ is continuous and completely continuous.}$$

Thus in addition to showing that N is continuous, we need to prove that for each bounded set $\Omega \subseteq C[0,1]$, $N(\Omega)$ is relatively compact in $C[0,1]$. Compactness criteria for $C[0,1]$ in the form of the Arzéla-Ascoli theorem are therefore essential to the existence principle.

Theorem 1.2.4. Arzéla-Ascoli Theorem.
Let $\Omega \subseteq C[0,1]$. If Ω is uniformly bounded and equicontinuous in $C[0,1]$, then Ω is relatively compact in $C[0,1]$.

Before presenting our first existence principle for (1.2.8) let us recall the following definition:

Definition 1.2.1. A function $f : [0,1] \times \mathbf{R} \to \mathbf{R}$ is a L^q*-Carathéodory function* if the following conditions hold:

- (i) the map $t \mapsto f(t,y)$ is measurable for all $y \in \mathbf{R}$,
- (ii) the map $y \mapsto f(t,y)$ is continuous for almost all $t \in [0,1]$,
- (iii) for any $r > 0$, there exists $\mu_r \in L^q[0,1]$ such that $|y| \le r$ implies that $|f(t,y)| \le \mu_r(t)$ for almost all $t \in [0,1]$.

A function $f : [0,1] \times \mathbf{R} \to \mathbf{R}$ is a *Carathéodory function* if conditions (i) and (ii) hold.

Theorem 1.2.5. *Let $1 \le p \le \infty$ be a constant and q be such that $1/p + 1/q = 1$. Assume that*

$$g \in C[0,1], \tag{1.2.9}$$

$$f : [0,1] \times \mathbf{R} \to \mathbf{R} \textit{ is a } L^q\textit{-Carathéodory function,} \tag{1.2.10}$$

$$k_t(s) := k(t,s) \in L^p[0,1] \textit{ for each } t \in [0,1] \tag{1.2.11}$$

and

$$\textit{the map } t \mapsto k_t \textit{ is continuous from } [0,1] \to L^p[0,1] \tag{1.2.12}$$

hold. In addition, suppose that there exists a constant $M > 0$, independent of λ, with $|y|_0 \neq M$ for any solution $y \in C[0,1]$ of

$$(1.2.13)_\lambda \qquad y(t) = \lambda\left(g(t) + \int_0^1 k(t,s)f(s,y(s))\,ds\right),\ t \in [0,1],$$

for each $\lambda \in (0,1)$. Then (1.2.8) has at least one solution $y \in C[0,1]$.

Notation. To avoid repetition in the proof of this theorem, for $p = \infty$ we will use $\left(\int_0^1 |u(t)|^p\,dt\right)^{\frac{1}{p}}$ to denote $||u||_\infty$.

PROOF: Define the operator N by

$$Ny(t) := g(t) + \int_0^1 k(t,s)f(s,y(s))\,ds,\ t \in [0,1].$$

To verify that $N : C[0,1] \to C[0,1]$ is well-defined, note that for any $y \in C[0,1]$, there exists $r > 0$ such that $|y|_0 \le r$. Since f is L^q-Carathéodory, there exists $\mu_r \in L^q[0,1]$ such that $|f(s,y)| \le \mu_r(s)$, a.e. $s \in [0,1]$. Therefore for any $t_1, t_2 \in [0,1]$, we see from the hypotheses that

$$(1.2.14) \qquad |Ny(t_1) - Ny(t_2)| \le |g(t_1) - g(t_2)| + \left(\int_0^1 |k_{t_1}(s) - k_{t_2}(s)|^p\,ds\right)^{\frac{1}{p}} ||\mu_r||_q \to 0 \text{ as } t_1 \to t_2.$$

Consequently $N : C[0,1] \to C[0,1]$.

As mentioned above, to apply the nonlinear alternative we need to show that

$$N : C[0,1] \to C[0,1] \text{ is continuous and completely continuous.}$$

We first show the continuity of the operator, that is, if $y_n \to y$ in $C[0,1]$, then $Ny_n \to Ny$ in $C[0,1]$. There exists $r > 0$ and $\mu_r \in L^q[0,1]$ with $|y_n|_0$, $|y|_0 \le r$ and $|f(s,y_n(s)|$, $|f(s,y(s))| \le \mu_r(s)$, a.e. $s \in [0,1]$. By an argument similar to the one used to derive (1.2.14), we can show that for any $\epsilon > 0$, there exists $\delta > 0$ such that for $t_1, t_2 \in [0,1]$, $|t_1 - t_2| < \delta$ implies

$$(1.2.15) \qquad |Ny_n(t_1) - Ny_n(t_2)| < \frac{\epsilon}{3} \text{ and } |Ny(t_1) - Ny(t_2)| < \frac{\epsilon}{3}.$$

Also $Ny_n(t) \to Ny(t)$ pointwise on $[0,1]$, since

$$|Ny_n(t) - Ny(t)| \le \sup_{t\in[0,1]} ||k_t||_p \left(\int_0^1 |f(s,y_n(s)) - f(s,y(s))|^q\,ds\right)^{\frac{1}{q}}$$

and hence by (1.2.10) (and the Lebesgue dominated convergence theorem if $1 \leq q < \infty$) we have

$$|Ny_n(t) - Ny(t)| \to 0 \text{ as } n \to \infty. \tag{1.2.16}$$

Combining (1.2.15) and (1.2.16), and using the fact that $[0,1]$ is compact, yields a constant $C > 0$ such that

$$\text{for all } n \geq C, \ |Ny_n(t) - Ny(t)| < \epsilon, \ \text{for all } t \in [0,1].$$

We have thus shown that $N : C[0,1] \to C[0,1]$ is continuous.

We now use the Arzéla-Ascoli theorem to show that $N : C[0,1] \to C[0,1]$ is completely continuous. Let Ω be a bounded set in $C[0,1]$ and let $r > 0$ be such that $|y|_0 \leq r$ for all $y \in \Omega$. Clearly $N(\Omega)$ is uniformly bounded since there exists $\mu_r \in L^q[0,1]$ such that $|f(s, y(s))| \leq \mu_r(s)$, a.e. $s \in [0,1]$, $y \in \Omega$ and therefore

$$|Ny|_0 \leq |g|_0 + K||\mu_r||_q \text{ for all } y \in \Omega,$$

where $K := \sup_{t\in[0,1]} ||k_t||_p$ exists by (1.2.12). In addition using an argument similar to the one used to obtain (1.2.14), one can see that $N(\Omega)$ is equicontinuous. It follows from the Arzéla-Ascoli theorem that $N(\Omega)$ is relatively compact in $C[0,1]$.

The proof now follows from the nonlinear alternative with $U := \{y \in C[0,1] : |y|_0 < M\}$ and $C = E = C[0,1]$. □

As was discussed earlier in relation to the general operator equation (1.2.1), one instance where it is possible to find an $M > 0$ such that any solution $y \in C[0,1]$ of $(1.2.13)_\lambda$ satisfies $|y|_0 \neq M$, is when f satisfies a certain growth condition. For example suppose that $f : [0,1] \times \mathbf{R} \to \mathbf{R}$ is such that

$$\begin{cases} \text{there exists } \psi : [0,\infty) \to [0,\infty), \text{ with } \psi \in C[0,\infty) \\ \text{such that } ||Fy||_q \leq \psi(|y|_0) \text{ for all } y \in C[0,1] \end{cases} \tag{1.2.17}$$

is true. Then any solution $y \in C[0,1]$ of $(1.2.13)_\lambda$ satisfies

$$|y|_0 \leq |g|_0 + \sup_{t\in[0,1]} ||k_t||_p||Fy||_q \leq |g|_0 + K\psi(|y|_0).$$

Thus if

$$\text{there exists } M > 0 \text{ such that } \frac{M}{|g|_0 + K\psi(M)} > 1, \tag{1.2.18}$$

holds, then clearly any solution $y \in (1.2.13)_\lambda$ must satisfy $|y|_0 \neq M$.

We have therefore proved the following existence result for (1.2.8).

Theorem 1.2.6. *Let* $1 \le p \le \infty$ *be a constant and* q *be such that* $1/p + 1/q = 1$. *Assume that* (1.2.9) – (1.2.12) *hold in addition to* (1.2.17) *and* (1.2.18). *Then* (1.2.8) *has at least one solution* $y \in C[0,1]$.

Example 1.2.1. Let $f(s,y) = |y|^n$, $n \ge 0$. Then for $y \in C[0,1]$,

$$\left(\int_0^1 |f(s,y(s))|^q\,ds\right)^{\frac{1}{q}} = \left(\int_0^1 |y(s)|^{nq}\,ds\right)^{\frac{1}{q}} \le |y|_0^n.$$

Therefore if we define $\psi(t) := t^n$, $t \ge 0$, then we see that this choice of f satisfies (1.2.17).

Note that for $0 \le n < 1$, the inequality in (1.2.18) is satisfied if we choose $M > 0$ "large enough," while for $n \ge 1$, (1.2.18) holds if K is "sufficiently small."

Our second main result discusses the conditions required on g, k and f to ensure the existence of a solution $y \in L^1[0,1]$ of (1.2.8) (and for completeness we also discuss the case when the solution lies in $L^p[0,1]$, $1 < p < \infty$). Banas [28,29] and Emmanuele [59,60] have examined this type of problem extensively over the last fifteen years or so. Their analysis relies on the notion of measures of weak noncompactness and the Schauder fixed point theorem. However, once again we will apply the nonlinear alternative to obtain an existence principle for (1.2.8), and in showing that the integral operator N is such that

$$N : L^p[0,1] \to L^p[0,1] \text{ is completely continuous,}$$

we will use the older compactness results of Riesz and Kolmogorov (see [1,34,47,57,85]) which are stated below.

Theorem 1.2.7. Riesz Compactness Criteria.
Let $\Omega \subseteq L^p[0,1]$, $1 \le p < \infty$. *Then* Ω *is relatively compact in* $L^p[0,1]$ *if and only if the following conditions hold:*

(*i*) Ω *is bounded in* $L^p[0,1]$,

(*ii*) $\int_0^1 |u(t+h) - u(t)|^p\,dt \to 0$ *as* $h \to 0$ *uniformly w.r.t.* $u \in \Omega$.

Theorem 1.2.8. Kolmogorov Compactness Criteria.
Let $\Omega \subseteq L^p[0,1]$, $1 \le p < \infty$. *Then* Ω *is relatively compact in* $L^p[0,1]$ *if and only if the following conditions hold:*

(*i*) *Ω is bounded in $L^p[0,1]$,*
(*ii*) *$u_h \to u$ (convergence in $L^p[0,1]$) as $h \to 0$ uniformly w.r.t. $u \in \Omega$;*
here

$$u_h(t) = \frac{1}{h}\int_t^{t+h} u(s)\, ds.$$

Remark 1.2.3. In both of the above theorems, it is agreed that $u \in \Omega$ is extended to an interval $(a,b) \supseteq [0,1]$ by letting $u(t) = 0$ outside $[0,1]$.

Finally before presenting the existence principle for (1.2.8), for completeness we state the following well known result of Krasnoselskii [89] which will also be required in the proof:

Theorem 1.2.9. *Let $f : [0,1] \times \mathbf{R} \to \mathbf{R}$ be a Carathéodory function such that $y \in L^{p_1}[0,1]$ implies that $f(t,y) \in L^{p_2}[0,1]$ ($p_1, p_2 \geq 1$). Then the operator $F : L^{p_1}[0,1] \to L^{p_2}[0,1]$ defined by $Fy(t) := f(t,y(t))$ is continuous and bounded. In particular there exists $a_1 \in L^{p_2}[0,1]$ and $a_2 > 0$ such that*

$$|f(t,y)| \leq a_1(t) + a_2|y|^{\frac{p_1}{p_2}}.$$

Theorem 1.2.10. *Suppose the following conditions are satisfied:*

(1.2.19) $$g \in L^1[0,1],$$

(1.2.20) $k : [0,1] \times [0,1] \to \mathbf{R}$ *is measurable and* $\left\| \int_0^1 |k(t,s)|\, dt \right\|_\infty < \infty,$

(1.2.21) $$\lim_{h \to 0} \int_0^1 \left\| \frac{1}{h}\int_t^{t+h} |k(x,s) - k(t,s)|\, dx \right\|_\infty dt = 0$$

and

(1.2.22) $$\begin{cases} f : [0,1] \times \mathbf{R} \to \mathbf{R} \text{ is a Carathéodory function and} \\ \text{there exists } a \in L^1[0,1] \text{ and a constant } b \geq 0 \text{ with} \\ |f(t,u)| \leq a(t) + b|u|, \text{ for a.e. } t \in [0,1] \text{ and all } u \in \mathbf{R}. \end{cases}$$

In addition assume there is a constant $M_0 > 0$, independent of λ, with

(1.2.23) $$\|y\|_1 \neq M_0$$

for any solution $y \in L^1[0,1]$ *of*

$$(1.2.24)_\lambda \qquad y(t) = \lambda\left(g(t) + \int_0^1 k(t,s)\, f(s, y(s))\, ds\right), \quad a.e. \quad t \in [0,1]$$

for each $\lambda \in (0,1)$. *Then* (1.2.8) *has at least one solution* $y \in L^1[0,1]$.

PROOF: Define the operators

$$K : L^1[0,1] \to L^1[0,1] \text{ by } Ku(t) := g(t) + \int_0^1 k(t,s)u(s)\, ds,$$

$$K_0 : L^1[0,1] \to L^1[0,1] \text{ by } K_0u(t) := \int_0^1 k(t,s)u(s)\, ds$$

and

$$F : L^1[0,1] \to L^1[0,1] \text{ by } Fu(t) := f(t, u(t)).$$

A solution of (1.2.8) is a fixed point of the operator $N := KF : L^1[0,1] \to L^1[0,1]$. In addition, finding a $y \in L^1[0,1]$ which satisfies $(1.2.24)_\lambda$ is equivalent to solving the fixed point problem $y = \lambda N y$. The result of Krasnoselskii stated in Theorem 1.2.9 implies (since (1.2.22) holds) that

$$F : L^1[0,1] \to L^1[0,1] \text{ is continuous and bounded.}$$

We now show that $K : L^1[0,1] \to L^1[0,1]$ is continuous and completely continuous.

The fact that K is continuous will follow once we show the linear operator $K_0 : L^1[0,1] \to L^1[0,1]$ is continuous. Let $u \in L^1[0,1]$. Then

$$\begin{aligned} \int_0^1 \int_0^1 |k(t,s)||u(s)|\, ds\, dt &= \int_0^1 |u(s)| \int_0^1 |k(t,s)|\, dt\, ds \\ &\leq \|u\|_1 \left\| \int_0^1 |k(t,s)|\, dt \right\|_\infty \end{aligned}$$

and therefore

$$\|K_0\, u\|_1 \leq \|u\|_1 \left(\left\| \int_0^1 |k(t,s)|\, dt \right\|_\infty \right).$$

Hence $K_0 : L^1[0,1] \to L^1[0,1]$ (and $K : L^1[0,1] \to L^1[0,1]$) is continuous.

To show that $K : L^1[0,1] \to L^1[0,1]$ is completely continuous we apply Theorem 1.2.8 – the Kolmogorov compactness criteria. Let Ω be a bounded subset of $L^1[0,1]$, that is, there exists $M > 0$ with $\|v\|_1 \le M$ for all $v \in \Omega$. Let $u \in \Omega$. Then

$$\begin{aligned}
\|K\,u\|_1 &= \int_0^1 \left| g(t) + \int_0^1 k(t,s)u(s)\,ds \right| dt \\
&\le \int_0^1 |g(t)|\,dt + \int_0^1 \int_0^1 |k(t,s)||u(s)|\,ds\,dt \\
&\le \int_0^1 |g(t)|\,dt + \|u\|_1 \left\| \int_0^1 |k(t,s)|\,dt \right\|_\infty \\
&\le \int_0^1 |g(t)|\,dt + M \left\| \int_0^1 |k(t,s)|\,dt \right\|_\infty
\end{aligned}$$

and therefore $K(\Omega)$ is bounded in $L^1[0,1]$. Next we will show that $(Ku)_h \to Ku$ (convergence in $L^1[0,1]$) as $h \to 0$, uniformly with respect to $u \in \Omega$. To see this notice

$$\begin{aligned}
\int_0^1 |(Ku)_h(t) \quad &- \quad Ku(t)|\,dt = \int_0^1 \left| \frac{1}{h} \int_t^{t+h} Ku(x)\,dx - Ku(t) \right| dt \\
&= \int_0^1 \left| \frac{1}{h} \int_t^{t+h} [Ku(x) - Ku(t)]\,dx \right| dt \\
&\le \int_0^1 \frac{1}{h} \int_t^{t+h} |g(x) - g(t)|\,dx\,dt \\
&+ \int_0^1 \frac{1}{h} \int_t^{t+h} \int_0^1 |k(x,s) - k(t,s)||u(s)|\,ds\,dx\,dt \\
&= \int_0^1 \frac{1}{h} \int_t^{t+h} |g(x) - g(t)|\,dx\,dt \\
&+ \int_0^1 \int_0^1 |u(s)| \left(\frac{1}{h} \int_t^{t+h} |k(x,s) - k(t,s)|\,dx \right) ds\,dt \\
&\le \int_0^1 \frac{1}{h} \int_t^{t+h} |g(x) - g(t)|\,dx\,dt \\
&+ \|u\|_1 \int_0^1 \left\| \frac{1}{h} \int_t^{t+h} |k(x,s) - k(t,s)|\,dx \right\|_\infty dt \\
&\le \int_0^1 \frac{1}{h} \int_t^{t+h} |g(x) - g(t)|\,dx\,dt \\
&+ M \int_0^1 \left\| \frac{1}{h} \int_t^{t+h} |k(x,s) - k(t,s)|\,dx \right\|_\infty dt.
\end{aligned}$$

From [153,Chpt.4], since $g \in L^1[0,1]$, we have

$$\lim_{h\to 0} \frac{1}{h}\int_t^{t+h} |g(x) - g(t)|\,dx = 0 \text{ for a.e. } t \in [0,1]$$

and consequently

$$\lim_{h\to 0}\int_0^1 \frac{1}{h}\int_t^{t+h} |g(x) - g(t)|\,dx\,dt = 0, \tag{1.2.25}$$

since

$$\begin{aligned}\int_0^1 \frac{1}{h}\int_t^{t+h} |g(x) - g(t)|\,dx\,dt &= \int_0^h \frac{1}{h}\int_0^1 |g(t+x) - g(t)|\,dt\,dx \\ &\le \sup_{0<s<h} \|g(\cdot + s) - g(\cdot)\|_1.\end{aligned}$$

Now (1.2.21) and (1.2.25) imply that

$$(Ku)_h \to Ku \text{ (convergence in } L^1[0,1]) \text{ as } h \to 0,$$

uniformly with respect to $u \in \Omega$. Consequently

$$K : L^1[0,1] \to L^1[0,1] \text{ is completely continuous.}$$

As a result $N = KF : L^1[0,1] \to L^1[0,1]$ is continuous and completely continuous.

Set

$$U = \left\{u \in L^1[0,1] : \|u\|_1 < M_0\right\}, \quad C = E = L^1[0,1].$$

Now the nonlinear alternative implies that N has a fixed point (notice $(A2)$ in does not occur since (1.2.23) holds). □

Next we discuss the case when solutions to (1.2.8) lie in $L^p[0,1]$, $1 \le p < \infty$.

Theorem 1.2.11. *Let $1 \le p < \infty$ and $1 < q < \infty$. Suppose the following conditions are satisfied:*

$$g \in L^p[0,1], \tag{1.2.26}$$

$$k : [0,1] \times [0,1] \to \mathbf{R} \textit{ is measurable,} \tag{1.2.27}$$

$$\int_0^1 \left(\int_0^1 |k(t,s)|^r\,ds\right)^{\frac{p}{r}} dt < \infty; \textit{ here } r \textit{ is the conjugate of } q \tag{1.2.28}$$

and

$$(1.2.29)\qquad \begin{cases} f:[0,1]\times\mathbf{R}\to\mathbf{R} \text{ is a Carathéodory function and} \\ \text{there exists } a\in L^q[0,1], \text{ and a constant } b\geq 0 \text{ with} \\ |f(t,u)|\leq a(t)+b|u|^{\frac{p}{q}} \text{ for a.e. } t\in[0,1] \text{ and all } u\in\mathbf{R}. \end{cases}$$

In addition assume there is a constant $M_0>0$, independent of λ, with

$$(1.2.30)\qquad \|y\|_p\neq M_0$$

for any solution $y\in L^p[0,1]$ of $(1.2.24)_\lambda$ for each $\lambda\in(0,1)$. Then (1.2.8) has at least one solution $y\in L^p[0,1]$.

Remark 1.2.4. The case $q=1$ is discussed in Theorem 1.2.12. Of course the case $p=q=1$ has already discussed in Theorem 1.2.10.

PROOF: Let $K, K_0 : L^q[0,1]\to L^p[0,1]$ and $F: L^p[0,1]\to L^q[0,1]$ be given as in Theorem 1.2.10. A solution of (1.2.8) is a fixed point of the operator $N=KF: L^p[0,1]\to L^p[0,1]$. Theorem 1.2.9 implies

$$F: L^p[0,1]\to L^q[0,1] \text{ is continuous and bounded.}$$

Essentially the same reasoning as in Theorem 1.2.10 establishes the result once we show that

$$K: L^q[0,1]\to L^p[0,1] \text{ is continuous and completely continuous.}$$

Let $u\in L^q[0,1]$. Then

$$\int_0^1\left(\int_0^1 |k(t,s)||u(s)|\,ds\right)^p dt \leq \|u\|_q^p\int_0^1\left(\int_0^1 |k(t,s)|^r\,ds\right)^{\frac{p}{r}} dt$$

and therefore

$$\|K_0u\|_p\leq\|u\|_q\left(\int_0^1\left(\int_0^1 |k(t,s)|^r\,ds\right)^{\frac{p}{r}} dt\right)^{\frac{1}{p}}.$$

Thus $K: L^q[0,1]\to L^p[0,1]$ is continuous.

To show $K: L^q[0,1]\to L^p[0,1]$ is completely continuous we apply Theorem 1.2.7 – the Riesz compactness criteria. Let Ω be a bounded subset of $L^q[0,1]$, that is, there exists $M>0$ with $\|v\|_q\leq M$ for all $v\in\Omega$. Let

$u \in \Omega$. Then

$$\begin{aligned} \|Ku\|_p^p &= \int_0^1 \left| g(t) + \int_0^1 k(t,s)u(s)\,ds \right|^p dt \\ &\le 2^{p-1} \int_0^1 |g(t)|^p\,dt + 2^{p-1} \int_0^1 \left(\int_0^1 |k(t,s)||u(s)|\,ds \right)^p dt \\ &\le 2^{p-1} \int_0^1 |g(t)|^p\,dt + 2^{p-1} \|u\|_q^p \int_0^1 \left(\int_0^1 |k(t,s)|^r\,ds \right)^{\frac{p}{r}} dt \end{aligned}$$

and therefore $K(\Omega)$ is bounded in $L^p[0,1]$. Next we show that

$$\int_0^1 |Ku(t+h) - Ku(t)|^p\,dt \to 0 \text{ as } h \to 0, \text{ uniformly with respect to } u \in \Omega.$$

To see this notice

$$\begin{aligned} \int_0^1 |Ku(t+h) &- Ku(t)|^p\,dt \le 2^{p-1} \int_0^1 |g(t+h) - g(t)|^p\,dt \\ &+ 2^{p-1} \int_0^1 \left(\int_0^1 |k(t+h,s) - k(t,s)||u(s)|\,ds \right)^p dt \\ &\le 2^{p-1} \int_0^1 |g(t+h) - g(t)|^p\,dt \\ &+ 2^{p-1} M^p \int_0^1 \left(\int_0^1 |k(t+h,s) - k(t,s)|^r\,ds \right)^{\frac{p}{r}} dt \\ &\to 0 \text{ as } h \to 0 \text{ (see [159,Chpt.12])}. \end{aligned}$$

Consequently $K : L^q[0,1] \to L^p[0,1]$ is completely continuous. □

Theorem 1.2.12. *Let $1 < p < \infty$. Suppose that (1.2.26) and (1.2.29) hold with $q = 1$, and assume that*

(1.2.31) $k : [0,1] \times [0,1] \to \mathbf{R}$ *is measurable and* $\left\| \int_0^1 |k(t,s)|^p\,dt \right\|_\infty < \infty$

and

(1.2.32) $$\lim_{h \to 0} \left\| \int_0^1 |k(t+h,s) - k(t,s)|^p\,dt \right\|_\infty = 0$$

are true. In addition assume there is a constant $M_0 > 0$, independent of λ, with (1.2.30) true for any solution $y \in L^p[0,1]$ of $(1.2.24)_\lambda$ for each $\lambda \in (0,1)$. Then (1.2.8) has at least one solution in $y \in L^p[0,1]$.

PROOF: Let $K, K_0 : L^1[0,1] \to L^p[0,1]$ and $F : L^p[0,1] \to L^1[0,1]$ be given as in Theorem 1.2.10. Essentially the same reasoning as in Theorem 1.2.10 establishes the result once we show that $K : L^1[0,1] \to L^p[0,1]$ is continuous and completely continuous. Let $u \in L^1[0,1]$ and let m be the conjugate of p. Then

$$\begin{aligned}
&\int_0^1 \left(\int_0^1 |k(t,s)||u(s)|\,ds\right)^p dt \\
&= \int_0^1 \left(\int_0^1 |k(t,s)||u(s)|^{\frac{1}{p}}|u(s)|^{\frac{1}{m}}\,ds\right)^p dt \\
&\leq \int_0^1 \int_0^1 |k(t,s)|^p|u(s)|\,ds \left(\int_0^1 |u(s)|\,ds\right)^{\frac{p}{m}} dt \\
&= \|u\|_1^{\frac{p}{m}} \int_0^1 \int_0^1 |k(t,s)|^p|u(s)|\,ds\,dt \\
&= \|u\|_1^{\frac{p}{m}} \int_0^1 |u(s)| \int_0^1 |k(t,s)|^p\,dt\,ds \\
&\leq \|u\|_1^{\frac{p}{m}} \|u\|_1 \left\|\int_0^1 |k(t,s)|^p\,dt\right\|_\infty \\
&= \|u\|_1^p \left\|\int_0^1 |k(t,s)|^p\,dt\right\|_\infty
\end{aligned}$$

and therefore

$$\|K_0 u\|_p \leq \|u\|_1 \left(\int_0^1 |k(t,s)|^p\,dt\|_\infty\right)^{\frac{1}{p}}.$$

Thus $K : L^1[0,1] \to L^p[0,1]$ is continuous.

To show that $K : L^1[0,1] \to L^p[0,1]$ is completely continuous we apply Theorem 1.2.7. Let Ω be a bounded subset of $L^1[0,1]$, that is, there exists $M > 0$ with $\|v\|_1 \leq M$ for all $v \in \Omega$. Let $u \in \Omega$. Then

$$\begin{aligned}
\|Ku\|_p^p &\leq 2^{p-1}\int_0^1 |g(t)|^p\,dt + 2^{p-1}\int_0^1 \left(\int_0^1 |k(t,s)||u(s)|\,ds\right)^p dt \\
&\leq 2^{p-1}\int_0^1 |g(t)|^p\,dt + 2^{p-1}\|u\|_1^p \left(\left\|\int_0^1 |k(t,s)|^p\,dt\right\|_\infty\right) \\
&\leq 2^{p-1}\int_0^1 |g(t)|^p\,dt + 2^{p-1}M^p \left(\left\|\int_0^1 |k(t,s)|^p\,dt\right\|_\infty\right)
\end{aligned}$$

and thus $K(\Omega)$ is bounded in $L^p[0,1]$. Also

$$\int_0^1 |Ku(t+h) \quad - \quad Ku(t)|^p\,dt \leq 2^{p-1}\int_0^1 |g(t+h) - g(t)|^p\,dt$$

$$
\begin{aligned}
&+\ 2^{p-1}\int_0^1\left(\int_0^1|k(t+h,s)-k(t,s)||u(s)|\,ds\right)^p dt\\
&\le\ 2^{p-1}\int_0^1|g(t+h)-g(t)|^p\,dt\\
&+\ 2^{p-1}M^p\left(\left\|\int_0^1|k(t+h,s)-k(t,s)|^p\,dt\right\|_\infty\right).
\end{aligned}
$$

Hence

$$\int_0^1|Ku(t+h)-Ku(t)|^p\,dt\to 0 \text{ as } h\to 0,$$

uniformly with respect to $u\in\Omega$. Consequently $K:L^1[0,1]\to L^p[0,1]$ is completely continuous. $\square$

To complete the discussion on the existence of $L^p[0,1]$ solutions of (1.2.8), we present one existence result as an application of Theorem 1.2.11.

Theorem 1.2.13. *Let* $1\le p<\infty$ *and* $1<q<\infty$. *Suppose* (1.2.26) – (1.2.29) *hold. In addition assume*

$$\sup_{x\in[0,\infty)}\left(\frac{x}{2^{p-1}||g||_p^p+2^{2(p-1)}[||a||_q^p+b^px^{\frac{p}{q}}]\tilde{K}}\right)>1, \tag{1.2.33}$$

holds where

$$\tilde{K}:=\int_0^1\left(\int_0^1|k(t,s)|^r\,ds\right)^{\frac{p}{r}}dt.$$

Then (1.2.8) *has at least one solution in* $y\in L^p[0,1]$.

Remark 1.2.5. If $q>p$ then (1.2.33) is satisfied since the left hand side of (1.2.33) is equal to infinity. On the other hand, if $q\le p$ then (1.2.33) is again satisfied provided b is "sufficiently small."

PROOF: Let $M_1>0$ satisfy

$$\frac{M_1}{2^{p-1}||g||_p^p+2^{2(p-1)}[||a||_q^p+b^pM_1^{\frac{p}{q}}]\tilde{K}}>1. \tag{1.2.34}$$

Let $y\in L^p[0,1]$ be any solution of $(1.2.24)_\lambda$ for $0<\lambda<1$. Then for a.e. $t\in[0,1]$ we have that

$$|y(t)|^p\le 2^{p-1}|g(t)|^p+2^{p-1}\left(\int_0^1|k(t,s)|[a(s)+b|y(s)|^{\frac{p}{q}}]\,ds\right)^p$$

and therefore

$$\begin{aligned}\int_0^1 |y|^p\,dt &\le 2^{p-1}\int_0^1 |g|^p\,dt + 2^{p-1}\left[2^{p-1}\int_0^1\left(\int_0^1 |k(t,s)||a(s)\,ds\right)^p dt\right.\\ &+ \left. 2^{p-1}\int_0^1\left(\int_0^1 |k(t,s)|b|y(s)|^{\frac{p}{q}}\,ds\right)^p dt\right]\\ &\le 2^{p-1}||g||_p^p + 2^{2(p-1)}||a||_q^p\tilde{K} + 2^{2(p-1)}b^p\left(\int_0^1 |y|^p\,ds\right)^{\frac{p}{q}}\tilde{K}.\end{aligned}$$

Thus

$$\text{(1.2.35)}\qquad \frac{\displaystyle\int_0^1 |y|^p\,dt}{2^{p-1}||g||_p^p + 2^{2(p-1)}\left[||a||_q^p + b^p\left(\displaystyle\int_0^1 |y|^p\,dt\right)^{\frac{p}{q}}\right]\tilde{K}} \le 1.$$

Let $M_0 = M_1^{\frac{1}{p}}$. Suppose $\|y\|_p = M_0$. Then (1.2.35) yields

$$\frac{M_1}{2^{p-1}||g||_p^p + 2^{2(p-1)}[||a||_q^p + b^p M_1^{\frac{p}{q}}]\tilde{K}} \le 1.$$

This contradicts (1.2.34). Thus any solution $y \in L^p[0,1]$ of $(1.2.24)_\lambda$ satisfies $\|y\|_p \ne M_0$. Theorem 1.2.11 now implies that (1.2.8) has a solution $y \in L^p[0,1]$. □

Thus far we have considered the general operator equation (1.2.1) when $N : E(I) \to E(I)$ is a Fredholm integral operator defined on a compact interval $I \subseteq \mathbf{R}$, and we have used the nonlinear alternative to obtain our existence principles. We now briefly consider what happens if the interval I is not compact. Suppose then that I is given by $I = [0,\infty)$ and N is the Fredholm integral operator defined by

$$\text{(1.2.36)}\qquad Ny(t) := g(t) + \int_0^\infty k(t,s)f(s,y(s))\,ds.$$

If we still wish to apply the nonlinear alternative, then the space $E(I)$ in which we want the solution to exist, must be a normed linear space and we must have compactness criteria for this space since we will need to ensure that

$$N : E(I) \to E(I) \text{ is completely continuous.}$$

Two such normed linear spaces are $L^p[0,\infty)$, $1 \le p < \infty$ and $C_l[0,\infty)$. Here $L^p[0,\infty)$ denotes the Banach space of L^p-integrable functions with $||u||_p := (\int_0^\infty |u(t)|^p\,dt)^{\frac{1}{p}}$. If we denote the set of bounded, continuous functions defined on $[0,\infty)$ by $BC[0,\infty)$ with norm $|u|_0 := \sup_{t\in[0,\infty)} |u(t)|$, then $C_l[0,\infty) \subseteq BC[0,\infty)$ is the space of all bounded, continuous functions on $[0,\infty)$ whose limit exists at ∞. We denote by $u(\infty) := \lim_{t\to\infty} u(t)$. We have the following compactness criteria for $L^p[0,\infty)$ and $C_l[0,\infty)$ respectively:

Theorem 1.2.14. Yosida [166,P.275].
Let S be the real line, $\mathbf{B}$ the σ-ring of Baire subsets B of S and $m(B) = \int_B dx$ the ordinary Lebesgue measure of B. Then a subset Ω of $L^p(S,\mathbf{B},m)$, $1 \le p < \infty$, is strongly relatively compact if and only if it satisfies the following conditions:

$$(i) \quad \sup_{x\in\Omega} ||x|| = \sup_{x\in\Omega} \left(\int_S |x(s)|^p\,ds \right)^{\frac{1}{p}} < \infty,$$

$$(ii) \quad \lim_{t\to 0} \int_S |x(t+s) - u(s)|^p\,ds = 0 \text{ uniformly in } x \in \Omega, \text{ and}$$

$$(iii) \quad \lim_{\alpha\uparrow\infty} \int_{|s|>\alpha} |x(s)|^p\,ds = 0 \text{ uniformly in } x \in \Omega.$$

Theorem 1.2.15. Corduneanu [48,P.62].
Let $\Omega \subseteq C_l[0,\infty)$. Then Ω is compact in $C_l[0,\infty)$ if the following conditions hold:

(i) *Ω is bounded in $C_l[0,\infty)$,*
(ii) *the functions belonging to Ω are equicontinuous on any compact interval of $[0,\infty)$, and*
(iii) *the functions from Ω are equiconvergent, that is, given $\epsilon > 0$, there corresponds $T(\epsilon) > 0$ such that $|f(t) - f(\infty)| < \epsilon$ for any $t \ge T(\epsilon)$ and $f \in \Omega$.*

For detailed existence principles which prove the existence of a solution $y \in L^p[0,\infty)$, $1 \le p < \infty$ or $y \in C_l[0,\infty)$ of $y(t) = Ny(t)$, $t \in [0,\infty)$ with N as defined in (1.2.36), we refer the reader to [137,Chpt.5].

We also observe with regard to the general operator equation (1.2.1), that the only example of an operator N that has been given so far has been a Fredholm integral operator. We therefore note that a special case of the Fredholm integral equation described in (1.2.8), occurs when the kernel $k : [0,1] \times [0,1] \to \mathbf{R}$ is given by

$$k(t,s) := \begin{cases} \tilde{k}(t,s), & 0 \le s \le t \le 1 \\ 0, & 0 \le t \le s \le 1. \end{cases}$$

In this case we obtain the Volterra integral equation

$$(1.2.37) \qquad y(t) = g(t) + \int_0^t \tilde{k}(t,s) f(s, y(s))\, ds, \text{ on } [0,1].$$

Since (1.2.37) can be viewed in this way (as being a special case of (1.2.8)), existence principles and results for the Volterra integral equation will follow as corollaries from the theorems outlined already in this section for the Fredholm integral equation. (For details see [137].) Similarly existence principles for the Volterra equation

$$(1.2.38) \qquad y(t) = g(t) + \int_0^t \tilde{k}(t,s) f(s, y(s))\, ds, \text{ on } [0,T),\ 0 \le T \le \infty,$$

may be obtained from those established for

$$(1.2.39) \qquad y(t) = g(t) + \int_0^T k(t,s) f(s, y(s))\, ds, \text{ on } [0,T),\ 0 \le T \le \infty.$$

However, studying (1.2.38) independently of (1.2.39) can lead in certain cases, to improved existence principles.

Let us consider (1.2.38) in its abstract form, that is,

$$(1.2.40) \qquad y(t) = Vy(t) \text{ on } [0,T) \text{ with } 0 \le T \le \infty,$$

where we assume that $V : E \to E$ is an *abstract Volterra operator.*

Definition 1.2.2. An operator $V : E \to F$ where $E = E(I)$ and $F = F(I)$ are function spaces and $I = [0,T]$ or $[0,T)$, is an *abstract Volterra operator* if for an arbitrary $T^\star \in I$, V satisfies the following condition:

$$x(t) = y(t),\ t \in [0,T^\star] \text{ implies that } Vx(t) = Vy(t),\ t \in [0,T^\star].$$

When considering spaces of measurable functions, the relevant equalities are replaced by equalities almost everywhere.

We want to consider (1.2.40) with $V : E \to E$ a Volterra operator, firstly when $E := C[0,T)$, and secondly when $E := L^p_{loc}[0,T)$. $C[0,T)$ denotes the Fréchet space of continuous functions on $[0,T)$. For $y \in C[0,T)$ we define for every $m \in \{1,2,\dots\}$, the seminorm $\rho_{0,m}(y)$ by

$$\rho_{0,m}(y) := \sup_{t\in[0,t_m]} |y(t)| \text{ where } t_m \uparrow T.$$

A subset Ω of $C[0,T)$ is compact if and only if it compact on each subinterval of $[0,T)$. Using the Arzéla-Ascoli compactness criteria, we see that

$\Omega \subseteq C[0,T)$ is compact if and only if Ω is uniformly bounded and equicontinuous on each compact interval of $[0,T)$.

Similarly $L^p_{loc}[0,T)$, $1 \le p \le \infty$, is a Fréchet space, and denotes the space of functions that are L^p-integrable on each compact subinterval of $[0,T)$. For $y \in L^p_{loc}[0,T)$, the seminorm $\rho_{p,m}(y)$ is given by

$$\rho_{p,m}(y) := \left(\int_0^{t_m} |y(t)|^p \, dt \right)^{\frac{1}{p}},$$

for $1 \le p < \infty$, while for $p = \infty$,

$$\rho_{p,m}(y) := ess - sup_{t \in [0,t_m]} |y(t)|;$$

where again $t_m \in [0,T)$ is such that $t_m \uparrow T$.

For $V : C[0,T) \to C[0,T)$ we have the following existence principle for (1.2.40).

Theorem 1.2.16. *Let $0 \le T \le \infty$ and assume that*

$$(1.2.41) \qquad \begin{cases} V : C[0,T) \to C[0,T) \text{ is a continuous and} \\ \text{completely continuous, abstract Volterra operator} \end{cases}$$

is true. Let $0 < t_1 < t_2 < \ldots < t_n < \ldots$, with $t_n \uparrow T$. Suppose for each $n = 1, 2, \ldots$, that there exists $y_n \in C[0,t_n]$ that solves

$$y_n(t) = V y_n(t), \ t \in [0, t_n],$$

and that there are bounded sets $B_k \subseteq \mathbf{R}$, for $k = 1, 2, \ldots$, such that $n \ge k$ implies $y_n(t) \in B_k$ for each $t \in [0, t_k]$. Then (1.2.40) has a solution $y \in C[0,T)$ such that $y(t) \in B_k$ for each $t \in [0,t_k]$.

PROOF: We first show that

$$(1.2.42) \qquad \begin{cases} \text{for each } k = 1, 2, \ldots, \text{ the set } \{y_n\}_{n \ge k} \\ \text{is relatively compact in } C[0, t_k] \end{cases}$$

is true. Clearly $\{y_n\}_{n \ge k}$ is bounded by B_k on $[0,t_k]$, and therefore by (1.2.41) we see that $\{V y_n\}_{n \ge k}$ is relatively compact in $C[0,t_k]$. Thus by definition of y_n and the fact that V is a Volterra operator we have that (1.2.42) is true.

Now (1.2.42) and the Arzéla-Ascoli theorem yield a subsequence N_1 of $\{1, 2, \ldots\}$, and a function $z_1 \in C[0,t_1]$ such that $y_n \to z_1$ in $C[0,t_1]$, as $n \to \infty$ in N_1. Let $N_1^\star := N_1 \backslash \{1\}$. Then once again, (1.2.42) and the Arzéla-Ascoli theorem yield a subsequence N_2 of $N_1^\star$, and a function $z_2 \in C[0,t_2]$

such that $y_n \to z_2$ in $C[0, t_2]$, as $n \to \infty$ in N_2. Note that $z_2 = z_1$ on $[0, t_1]$ since $N_2 \subseteq N_1$. Proceed inductively to obtain subsequences of integers

$$N_1 \supseteq N_2 \supseteq \ldots \supseteq N_k \supseteq \ldots, \quad N_k \subseteq \{k, k+1, \ldots\},$$

and functions $z_k \in C[0, t_k]$ with

$$y_n \to z_k \text{ uniformly on } [0, t_k] \text{ as } n \to \infty \text{ in } N_k,$$

and $z_{k+1} = z_k$ on $[0, t_k]$ for $k = 1, 2, \ldots$.

Define a function $y : [0, T) \to \mathbf{R}$ by $y(t) := z_k(t)$ on $[0, t_k]$. Fix $t \in [0, T)$ and choose and fix k with $t \in [0, t_k]$. Then

$$y_n(t) = V y_n(t), \quad n \geq k.$$

Letting $n \to \infty$ in N_k, we see that $z_k(t) = V z_k(t)$ since $V : C[0, T) \to C[0, T)$ is continuous. Thus $y \in C[0, T)$ solves (1.2.40). □

An analogous result exists for (1.2.40) when $V : L^p_{loc}[0, T) \to L^p_{loc}[0, T)$. We just state the result since the proof is similar to that of Theorem 1.2.16.

Theorem 1.2.17. *Let $0 \leq T \leq \infty$ and $1 \leq p < \infty$. Assume that*

$$(1.2.43) \qquad \begin{cases} V : L^p_{loc}[0, T) \to L^p_{loc}[0, T) \text{ is a continuous and} \\ \text{completely continuous, abstract Volterra operator} \end{cases}$$

is true. Let $0 < t_1 < t_2 < \ldots < t_n < \ldots$, with $t_n \uparrow T$. Suppose for each $n = 1, 2, \ldots$, that there exists $y_n \in L^p[0, t_n]$ that solves

$$y_n(t) = V y_n(t), \quad a.e.\ t \in [0, t_n],$$

and that there are constants $C_k \in [0, \infty)$ for $k = 1, 2, \ldots$, such that $n \geq k$ implies that

$$\rho_{p,k}(y_n) = \left(\int_0^{t_k} |y_n(t)|^p \, dt \right)^{\frac{1}{p}} \leq C_k.$$

Then (1.2.40) has a solution $y \in L^p_{loc}[0, T)$.

To conclude, we note that if the operator V is given by

$$V y(t) := g(t) + \int_0^t \tilde{k}(t, s) f(s, y(s)) \, ds \text{ on } [0, T),$$

then the conditions required on g, k and f to ensure that either (1.2.41) or (1.2.43) hold, are explicitly described in [137,Chpt.5]. In addition, existence principles which guarantee either a solution $y \in C[0,t_n]$ or $y \in L^p[0,t_n]$ of $y(t) = Vy(t)$, $t \in [0,t_n]$ can also be found in [137,Chpt.4]. (We will also present similar results in Chapter 6, Section 6.4.) Thus one can show that under certain conditions, the Volterra integral equation (1.2.38) has either a solution $y \in C[0,T)$ or $y \in L^p[0,T)$.

1.3. Functional-differential equations.

In this section we consider the functional-differential equation

$$\begin{cases} y'(t) = Vy(t) \\ y(0) = y_0 \end{cases} \tag{1.3.1}$$

firstly on the compact interval $[0,T]$, and secondly on the half-open interval $[0,T)$. In particular, we consider the cases when the operator V is such that $V : C[0,T] \to L^p[0,T]$ and $V : C[0,T) \to L^p_{loc}[0,T)$ respectively.

Let us first consider (1.3.1) on the compact interval $[0,T]$, that is,

$$\begin{cases} y'(t) = Vy(t), \text{ a.e. } t \in [0,T] \\ y(0) = y_0. \end{cases} \tag{1.3.2}$$

We assume that $V : C[0,T] \to L^p[0,T]$, $0 \le T < \infty$, and we apply the nonlinear alternative once again to prove the existence of a solution $y \in AC[0,T]$ of (1.3.2).

Notation. $AC[0,T]$ denotes the space of absolutely continuous functions on $[0,T]$, and is such that for any $y \in AC[0,T]$, y is continuous and y' is integrable on $[0,T]$. $AC_{loc}[0,T)$, where $0 \le T \le \infty$, in turn represents the space of functions that are absolutely continuous on each compact subinterval of $[0,T)$.

Theorem 1.3.1. *Let $1 \le p \le \infty$. Suppose that*

$$V : C[0,T] \to L^p[0,T] \text{ is a continuous operator} \tag{1.3.3}$$

and

$$\begin{cases} \text{for any constant } r > 0, \text{ there exists } \mu_r \in L^p[0,T], \\ \text{such that for any } y \in C[0,T], \text{ with } |y|_0 \le r, \\ \text{we have } |Vy(t)| \le \mu_r(t), \text{ for a.e. } t \in [0,T] \end{cases} \tag{1.3.4}$$

hold. In addition, suppose there exists a constant $M > 0$, independent of λ, with

$$|y_0| < M \text{ and } |y|_0 \ne M,$$

for any solution $y \in AC[0,T]$ *of*

$$(1.3.5)_\lambda \qquad \begin{cases} y'(t) = \lambda V y(t), & a.e.\ t \in [0,T] \\ y(0) = y_0, \end{cases}$$

for each $\lambda \in (0,1)$. *Then* (1.3.2) *has at least one solution* $y \in AC[0,T]$.

PROOF: From (1.3.3), solving $(1.3.5)_\lambda$ is equivalent to finding $y \in C[0,T]$ that satisfies

$$(1.3.6) \qquad y(t) = y_0 + \lambda \left(\int_0^t V y(s)\, ds \right), \quad t \in [0,T].$$

Rewrite (1.3.6) as

$$y(t) = (1-\lambda) y_0 + \lambda \left(y_0 + \int_0^t V y(s)\, ds \right), \quad t \in [0,T],$$

and define the operator $S : C[0,T] \to C[0,T]$ by

$$Sy(t) := y_0 + \int_0^t V y(s)\, ds, \quad t \in [0,T].$$

Consequently $(1.3.5)_\lambda$ is equivalent to the fixed point problem

$$y = (1-\lambda) y_0 + \lambda S y.$$

To apply the nonlinear alternative (Theorem 1.2.1) we must ensure that

$$S : C[0,T] \to C[0,T] \text{ is continuous and completely continuous.}$$

To illustrate the continuity of S, let $y_n \to y$ in $C[0,T]$. Then from (1.3.3), with q the conjugate of p, we see that

$$|Sy_n - Sy|_0 \le T^{\frac{1}{q}} \left(\int_0^T |V y_n(s) - V y(s)|^p\, ds \right)^{\frac{1}{p}} \to 0, \text{ if } 1 \le p < \infty,$$

and

$$|Sy_n - Sy|_0 \le T ess - sup_{t \in [0,T]} |V y_n(t) - V y(t)| \to 0, \text{ if } p = \infty.$$

Thus $S : C[0,T] \to C[0,T]$ is continuous.

The complete continuity of N will follow from the Arzéla-Ascoli theorem (Theorem 1.2.4). Let $\Omega \subseteq C[0,T]$ be bounded, that is, there exists $r > 0$ such that $|y|_0 \le r$ for all $y \in \Omega$. From (1.3.4) we see that for $y \in \Omega$,

$$|Sy|_0 \le |y_0| + \int_0^T \mu_r(s)\, ds \equiv M_0 < \infty,$$

hence $S(\Omega)$ is uniformly bounded. Also for $y \in \Omega$ and $t_1, t_2 \in [0,T]$ with $t_2 < t_1$, we obtain

$$|Sy(t_1) - Sy(t_2)| \leq \int_{t_2}^{t_1} |Vy(s)|\, ds \leq \int_{t_2}^{t_1} \mu_r(s)\, ds \to 0 \text{ as } t_2 \to t_1.$$

Thus $S(\Omega)$ is equicontinuous on $[0,T]$. The Arzéla-Ascoli theorem now guarantees the complete continuity of $N : C[0,T] \to C[0,T]$.

Let $U := \{y \in C[0,T] : |y|_0 < M\}$, $C = E = C[0,T]$ and $p^\star = y_0$. The nonlinear alternative ensures that V has a fixed point in U, that is, (1.3.2) has a solution $y \in C[0,T]$. In particular note that $y \in AC[0,T]$. □

A special case of (1.3.2) occurs when

$$Ny(t) := f(t, y(t), Ky(t)), \text{ a.e. } t \in [0,T],$$

where $f : [0,T] \times \mathbf{R} \times \mathbf{R} \to \mathbf{R}$ and K is a Fredholm or Volterra integral operator. We then have the integro-differential equation

$$\text{(1.3.7)} \qquad \begin{cases} y'(t) = f(t, y(t), Ky(t)), \text{ a.e. } t \in [0,T] \\ y(0) = y_0. \end{cases}$$

For detailed existence principles on (1.3.7) we refer the reader to [137, Chapter2].

We now concentrate on (1.3.1) defined on the half-open interval $[0,T)$, where $0 \leq T \leq \infty$, that is, we examine the functional-differential equation

$$\text{(1.3.8)} \qquad \begin{cases} y'(t) = Vy(t), \text{ a.e. } t \in [0,T) \\ y(0) = y_0. \end{cases}$$

Here we assume that the operator $V : C[0,T) \to L^p_{loc}[0,T)$. Our first existence principle follows from the Schauder-Tychonoff fixed point theorem which we state for completeness.

Theorem 1.3.2 The Schauder-Tychonoff Fixed Point Theorem.
Let D be a closed, convex subset of a locally convex Hausdorff space E. Assume that $N : D \to D$ is continuous and $N(D)$ is relatively compact in E. Then N has at least one fixed point in D.

Theorem 1.3.3. *Let $1 \leq p \leq \infty$ and D be a closed, convex subset of $C[0,T)$. Assume that*

$$\text{(1.3.9)} \qquad V : C[0,T) \to L^p_{loc}[0,T) \text{ is a continuous operator,}$$

$$\text{(1.3.10)} \qquad \begin{cases} \text{for all } y \in D \text{ there exists } \mu \in L^p_{loc}[0,T), \\ \text{with } |Vy(t)| \le \mu(t) \text{ for a.e. } t \in [0,T) \end{cases}$$

and

$$\text{(1.3.11)} \qquad \begin{cases} S : D \to D \text{ where } S : C[0,T) \to C[0,T) \\ \text{is given by } Sy(t) := y_0 + \int_0^t Vy(s)\,ds \end{cases}$$

hold. Then (1.3.8) *has at least one solution* $y \in AC_{loc}[0,T)$.

PROOF: A solution $y \in AC_{loc}[0,T)$ of (1.3.8) is a fixed point of the operator $S : C[0,T) \to C[0,T)$. It is easy to show that $S : D \to D$ is continuous, that is, $\rho_{0,m}(y_n) \to \rho_{0,m}(y)$ implies that $\rho_{0,m}(Sy_n - Sy) \to 0$ for each m. In addition one can show that $S(D)$ is relatively compact in $C[0,T)$ by proving, using the Arzéla-Ascoli Theorem, that $S(D)$ is uniformly bounded and equicontinuous on each compact subinterval of $[0,T)$. We omit the detail as it is similar to that in the proof of Theorem 1.3.1. The result now follows from the Schauder-Tychonoff theorem. □

As in the case of the functional equations discussed in Section 1.2, very often one can obtain an improved existence theory if it is assumed that the operator in question – in this case $V : C[0,T) \to L^p_{loc}[0,T)$ – is an abstract Volterra operator. In our second existence principle for (1.3.8) we assume that $V : C[0,T) \to L^p_{loc}[0,T)$ is an abstract Volterra operator and use an idea of Lee and O'Regan [94] to prove the existence of a solution $y \in AC_{loc}[0,T)$.

Theorem 1.3.4. *Let* $1 \le p \le \infty$ *and suppose that* $0 < t_1 < t_2 < \ldots < t_n < \ldots$, *with* $t_n \uparrow T$. *Assume that*

$$\text{(1.3.12)} \qquad \begin{cases} V : C[0,T) \to L^p_{loc}[0,T) \text{ is a continuous,} \\ \text{abstract Volterra operator,} \end{cases}$$

$$\text{(1.3.13)} \qquad \begin{cases} \text{for any constant } r_k > 0, \text{ there exists } \mu_{r_k} \in L^p[0,t_k], \\ \text{such that for any } y \in C[0,t_k], \text{ with } \rho_{0,k}(y) \le r_k, \\ \text{we have } |Vy(t)| \le \mu_{r_k}(t), \text{ for a.e. } t \in [0,t_k] \end{cases}$$

and

$$(1.3.14)\quad \begin{cases} \textit{suppose for each } n = 1, 2, \ldots, \textit{ there exists } y_n \in AC[0, t_n], \\ \textit{that solves} \\ \begin{cases} y_n'(t) = V y_n(t), \ \textit{a.e. } t \in [0, t_n] \\ y_n(0) = y_0 \end{cases} \\ \textit{and that there are bounded sets } B_k \subseteq \mathbf{R} \textit{ for } k = 1, 2, \ldots, \\ \textit{such that } n \geq k \textit{ implies } y_n(t) \in B_k \textit{ for } t \in [0, t_k] \end{cases}$$

hold. Then (1.3.8) *has a solution* $y \in AC_{loc}[0, T)$ *such that* $y_n(t) \in \overline{B}_k$ *on* $[0, t_k]$ *for each* k.

PROOF: The proof is similar to the proof of Theorem 1.2.16 in Section 1.2 but we include some of the details for completeness. It is easy to show from (1.3.13) and (1.3.14) that we have

$$(1.3.15)\quad \begin{cases} \text{for each } k \in \{1, 2, \ldots\}, \text{ the sequence } \{y_n\}_{n \geq k} \\ \text{is uniformly bounded and equicontinuous on } [0, t_k]. \end{cases}$$

This and the Arzéla-Ascoli theorem yield a subsequence N_1 of $\{1, 2, \ldots\}$ and a function $z_1 \in C[0, t_1]$ such that $y_n \to z_1$ uniformly on $[0, t_1]$ as $n \to \infty$ in N_1. Let $N_2^\star = N_1 \backslash \{1\}$. Then once again, (1.3.15) and the Arzéla-Ascoli theorem yield a subsequence N_2 of $N_2^\star$ and a function $z_2 \in C[0, t_2]$ such that $y_n \to z_2$ uniformly on $[0, t_2]$ as $n \to \infty$ in N_2. Proceed inductively to obtain a subsequence of integers

$$N_1 \supseteq N_2 \supseteq \ldots \supseteq N_k \supseteq \ldots, N_k \subseteq \{k, k+1, \ldots\},$$

and functions $z_k \in C[0, t_k]$ with

$$y_n \to z_k \text{ uniformly on } [0, t_k] \text{ as } n \to \infty \text{ in } N_k,$$

and $z_{k+1} = z_k$ on $[0, t_k]$ for $k = 1, 2, \ldots$.

Define a function $y : [0, T) \to \mathbf{R}$ by $y(t) = z_k(t)$ for $t \in [0, t_k]$. Fix $t \in [0, T)$ and choose and fix k with $t \in [0, t_k]$. Now (1.3.12) and (1.3.14) along with the Lebesgue dominated convergence theorem imply that

$$z_k(t) = y_0 + \int_0^t V z_k(s)\, ds,$$

and therefore

$$y(t) = y_0 + \int_0^t V y(s)\, ds.$$

Thus $y \in AC_{loc}[0,T)$ solves (1.3.8). □

1.4. Singular functional-differential equations.

Functional-differential equations of the form discussed in Section 1.3, namely,

$$(1.4.1) \qquad \begin{cases} y'(t) = Vy(t), \text{ a.e. } t \in I \\ y(0) = y_0, \end{cases}$$

where V is an operator and $I = [0,T]$ or $[0,T)$. have received growing attention in the literature over the past thirty years [46,51]. In the previous section we presented some very general, well-known existence principles for (1.4.1) when the operator V satisfies

$$(1.4.2) \qquad V : C[0,T] \to L^p[0,T], \ 1 \le p \le \infty$$

or

$$(1.4.3) \qquad V : C[0,T) \to L^p_{loc}[0,T), \ 1 \le p \le \infty, \ 0 \le T \le \infty,$$

and a solution $y \in AC[0,T]$ or $y \in AC_{loc}[0,T)$ respectively, of (1.4.1) is sought. Indeed, in much of the literature the operator V in (1.4.1) is assumed to be an abstract Volterra operator (see Definition 1.2.2).

Two classic examples of Volterra operators are

$$(1.4.4) \qquad Vy(t) = f(t, y(t))$$

– a Niemytzki operator (a very special type of Volterra operator), and

$$(1.4.5) \qquad Vy(t) = g(t) + \int_0^t k(t,s) f(s, y(s))\, ds$$

– a Volterra integral operator. With V as described in (1.4.4), we see that (1.4.1) reduces to the Cauchy problem, while if V is as in (1.4.5), we have that (1.4.1) is an integrodifferential equation of the kind studied in [104,137].

The underlying spaces given in (1.4.2) and (1.4.3) are typical of those considered in the study of (1.4.1) in general, and indeed in the study of (1.4.1) when V is as described in (1.4.4) or (1.4.5). These spaces in turn encourage us to find solutions that lie in either $AC[0,T]$ or $AC_{loc}[0,T)$. In this section we wish to examine what happens if the operator V is not as "well-behaved" as the V's described in the examples above, implying in turn that the underlying spaces are not "as nice" as $C[0,T]$, $C[0,T)$, $L[0,T]$ or $L_{loc}[0,T)$.

Formally speaking, we wish to examine a variation of (1.4.1) namely

$$\begin{cases} y'(t) = Vy(t), \text{ a.e. } t \in I \\ \lim_{t\to 0^+} h(t)y(t) = a, \end{cases} \tag{1.4.6}$$

where $I = (0,T]$ or $(0,T)$, and V is a *singular* or *singular Volterra* operator.

Definition 1.4.1. Let $E[0,T]$ be a normed space with norm given by $|.|_E$. An operator $V : E[0,T] \to L[0,T]$ is a *Carathéodory operator* if it satisfies the following conditions:

(*i*) V is continuous,

(*ii*) for any $r > 0$, there exists $\mu_r \in L[0,T]$ such that for any $y \in E[0,T]$ with $|y|_E \le r$, we have $|Vy(t)| \le \mu_r(t)$, for a.e. $t \in [0,T]$.

Definition 1.4.2. Let $I = [0,T]$ with one or both endpoints excluded. An operator $V : E[0,T] \to L_{loc}(I)$ is a *local Carathéodory operator* if it satisfies the following conditions:

(*i*) V is continuous,

(*ii*) for any $r > 0$, there exists $\mu_r \in L_{loc}(I)$ such that for all $y \in E[0,T]$ with $|y|_E \le r$, we have $|Vy(t)| \le \mu_r(t)$, for a.e. $t \in I$.

Definition 1.4.3. Let μ_r be as described in either of the two definitions above. If

$$\int_0^T \mu_r(t)\,dt < \infty,$$

then V is a *regular operator.* Otherwise we say that V is a *singular operator.* (Note that the operator V described in Definition 1.4.1 is a regular operator.)

The main functional-differential equation of interest to us in this section is (1.4.6). We assume that $h \in C[0,T]$, $a \in \mathbf{R}$ and that the operator V satisfies

$$V : E_h[0,T] \to L_{loc}(0,T], \tag{1.4.7}$$

where $E_h[0,T]$ will be described shortly. Equations of the form (1.4.6) are discussed in [86] and solutions $y \in C[0,T]$ are sought. However, prompted by the initial condition in (1.4.6) we define

$$E_h[0,T] = \{y \in C(0,T] : hy \in C[0,T]\} \tag{1.4.8}$$

and seek a solution y of (1.4.6) which resides in this weighted Banach space $E_h[0,T]$, with norm given by

$$|y|_{0,h} = |hy|_0 = \sup_{t\in[0,T]} |h(t)y(t)|.$$

For V a Niemytzki operator as in (1.4.4) there is no difficulty in having V operate on $E_h[0,T]$ (or indeed $C(0,T]$), but if V is a Volterra integral operator as described in (1.4.5) and we insist that (1.4.7) is satisfied, then obviously a strong dependence of V on h is implied. In general, for an abstract operator V to satisfy (1.4.7), the reliance of V on h is inherent.

However before considering (1.4.6) we will first examine the functional-differential equation

$$\text{(1.4.9)} \qquad \begin{cases} (h(t)y(t))' = Ny(t), \text{ a.e. } t \in [0,T] \\ \lim_{t\to 0^+} h(t)y(t) = a, \end{cases}$$

where $N : E_h[0,T] \to L[0,T]$ is a regular operator. An existence principle follows easily for (1.4.9) by incorporating the weighted space $E_h[0,T]$ with a result from [104,138].

Theorem 1.4.1. *Suppose*

$$\text{(1.4.10)} \qquad h \in C[0,T] \textit{ with } h(t) \neq 0 \textit{ for } t \in (0,T]$$

and

$$\text{(1.4.11)} \qquad N : E_h[0,T] \to L[0,T] \textit{ is a Carathéodory operator}$$

hold. In addition, suppose that there exists a constant $M > 0$, independent of λ, with

$$|a| < M \textit{ and } |y|_{0,h} = |hy|_0 \neq M,$$

for any solution $y \in E_h[0,T]$ of

$$\text{(1.4.12)}_\lambda \qquad \begin{cases} (h(t)y(t))' = \lambda Ny(t), \text{ a.e. } t \in [0,T] \\ \lim_{t\to 0^+} h(t)y(t) = a, \end{cases}$$

for each $\lambda \in (0,1)$. Then (1.4.9) *has at least one solution $y \in C(0,T]$ with $hy \in AC[0,T]$.*

PROOF: By (1.4.11), solving $(1.4.12)_\lambda$ is equivalent to finding $y \in E_h[0,T]$ that satisfies

$$h(t)y(t) = a + \lambda \int_0^t Ny(s)\,ds, \quad t \in [0,T],$$

which on rewriting becomes

$$y(t) = (1-\lambda)\frac{a}{h(t)} + \lambda\left(\frac{a}{h(t)} + \frac{1}{h(t)}\int_0^t Ny(s)\,ds,\right), \quad t \in (0,T].$$

Defining

$$\tilde{N}y(t) = \frac{a}{h(t)} + \frac{1}{h(t)}\int_0^t Ny(s)\,ds, \quad t \in (0,T],$$

it is easy to check that $\tilde{N} : E_h[0,T] \to E_h[0,T]$ is well-defined.

To apply the nonlinear alternative we must verify the continuity and complete continuity of $\tilde{N} : E_h[0,T] \to E_h[0,T]$. Since by (1.4.11), $N : E_h[0,T] \to L[0,T]$ is a continuous operator, it follows that if $y_n \to y$ in $E_h[0,T]$, then

$$\begin{aligned} |h(t)\tilde{N}y_n(t) - h(t)\tilde{N}y(t)| &\leq \int_0^t |Ny_n(s) - Ny(s)|\,ds \\ &\leq \int_0^T |Ny_n(s) - Ny(s)|\,ds \to 0, \end{aligned}$$

thus illustrating the continuity of $\tilde{N} : E_h[0,T] \to E_h[0,T]$.

We now show the complete continuity of $\tilde{N} : E_h[0,T] \to E_h[0,T]$. Let $\Omega \subseteq E_h[0,T]$ be bounded, that is, there exists $r > 0$ such that $|y|_{0,h} = |hy|_0 \leq r$ for each $y \in \Omega$. For $y \in \Omega$,

$$|\tilde{N}y|_{0,h} = |h\tilde{N}y|_0 \leq |a| + \int_0^T \mu_r(s)\,ds \equiv M_0 < \infty$$

implying that $\tilde{N}\Omega$ is uniformly bounded in $E_h[0,T]$. Also for $y \in \Omega$ and t, $t^\star \in [0,T]$, $t < t^\star$,

$$|h(t^\star)\tilde{N}y(t^\star) - h(t)\tilde{N}y(t)| \leq \int_t^{t^\star} |Ny(s)|\,ds \leq \int_t^{t^\star} \mu_r(s)\,ds \to 0 \text{ as } t \to t^\star.$$

Consequently by the Arzéla-Ascoli theorem, $\tilde{N} : E_h[0,T] \to E_h[0,T]$ is completely continuous.

Applying the nonlinear alternative with $U = \{y \in E_h[0,T] : |y|_{0,h} < M\}$, $C = E = E_h[0,T]$, $G = \tilde{N}$ and $p^\star = a/h$, we notice that (ii) cannot occur. Consequently, $\tilde{N}$ has a fixed point in $E_h[0,T]$ or equivalently (1.4.9) has a solution $y \in E_h[0,T]$ with $hy \in AC[0,T]$. □

With Theorem 1.4.1 established we are now in a better position to examine (1.4.6). We note that finding a solution $y \in E_h[0,T]$ of (1.4.6) is equivalent (under appropriate conditions) to finding a solution $y \in E_h[0,T]$ of

$$(1.4.13) \qquad \begin{cases} (h(t)y(t))' = h'(t)y(t) + h(t)Vy(t), \text{ a.e. } t \in (0,T] \\ \lim_{t\to 0^+} h(t)y(t) = a. \end{cases}$$

Therefore defining

$$(1.4.14) \qquad Ny(t) := h'(t)y(t) + h(t)Vy(t),$$

we see that we are able, in certain circumstances, to write (1.4.6) in the form of (1.4.9). In particular, if h and V are such that

$$(1.4.15) \qquad N : E_h[0,T] \to L[0,T],$$

and the hypotheses of Theorem 1.4.1 are satisfied, then it will follow that (1.4.6) has a solution $y \in E_h[0,T]$ with $hy \in AC[0,T]$. We now state this formally and illustrate with an example.

Theorem 1.4.2. *Suppose that*

$$(1.4.16) \qquad h \in AC[0,T] \text{ and } h(t) \neq 0 \text{ for } t \in (0,T]$$

and

$$(1.4.17) \qquad \begin{cases} \text{the operator } V : E_h[0,T] \to L_{loc}(0,T] \text{ is} \\ \text{such that the operator } N, \text{ defined by} \\ Ny(t) = h'(t)y(t) + h(t)Vy(t), \\ \text{satisfies } N : E_h[0,T] \to L[0,T], \\ \text{and is a Carathéodory operator} \end{cases}$$

hold. In addition, suppose that there exists a constant $M > 0$, independent of λ, with

$$|a| < M \text{ and } |y|_{0,h} = |hy|_0 \neq M,$$

for any solution $y \in E_h[0,T]$ of

$$(1.4.18)_\lambda \qquad \begin{cases} (h(t)y(t))' = \lambda\,(h'(t)y(t) + h(t)Vy(t)), \text{ a.e. } t \in [0,T] \\ \lim_{t\to 0^+} h(t)y(t) = a, \end{cases}$$

for each $\lambda \in (0,1)$. Then (1.4.6) has at least one solution $y \in C(0,T]$ with $hy \in AC[0,T]$.

PROOF: The proof follows directly from Theorem 1.4.1. □

In many of the examples that we wish to consider, the weight function is given by $h(t) = t^n$, $n > 0$. We now examine one such example which satisfies the hypotheses of Theorem 1.4.2 and which has also been discussed in [86].

Example 1.4.1. Consider

$$\text{(1.4.19)} \qquad \begin{cases} y'(t) = Vy(t) := -\frac{n}{t}y(t) + mt^{m-1-n}, & t \in (0,T] \\ \lim_{t\to 0^+} t^n y(t) = 0, \end{cases}$$

where $n > 0$, $m \in (0,n)$ and V is a singular operator. This problem has a unique solution $y(t) = t^{m-n}$, which of course lies in $E_h[0,T]$, where $h(t) = t^n$. Here

$$\begin{aligned} Ny(t) &= h'(t)y(t) + h(t)Vy(t) \\ &= nt^{n-1}y(t) + t^n\left(-\frac{n}{t}y(t) + mt^{m-1-n}\right) = mt^{m-1} \end{aligned}$$

which clearly satisfies (1.4.17).

Further examples will be discussed after we make the following observation. We note that in the proof of Theorem 1.4.1, we were able to integrate the functional-differential equation (1.4.9), and hence write it as a functional equation of the form

$$\text{(1.4.20)} \qquad h(t)y(t) = a + N^\star y(t), \quad t \in [0,T],$$

where $N^\star : E_h[0,T] \to C[0,T]$, or alternatively,

$$\text{(1.4.21)} \qquad y(t) = \tilde{N}y(t) := \frac{a}{h(t)} + \frac{1}{h(t)}N^\star y(t), \quad t \in (0,T],$$

where $\tilde{N} : E_h[0,T] \to E_h[0,T]$. It transpires that for certain operators V, it is more beneficial to rewrite (1.4.6) directly in the form (1.4.20), and then use an existence principle for (1.4.20) to establish the existence of a solution of (1.4.6). We will better illustrate this point in the next example. However, we first state the following existence principle for (1.4.20) :

Theorem 1.4.3. *Suppose* (1.4.10) *and*

$$\text{(1.4.22)} \qquad \begin{cases} N^\star : E_h[0,T] \to C[0,T] \text{ is a continuous} \\ \text{and completely continuous operator} \end{cases}$$

hold. In addition, suppose that there exists a constant $M > 0$, *independent of* λ, *with* $|a| < M$ *and* $|y|_{0,h} = |hy|_0 \neq M$, *for any solution* $y \in E_h[0,T]$ *of*

$$h(t)y(t) = a + \lambda N^\star y(t), \quad t \in [0,T],$$

for each $\lambda \in (0,1)$. *Then* (1.4.20) *has at least one solution* $y \in C(0,T]$ *with* $hy \in C[0,T]$.

PROOF: It is easy to see that $\tilde{N} := \frac{a}{h(t)} + \frac{1}{h(t)} N^\star : E_h[0,T] \to E_h[0,T]$ is a continuous and completely continuous operator. The proof now follows by using the nonlinear alternative in a similar fashion to that described in the proof of Theorem 1.4.1. □

Remark 1.4.1. The solution $y \in C(0,T]$ of (1.4.20) guaranteed by Theorem 1.4.3 satisfies $hy \in C[0,T]$. However, in this case it is not necessarily true that $(hy)' \in L[0,T]$, that is, unlike the solution obtained for (1.4.9) in Theorem 1.4.1, here hy may not belong to $AC[0,T]$.

We now present two applications of Theorem 1.4.3.

Example 1.4.2. Consider

$$\text{(1.4.23)} \quad \begin{cases} y'(t) = Vy(t) := -\dfrac{1}{t^3} \displaystyle\int_0^t [sy(s) + As|sy(s)|^{\frac{1}{n}}]\, ds, & t \in (0,T] \\ \lim_{t\to 0+} ty(t) = a, \end{cases}$$

where $1 < n \in \mathbf{N}$, $A, a \in \mathbf{R}$ and V is a singular Volterra operator. If y is a solution of (1.4.23), then by differentiating we see that

$$y''(t) + \frac{3y'(t)}{t} + \frac{y(t)}{t^2} = -\frac{A}{t^2} |ty(t)|^{\frac{1}{n}}.$$

Using the method of variation of parameters, we obtain

$$y(t) = \frac{C_1}{t} + \frac{C_2 \ln t}{t} + \frac{A}{t} \int_0^t \ln s\, |sy(s)|^{\frac{1}{n}}\, ds - \frac{A \ln t}{t} \int_0^t |sy(s)|^{\frac{1}{n}}\, ds, \quad t \in (0,T],$$

that is,

$$\text{(1.4.24)} \quad ty(t) = C_1 + C_2 \ln t + A \int_0^t \ln s\, |sy(s)|^{\frac{1}{n}}\, ds - A \ln t \int_0^t |sy(s)|^{\frac{1}{n}}\, ds, \quad t \in (0,T].$$

We seek a solution $y \in E_t[0,T] = \{y \in C(0,T] : ty \in C[0,T]\}$ of (1.4.24), and therefore we let $C_2 = 0$. Also since the initial condition in (1.4.23) must be satisfied and since for $y \in E_t[0,T]$, $\lim_{t\to 0+} \int_0^t \ln s |sy(s)|^{\frac{1}{n}}\, ds = \lim_{t\to 0+} \ln t \int_0^t |sy(s)|^{\frac{1}{n}}\, ds = 0$, we choose $C_1 = a$ and can rewrite (1.4.24) as

$$\text{(1.4.25)} \qquad ty(t) = a - A \int_0^t \frac{1}{s} \int_0^s |zy(z)|^{\frac{1}{n}}\, dz\, ds, \quad t \in [0,T].$$

Defining

$$N^\star y(t) := -A\int_0^t \frac{1}{s}\int_0^s |zy(z)|^{\frac{1}{n}}\,dz\,ds, \quad t\in[0,T],$$

we obtain

$$ty(t) = a + N^\star y(t), \quad t\in[0,T],$$

and consequently (1.4.23) is written in the form (1.4.20).

It is easy to check that $N^\star : E_t[0,T] \to C[0,T]$ is a continuous operator. To check the complete continuity of $N^\star$, let Ω be a bounded set in $E_t[0,T]$, that is, there exists $r > 0$ such that $|y|_{0,t} = |ty|_0 \le r$, for all $y\in\Omega$. Clearly $N^\star\Omega$ is uniformly bounded since for $y\in\Omega$,

$$|N^\star y|_0 \le |A|\int_0^t \frac{1}{s}\int_0^s |zy(z)|^{\frac{1}{n}}\,dz\,ds \le |A|Tr^{\frac{1}{n}}.$$

Letting t_1, $t_2\in[0,T]$, with $t_2 < t_1$, we have that

$$\begin{aligned}|N^\star y(t_1) - N^\star y(t_2)| &\le |A|\int_{t_2}^{t_1} \frac{1}{s}\int_0^s |zy(z)|^{\frac{1}{n}}\,dz\,ds\\ &\le |A|(t_1-t_2)r^{\frac{1}{n}} \to 0 \text{ as } t_1\to t_2,\end{aligned}$$

thus illustrating the equicontinuity of $N^\star\Omega$. The complete continuity of $N^\star : E_t[0,T] \to C[0,T]$ now follows from the Arzéla-Ascoli theorem. In summary, we have that $N^\star : E_t[0,T] \to C[0,T]$ is a continuous and completely continuous operator and thus satisfies (1.4.22) of Theorem 1.4.3.

The existence of a solution $y\in E_t[0,T]$ of (1.4.23) will now follow from Theorem 1.4.3 if we show that there exists an $M>0$, such that any solution $y\in E_t[0,T]$ of

$$\text{(1.4.26)}\quad ty(t) = a + \lambda N^\star y(t) = a - \lambda A\int_0^t \frac{1}{s}\int_0^s |zy(z)|^{\frac{1}{n}}\,dz\,ds, \quad t\in[0,T]$$

satisfies $|ty|_0 \ne M$. Since

$$|ty|_0 \le |a| + |A|T|ty|_0^{\frac{1}{n}},$$

the existence of such an M is clear (that is, one can find $M>0$ such that $M/(|a| + |A|TM^{\frac{1}{n}}) > 1$).

Note that if $A = 0$ in (1.4.23), then $y = a/t$ is a unique solution of the problem and clearly $y\in E_t[0,T]$.

Another result, this time concerning a singular boundary value problem, different to the type discussed in [84] but related to Example 1.4.2, is given in the following example.

Example 1.4.3. Consider

$$(1.4.47)\qquad \begin{cases} y''(t) + \dfrac{3y'(t)}{t} + \dfrac{y(t)}{t^2} = -\dfrac{A}{t^2}|ty(t)|^{\frac{1}{n}}, & t \in [0,1], \\ \lim_{t\to 0^+} ty(t) \text{ exists }, y(1) = 1, \end{cases}$$

where $1 < n \in \mathbf{N}$, and $A \in \mathbf{R}$. Notice that

$$\begin{aligned} ty(t) &= 1 - A\int_t^1 \ln s |sy(s)|^{\frac{1}{n}}\, ds - A\ln t \int_0^t |sy(s)|^{\frac{1}{n}}\, ds \\ &= 1 + A\int_t^1 \frac{1}{s}\int_0^s |zy(z)|^{\frac{1}{n}}\, dz\, ds, \quad t \in [0,1], \end{aligned}$$

or if let

$$N^\star y(t) := A\int_t^1 \frac{1}{s}\int_0^s |zy(z)|^{\frac{1}{n}}\, dz\, ds, \quad t \in [0,1],$$

we see that

$$(1.4.28)\qquad ty(t) = 1 + N^\star y(t), \quad t \in [0,1].$$

By essentially the same reasoning as in Example 1.4.2, we obtain that (1.4.28) has a solution $y \in C(0,1]$ with $ty \in C[0,1]$. Hence (1.4.27) has a solution $y \in E_t[0,1]$. (Unlike Example 1.4.2, the $N^\star : E_t[0,1] \to C[0,1]$ described above is not a Volterra operator. However, Example 1.4.2 and the example above, follow from Theorem 1.4.3 which doesn't discriminate against operators which are not Volterra.)

To complete the note, we remark that one can also consider a variation of (1.4.6) defined on the open interval $(0,T)$, $0 \le T \le \infty$, namely

$$(1.4.29)\qquad \begin{cases} y'(t) = Vy(t), \text{ a.e. } t \in (0,T) \\ \lim_{t\to 0^+} h(t)y(t) = a \end{cases}$$

where

$$V : E_h[0,T) \to L_{loc}(0,T).$$

Here $E_h[0,T)$ is the Fréchet space

$$E_h[0,T) = \{y \in C(0,T) : hy \in C[0,T)\},$$

where for every $m \in \{1,2,\ldots\}$, the seminorm $\rho_{m,h}(y)$ is defined by

$$\rho_{m,h}(y) = \sup_{t\in[0,t_m]} |h(t)y(t)|, \text{ where } t_m \uparrow T.$$

As in tackling (1.4.6), when we examined (1.4.9) first, here it helps to first consider the somewhat simpler problem

$$(1.4.30)\qquad \begin{cases} (h(t)y(t))' = Ny(t), \text{ a.e. } t \in [0,T) \\ \lim_{t\to 0+} h(t)y(t) = a, \end{cases}$$

where $N : E_h[0,T) \to L_{loc}[0,T)$.

We just state the following two existence principles for (1.4.30), since the proofs follow by modifying the ideas presented in [104,137] to take into consideration the space $E_h[0,T)$.

Theorem 1.4.4. *Let K be a closed, convex subset of $E_h[0,T)$, with*

$$(1.4.31)\qquad h \in C[0,T) \textit{ with } h(t) \neq 0 \textit{ for } t \in (0,T),$$

$$(1.4.32)\qquad N : E_h[0,T) \to L_{loc}[0,T) \textit{ a continuous operator,}$$

$$(1.4.33)\qquad \begin{cases} \textit{for all } y \in K, \textit{ there exists } a \in L_{loc}[0,T) \\ \textit{with } |Ny(t)| \leq a(t) \textit{ almost everywhere on } [0,T) \end{cases}$$

and

$$(1.4.34)\qquad \begin{cases} \tilde{N} : K \to K \textit{ where } \tilde{N} : E_h[0,T) \to E_h[0,T) \textit{ is the operator} \\ \textit{defined by } \tilde{N}y(t) = \dfrac{a}{h(t)} + \dfrac{1}{h(t)}\displaystyle\int_0^t Ny(s)\,ds, \quad t \in (0,T) \end{cases}$$

hold. Then (1.4.30) has at least one solution $y \in C(0,T)$ with $hy \in AC_{loc}[0,T)$.

Theorem 1.4.5. *Let $0 < t_1 < t_2 < \ldots < t_n < \ldots$, with $t_n \uparrow T$. Suppose (1.4.31),*

$$(1.4.35)\qquad N : E_h[0,T) \to L_{loc}[0,T) \textit{ is a continuous Volterra operator,}$$

$$(1.4.36)\qquad \begin{cases} \textit{for any constant } A_k > 0, \textit{ there exists } a_{A_k} \in L[0,t_k] \\ \textit{such that for any } y \in E_h[0,t_k], \textit{ with } \rho_{k,h}(y) \leq A_k, \\ \textit{we have } |Ny(t)| \leq a_{A_k}(t), \textit{ a.e. } t \in [0,t_k] \end{cases}$$

and

$$(1.4.37)\qquad \begin{cases} \textit{suppose for each } n = 1,2,\ldots, \textit{ there exists } y_n \in E_h[0,t_n] \\ \textit{that solves} \\ \quad \begin{cases} y_n'(t) = Ny_n(t), \textit{ a.e. } t \in [0,t_n] \\ \lim_{t\to 0+} h(t)y_n(t) = a \end{cases} \\ \textit{and that there are bounded sets } B_k \subseteq \mathbf{R} \textit{ for } k = 1,2,\ldots, \\ \textit{such that } n \geq k \textit{ implies } h(t)y_n(t) \in B_k \textit{ for } t \in [0,t_k] \end{cases}$$

hold. Then (1.4.30) has at least one solution $y \in C(0,T)$ *with* $hy \in AC_{loc}[0,T)$ *and* $h(t)y(t) \in \overline{B_k}$ *on* $[0, t_k]$, *for each* k.

Remark 1.4.2. Theorem 1.4.5 is the first theorem in this section where we have specifically requested that the operator V be of Volterra type.

It is now clear that if one can rewrite (1.4.29) in form of (1.4.30), that is,

$$\begin{cases} (h(t)y(t))' = Ny(t) := h'(t)y(t) + h(t)Vy(t), \text{ a.e. } t \in [0,T) \\ \lim_{t\to 0^+} h(t)y(t) = a, \end{cases}$$

where h and N satisfy the hypotheses of either Theorem 1.4.4 or Theorem 1.4.5, then (1.4.29) has at least one solution $y \in C(0,T)$ with $hy \in AC_{loc}[0,T)$. We omit the detail.

Remark 1.4.3. As in the first half of this section, existence principles can also be presented for

$$h(t)y(t) = a + N^\star y(t), \quad t \in [0,T), \tag{1.4.38}$$

and additional results then discussed for (1.4.29).

CHAPTER 2

MULTIPLE NONNEGATIVE SOLUTIONS OF NONLINEAR INTEGRAL EQUATIONS

2.1. Introduction.

This chapter is dedicated to proving the existence of multiple, nonnegative, continuous or L^p solutions of certain classes of nonlinear, Fredholm integral equations defined on either a compact or semi-infinite interval.

In Section 2.2 we consider the nonlinear Fredholm integral equation

$$y(t) = h(t) + \int_0^T k(t,s) f(y(s))\, ds, \quad t \in [0,T]. \tag{2.1.1}$$

Two existence results are presented for (2.1.1) which both guarantee the existence of *at least one* nonnegative solution $y \in C[0,T]$ of (2.1.1). The first result is proved using Krasnoselskii's fixed point theorem, while the proof of the second relies on the nonlinear alternative. These results are then combined to provide a *multiple solutions* existence result for (2.1.1).

The integral equation

$$y(t) = h(t) + \int_0^\infty k(t,s) f(y(s))\, ds, \quad t \in [0,\infty), \tag{2.1.2}$$

is studied in Section 2.3. Analogue results of those discussed in the previous section are presented, where now the existence of one or more nonnegative solutions $y \in BC[0,\infty)$ of (2.1.2), such that $\lim_{t\to\infty} y(t)$ exists, is established. A more general theorem which ensures the existence of at least two nonnegative solutions $y_1, y_2 \in BC[0,\infty)$ of (2.1.2) (and relies on the results already established for (2.1.1) along with an inductive argument using the Arzéla-Ascoli theorem) completes Section 2.3.

Finally in Section 2.4, we consider the Fredholm integral equation

$$y(t) = h(t) + \int_0^T k(t,s) f(s, y(s))\, ds, \ \text{a.e. } t \in [0,T], \tag{2.1.3}$$

and use Krasnoselskii's fixed point theorem on this occasion, to establish conditions under which (2.1.3) will have at least one positive solution $y \in L^p[0,T]$, $1 \le p < \infty$. In fact we first consider (2.1.3) when $h \equiv 0$, that is,

$$(2.1.4) \qquad y(t) = \int_0^T k(t,s) f(s, y(s))\, ds, \ \text{a.e. } t \in [0,T],$$

since the analysis is slightly more straight forward in this case.

We lastly consider the nonlinear integral equation on the half line, namely

$$(2.1.5) \qquad y(t) = h(t) + \int_0^\infty k(t,s) f(s, y(s))\, ds, \ \text{a.e. } t \in [0,\infty),$$

which can be dealt with in a similar manner to (2.1.3) and (2.1.4).

Throughout this chapter and the next we will make repeated use of Krasnoselskii's fixed point theorem which we now state.

Theorem 2.1.1. Krasnoselskii's Fixed Point Theorem.
Let E be a Banach space, and let $C \subset E$ be a cone in E. Assume Ω_1, Ω_2 are open subsets of E with $0 \in \Omega_1$, $\overline{\Omega}_1 \subset \Omega_2$, and let

$$K : C \cap (\overline{\Omega}_2 \backslash \Omega_1) \to C$$

be a completely continuous operator such that either

$$(i) \ ||Ku|| \le ||u||, \ u \in C \cap \partial\Omega_1 \text{ and } ||Ku|| \ge ||u||, \ u \in C \cap \partial\Omega_2$$

or

$$(ii) \ ||Ku|| \ge ||u||, \ u \in C \cap \partial\Omega_1 \text{ and } ||Ku|| \le ||u||, \ u \in C \cap \partial\Omega_2$$

is true. The K has a fixed point in $C \cap (\overline{\Omega}_2 \backslash \Omega_1)$.

2.2. Multiple nonnegative continuous solutions on a compact interval.

In this section we are interested in proving the existence of multiple solutions of the nonlinear integral equation

$$(2.2.1) \qquad y(t) = h(t) + \int_0^T k(t,s) f(y(s))\, ds, \ \ t \in [0,T],$$

where $0 \le T < \infty$. By a solution of (2.2.1) we mean $y \in C[0,T]$ such that y satisfies (2.2.1). In particular we want to establish the existence of

multiple nonnegative solutions of (2.2.1). However, before tackling this problem it is reasonable to first consider what conditions one requires on h, k and f in order for (2.2.1) to have at least **one nonnegative** solution $y \in C[0,T]$. The following theorem (whose proof relies on Krasnoselskii's fixed point theorem) is one such result:

Theorem 2.2.1. *Suppose*

$$0 \le k_t(s) = k(t,s) \in L^1[0,T] \text{ for each } t \in [0,T], \tag{2.2.2}$$

$$\text{the map } t \mapsto k_t \text{ is continuous from } [0,T] \text{ to } L^1[0,T], \tag{2.2.3}$$

$$\begin{cases} \text{there exists } 0 < M < 1,\ \kappa \in L^1[0,T] \text{ and} \\ \text{an interval } [a,b] \subseteq [0,T],\ a < b,\ \text{such that} \\ k(t,s) \ge M\kappa(s) \ge 0,\ t \in [a,b],\ \text{a.e. } s \in [0,T], \end{cases} \tag{2.2.4}$$

$$k(t,s) \le \kappa(s),\ t \in [0,T],\ \text{a.e. } s \in [0,T], \tag{2.2.5}$$

$$\begin{cases} h \in C[0,T] \text{ with } h(t) \ge 0,\ t \in [0,T] \text{ and} \\ \min_{t\in[a,b]} h(t) \ge M|h|_0 = M \sup_{t\in[0,T]} |h(t)|, \end{cases} \tag{2.2.6}$$

$$\begin{cases} f : \mathbf{R} \to \mathbf{R} \text{ is continuous and nondecreasing} \\ \text{with } f(u) > 0 \text{ for } u > 0, \end{cases} \tag{2.2.7}$$

$$\begin{cases} \text{there exists } \alpha > 0 \text{ such that } \dfrac{\alpha}{|h|_0 + K_1 f(\alpha)} > 1 \\ \left(\text{here } K_1 := \sup_{t\in[0,T]} \displaystyle\int_0^T k(t,s)\,ds > 0\right) \end{cases} \tag{2.2.8}$$

and

$$\begin{cases} \text{there exists } \beta > 0,\ \beta \ne \alpha \text{ and } t^\star \in [0,T] \text{ such that} \\ \dfrac{\beta}{h(t^\star) + f(M\beta)\displaystyle\int_a^b k(t^\star,s)\,ds} < 1 \end{cases} \tag{2.2.9}$$

hold. Then (2.2.1) *has at least one nonnegative solution* $y \in C[0,T]$ *and either*

(A) $0 < \alpha < |y|_0 < \beta$ *and* $y(t) \geq M\alpha,\ \ t \in [a,b]$ *if* $\alpha < \beta$

or

(B) $0 < \beta < |y|_0 < \alpha$ *and* $y(t) \geq M\beta,\ \ t \in [a,b]$ *if* $\beta < \alpha$

holds.

PROOF: Define the operator $K : C[0,T] \to C[0,T]$ by

$$Ky(t) := h(t) + \int_0^T k(t,s) f(y(s))\, ds, \quad t \in [0,T]; \tag{2.2.10}$$

the cone C by

$$C := \{y \in C[0,T] : y(t) \geq 0,\ t \in [0,T] \text{ and } \min_{t\in[a,b]} y(t) \geq M|y|_0\};$$

and let Ω_α and Ω_β be given by

$$\Omega_\alpha := \{y \in C[0,T] : |y|_0 < \alpha\}$$

and

$$\Omega_\beta := \{y \in C[0,T] : |y|_0 < \beta\}$$

respectively.

Conditions (2.2.2), (2.2.3), (2.2.6) and (2.2.7) imply that

$$K : C[0,T] \to C[0,T] \text{ is continuous and completely continuous}$$

(See Chapter 1). Also for $y \in C[0,T]$ with $y(t) \geq 0$, $t \in [0,T]$, we have from (2.2.2), (2.2.5), (2.2.6) and (2.2.7) that

$$\begin{cases} |Ky|_0 = \sup_{t\in[0,T]} \left(h(t) + \int_0^T k(t,s) f(y(s))\, ds \right) \\ \qquad \leq |h|_0 + \int_0^T \kappa(s) f(y(s))\, ds \end{cases} \tag{2.2.11}$$

holds. Therefore for $y \in C[0,T]$ with $y(t) \geq 0$, $t \in [0,T]$, this fact along with (2.2.4) and (2.2.6) yields

$$\begin{aligned} \min_{t\in[a,b]} Ky(t) &= \min_{t\in[a,b]} \left(h(t) + \int_0^T k(t,s) f(y(s))\, ds \right) \\ &\geq M \left(|h|_0 + \int_0^T \kappa(s) f(y(s))\, ds \right) \geq M|Ky|_0. \end{aligned}$$

Consequently we have that

$$K : C \to C \text{ is continuous and completely continuous.}$$

The desired result will now follow from Krasnoselskii's fixed point theorem if we show that

$$|Ky|_0 < |y|_0, \quad y \in C \cap \partial\Omega_\alpha \tag{2.2.12}$$

and

$$|Ky|_0 > |y|_0, \quad y \in C \cap \partial\Omega_\beta \tag{2.2.13}$$

are true.

We first show that (2.2.12) holds. Let $y \in C \cap \partial\Omega_\alpha$. In particular this means that $y(t) \geq 0$, $t \in [0,T]$ and $|y|_0 = \alpha$. Then from (2.2.2), (2.2.6), (2.2.7) and (2.2.8) we obtain

$$\begin{aligned}
\sup_{t\in[0,T]} |Ky(t)| &= \sup_{t\in[0,T]} \left(h(t) + \int_0^T k(t,s) f(y(s))\, ds \right) \\
&\leq \sup_{t\in[0,T]} h(t) + f(|y|_0) \sup_{t\in[0,T]} \int_0^T k(t,s)\, ds \\
&\leq |h|_0 + K_1 f(\alpha) < \alpha = |y|_0.
\end{aligned}$$

Therefore (2.2.12) is true.

We secondly show that (2.2.13) is true. Let $y \in C \cap \partial\Omega_\beta$. Then $y(t) \geq 0$, for $t \in [0,T]$, $|y|_0 = \beta$ and for $t \in [a,b]$, $M\beta \leq y(t) \leq \beta$. Now (2.2.2), (2.2.6), (2.2.7) and (2.2.9) imply that

$$\begin{aligned}
Ky(t^\star) &= h(t^\star) + \int_0^T k(t^\star,s) f(y(s))\, ds \\
&\geq h(t^\star) + \int_a^b k(t^\star,s) f(y(s))\, ds \\
&\geq h(t^\star) + f(M\beta) \int_a^b k(t^\star,s)\, ds > \beta = |y|_0.
\end{aligned}$$

Consequently (2.2.13) holds.

Hence by Krasnoselskii's fixed point theorem there exists $y \in C$ that solves (2.2.1). In particular, if $\alpha < \beta$ then $y \in C \cap (\overline{\Omega}_\beta \backslash \Omega_\alpha)$, while $y \in C \cap (\overline{\Omega}_\alpha \backslash \Omega_\beta)$ if $\beta < \alpha$. □

Remark 2.2.1. Condition (2.2.9) may be replaced with

$$(2.2.14)\qquad \begin{cases} \text{there exists } \beta > 0,\ \beta \neq \alpha, \text{ such that,} \\ \dfrac{\beta}{K_2 f(M\beta)} < 1 \text{ where } K_2 := \sup_{t\in[0,T]} \displaystyle\int_a^b k(t,s)\,ds, \end{cases}$$

since (2.2.13) also follows from (2.2.14). To see this note that for $y \in C \cap \partial\Omega_\beta$, (2.2.2), (2.2.6), (2.2.7) and (2.2.14) give

$$\begin{aligned} \sup_{t\in[0,T]} |Ky(t)| &= \sup_{t\in[0,T]} \left(h(t) + \int_0^T k(t,s) f(y(s))\,ds \right) \\ &\geq \sup_{t\in[0,T]} \int_a^b k(t,s) f(y(s))\,ds \geq K_2 f(M\beta) > \beta = |y|_0. \end{aligned}$$

Hence (2.2.14) implies that (2.2.13) is true. In particular if $|h|_0 = 0$, it is better to replace (2.2.9) with (2.2.14).

Remark 2.2.2. It is easy to check that the requirement that $f : \mathbf{R} \to \mathbf{R}$ (as described in (2.2.7)) be nondecreasing can be omitted, provided that (2.2.8) and (2.2.9) be replaced with

$$(2.2.15)\qquad \text{there exists } \alpha > 0 \text{ such that } \frac{\alpha}{|h|_0 + K_1 \sup_{z\in[0,\alpha]} f(z)} > 1$$

and

$$(2.2.16)\qquad \begin{cases} \text{there exists } \beta > 0,\ \beta \neq \alpha \text{ and } t^\star \in [0,T] \text{ such that} \\ \dfrac{\beta}{h(t^\star) + \inf_{z\in[M\beta,\beta]} f(z) \displaystyle\int_a^b k(t^\star,s)\,ds} < 1 \end{cases}$$

respectively.

Example 2.2.1. Consider the separable kernel

$$k(t,s) = k_1(t)k_2(s), \quad t \in [0,T], \text{ a.e. } s \in [0,T],$$

where $k_1 \in C[0,T]$ is such that $0 \leq k_1(t) < 1$, $t \in [0,T]$ with $k_1(t) > 0$ on some interval $[a,b] \subseteq [0,T]$, and $k_2 \in L^1[0,T]$ is such that $k_2(s) \geq 0$ a.e. $s \in [0,T]$. Then k satisfies (2.2.2) – (2.2.5) with $\kappa = k_2$ and $M = \min_{t\in[a,b]} k_1(t)$.

Extracting some of the hypotheses of Theorem 2.2.1 and applying the nonlinear alternative, we have the following existence result which also guarantees the existence of at least one nonnegative solution $y \in C[0,T]$ of (2.2.1). In this result however, it is possible that $y \equiv 0$ may be the solution, whereas in Theorem 2.2.1 we had that the guaranteed solution was positive on some interval $[a,b] \subseteq [0,T]$.

Theorem 2.2.2. *Suppose* (2.2.2), (2.2.3), (2.2.7), (2.2.8) *and*

$$h \in C[0,T] \textit{ with } h(t) \geq 0, \;\; t \in [0,T] \tag{2.2.17}$$

hold. Then (2.1.1) *has at least one nonnegative solution* $y \in C[0,T]$ *such that* $0 \leq y(t) < \alpha$, $t \in [0,T]$.

PROOF: Define

$$f^\star(y) = \begin{cases} f(y), & y \geq 0 \\ f(0), & y < 0 \end{cases}$$

and consider the integral equation

$$y(t) = h(t) + \int_0^T k(t,s) f^\star(y(s))\, ds, \;\; t \in [0,T]. \tag{2.2.18}$$

One can check that the operator $N^\star : C[0,T] \to C[0,T]$ given by

$$N^\star y(t) := h(t) + \int_0^T k(t,s) f^\star(y(s))\, ds, \;\; t \in [0,T]$$

is continuous and completely continuous (see Chapter 1). Let $y \in C[0,T]$ be a solution of

$$y(t) = \lambda \left(h(t) + \int_0^T k(t,s) f^\star(y(s))\, ds \right), \;\; t \in [0,T], \text{ some } \lambda \in (0,1).$$

Certainly $y(t) \geq 0$, $t \in [0,T]$. In addition, (2.2.7) implies that

$$|y|_0 = \sup_{t \in [0,T]} \left(h(t) + \int_0^T k(t,s) f(y(s))\, ds \right) \leq |h|_0 + K_1 f(|y|_0),$$

and consequently we have from (2.2.8) that $|y|_0 \neq \alpha$.

Applying the nonlinear alternative (Theorem 1.2.1) with $C = E = C[0,T]$, $U = \{y \in C[0,T] : |y|_0 < \alpha\}$ and $N = N^\star$, we see that (2.2.18) has at least one solution $y \in C[0,T]$ such that $0 \leq |y|_0 < \alpha$. However since $y(t) \geq 0$, $t \in [0,T]$, it is clear that y is also a solution of (2.2.1) and the result is proved. □

Remark 2.2.3. If h, k and f are such that the hypotheses of Theorem 2.2.2 hold and in addition $y \equiv 0$ is **not** a solution of (2.2.1), then Theorem 2.2.2 implies that (2.2.1) has at least one solution $y \in C[0,T]$ that satisfies $0 < |y|_0 < \alpha$.

Returning to Theorem 2.2.1 we note that conditions (2.2.8) and (2.2.9) indicate the central role played by the nonlinearity f in establishing the existence of at least one nonnegative solution $y \in C[0,T]$ of (2.2.1). By imposing further conditions of this type on f we now obtain the following multiple solutions result for (2.2.1) :

Theorem 2.2.3. *Suppose* (2.2.2) – (2.2.7),

$$\text{(2.2.19)} \qquad \begin{cases} \textit{there exists constants } \alpha_i > 0, \ i = 1, \ldots, n, \\ \textit{some } n \in \mathbf{N}, \textit{ such that for each } i \in \{1, \ldots, n\}, \\ \text{(2.2.8) } \textit{is satisfied with } \alpha = \alpha_i \end{cases}$$

and

$$\text{(2.2.20)} \qquad \begin{cases} \textit{there exists constants } \beta_j > 0 \textit{ and } t_j^\star \in [0,T], \\ j = 1, \ldots, m, \textit{ some } m \in \mathbf{N}, \textit{ such that for each} \\ j \in \{1, \ldots, m\}, \text{ (2.2.9) } \textit{is satisfied with } \beta = \beta_j \textit{ and } t^\star = t_j^\star \end{cases}$$

hold.

(I) If $m = n+1$ *and* $0 < \beta_1 < \alpha_1 < \ldots < \beta_n < \alpha_n < \beta_{n+1}$, *then* (2.2.1) *has at least* $2n$ *nonnegative solutions* $y_1, \ldots, y_{2n} \in C[0,T]$ *such that*

$$0 < \beta_1 < |y_1|_0 < \alpha_1 < \ldots < \alpha_n < |y_{2n}|_0 < \beta_{n+1}.$$

(II) If $m = n$ *and* $0 < \beta_1 < \alpha_1 < \ldots < \beta_n < \alpha_n$, *then* (2.2.1) *has at least* $2n-1$ *nonnegative solutions* $y_1, \ldots, y_{2n-1} \in C[0,T]$ *such that*

$$0 < \beta_1 < |y_1|_0 < \alpha_1 < \ldots < \beta_n < |y_{2n-1}|_0 < \alpha_n.$$

(III) If $n = m+1$ *and* $0 < \alpha_1 < \beta_1 < \ldots < \alpha_m < \beta_m < \alpha_{m+1}$, *then* (2.2.1) *has at least* $2m+1$ *nonnegative solutions* $y_0, \ldots, y_{2m} \in C[0,T]$ *such that*

$$0 \le |y_0|_0 < \alpha_1 < |y_1|_0 < \beta_1 < \ldots < \beta_m < |y_{2m}|_0 < \alpha_{m+1}.$$

(IV) If $m = n$ *and* $0 < \alpha_1 < \beta_1 < \ldots < \alpha_n < \beta_n$, *then* (2.2.1) *has at least* $2n$ *nonnegative solutions* $y_0, \ldots, y_{2n-1} \in C[0,T]$ *such that*

$$0 \le |y_0|_0 < \alpha_1 < |y_1|_0 < \beta_1 < \ldots < \alpha_n < |y_{2n-1}|_0 < \beta_n.$$

PROOF: In (III) and (IV), the existence of a solution $y_0 \in C[0,T]$ of (2.2.1) such that $0 \le |y_0|_0 < \alpha_1$, follows from Theorem 2.2.2. The remainder of the proof follows by repeated application of Theorem 2.2.1. We do not include the detail. □

Remark 2.2.4. Similar to Remark 2.2.1, we note that (2.2.20) may be replaced with

$$(2.2.21) \qquad \begin{cases} \text{there exists constants } \beta_j > 0,\ j = 1,\dots,m, \\ \text{some } m \in \mathbf{N}, \text{ such that for each } j \in \{1,\dots,m\}, \\ (2.2.14) \text{ is satisfied with } \beta = \beta_j. \end{cases}$$

Suppose now for simplicity that $|h|_0 = 0$. We give examples of functions f that satisfy conditions (2.2.8) and (2.2.14) (and hence (2.2.19) and (2.2.21) for various n and m respectively).

Example 2.2.2. Let $f(y) = y^m$, $0 \le m < 1$. Then f satisfies (2.2.8) and (2.2.14) with

$$\alpha > K_1^{\frac{1}{1-m}} \text{ and } \beta < (K_2 M^m)^{\frac{1}{1-m}}$$

respectively. Note that $\beta < \alpha$ since $K_2 \le K_1$ and $0 < M < 1$.

Example 2.2.3. Suppose $f(y) = y^n$, $n > 1$. Then f satisfies (2.2.8) and (2.2.14) with

$$\alpha < K_1^{\frac{1}{1-n}} \text{ and } \beta > (K_2 M^n)^{\frac{1}{1-n}}$$

respectively. In this case $\alpha < \beta$ since $\frac{1}{1-n} < 0$.

Example 2.2.4. Let $f(y) = 1 + y^m + y^n$, $0 \le m < 1 < n$. Since

$$\frac{y}{K_2(1 + (My)^m + (My)^n)} \to 0 \text{ as } y \to 0^+ \text{ and } y \to \infty,$$

there exists $0 < \tilde{\beta}_1 < \tilde{\beta}_2$ such that f satisfies (2.2.14) with both $\beta = \beta_1$ and $\beta = \beta_2$ where $\beta_1 \in (0, \tilde{\beta}_1)$ and $\beta_2 \in (\tilde{\beta}_2, \infty)$. In addition if

$$\sup_{y \in [0,\infty)} \frac{y}{K_1(1 + y^m + y^n)} > 1,$$

then there exists $\alpha > 0$ for which f satisfies (2.2.8). Note that we can choose α such that $0 < \beta_1 < \alpha < \beta_2$.

Remark 2.2.5. If f is such that

$$\frac{y}{|h|_0 + K_1 f(y)} \to \infty \text{ as } y \to 0^+,$$

then one can find $\alpha > 0$ "small enough" such that (2.2.8) is satisfied. Analogously, if

$$\frac{y}{|h|_0 + K_1 f(y)} \to \infty \text{ as } y \to \infty,$$

(2.2.8) is satisfied by choosing α "large enough." In a similar fashion if f is such that

$$\frac{y}{K_2 f(My)} \to 0, \text{ as } y \to 0^+ \text{ and/or } y \to \infty,$$

then one can choose β_1 "small enough" and/or β_2 "large enough" such that (2.2.14) is satisfied. This idea was illustrated in Example 2.2.4.

Remark 2.2.6. For applications of the results in this section we refer the reader to [63,64,97].

2.3. Multiple nonnegative continuous solutions on the half-line.

We now examine the nonlinear integral equation

$$y(t) = h(t) + \int_0^\infty k(t,s) f(y(s))\, ds, \quad t \in [0,\infty) \tag{2.3.1}$$

which we note is defined on the half-line $[0,\infty)$. Our first aim in this section is to extend Theorem 2.2.1 and Theorem 2.2.2 of Section 2.2 to obtain the existence of at least one nonnegative solution $y \in C[0,\infty)$ of (2.3.1) such that $\lim_{t\to\infty} y(t)$ exists. Recall the following notation from Chapter 1:

Notation. We denote by $BC[0,\infty)$ the normed space of bounded, continuous functions defined on $[0,\infty)$ with norm given by

$$|y|_0 := \sup_{t\in[0,\infty)} |y(t)|.$$

The space $C_l[0,\infty)$ is a subset of $BC[0,\infty)$ which consists of all $y \in BC[0,\infty)$ such that $\lim_{t\to\infty} y(t)$ exists. For convenience, we let $y(\infty) := \lim_{t\to\infty} y(t)$.

Theorem 2.3.1. *Suppose*

$$0 \le k_t(s) = k(t,s) \in L^1[0,\infty) \textit{ for each } t \in [0,\infty), \tag{2.3.2}$$

(2.3.3) $\quad$ *the map* $t \mapsto k_t$ *is continuous from* $[0,\infty)$ *to* $L^1[0,\infty)$,

(2.3.4) $$\begin{cases} \text{there exists } 0 < M < 1,\ \kappa \in L^1[0,\infty) \text{ and} \\ \text{an interval } [a,b] \subseteq [0,\infty),\ a < b, \text{ such that} \\ k(t,s) \geq M\kappa(s) \geq 0,\ t \in [a,b],\ \text{a.e. } s \in [0,\infty), \end{cases}$$

(2.3.5) $$k(t,s) \leq \kappa(s),\ t \in [0,\infty),\ \text{a.e. } s \in [0,\infty),$$

(2.3.6) *there exists* $\tilde{k} \in L^1[0,\infty)$ *such that* $k_t \to \tilde{k} \in L^1[0,\infty)$ *as* $t \to \infty$,

(2.3.7) $\quad h \in C_l[0,\infty)$ *with* $h(t) \geq 0,\ t \in [0,\infty)$ *and* $\min_{t\in[a,b]} h(t) \geq M|h|_0$,

(2.3.8) $$\begin{cases} f : \mathbf{R} \to \mathbf{R} \text{ is continuous and nondecreasing} \\ \text{with } f(u) > 0 \text{ for } u > 0, \end{cases}$$

(2.3.9) $$\begin{cases} \text{there exists } \alpha > 0 \text{ such that } \dfrac{\alpha}{|h|_0 + K_3 f(\alpha)} > 1 \\ \left(\text{here } K_3 := \sup_{t\in[0,\infty)} \displaystyle\int_0^\infty k(t,s)\,ds > 0\right) \end{cases}$$

and

(2.3.10) $$\begin{cases} \text{there exists } \beta > 0,\ \beta \neq \alpha \text{ and } t^\star \in [0,\infty) \text{ such that} \\ \dfrac{\beta}{h(t^\star) + f(M\beta)\displaystyle\int_a^b k(t^\star,s)\,ds} < 1 \end{cases}$$

hold. Then (2.3.1) *has at least one nonnegative solution* $y \in C_l[0,\infty)$ *and either*

(A) $0 < \alpha < |y|_0 < \beta$ *and* $y(t) \geq M\alpha,\ t \in [a,b]$ *if* $\alpha < \beta$

or

(B) $0 < \beta < |y|_0 < \alpha$ *and* $y(t) \geq M\beta,\ t \in [a,b]$ *if* $\beta < \alpha$

holds.

PROOF: Define the operator $K : C_l[0,\infty) \to C_l[0,\infty)$ by

$$(2.3.11) \qquad Ky(t) := h(t) + \int_0^\infty k(t,s) f(y(s))\, ds, \quad t \in [0,\infty);$$

the cone C by

$$C := \{y \in C_l[0,\infty) : y(t) \geq 0,\ t \in [0,\infty) \text{ and } \min_{t \in [a,b]} y(t) \geq M|y|_0\};$$

and let Ω_α and Ω_β be given by

$$\Omega_\alpha := \{y \in C_l[0,\infty) : |y|_0 < \alpha\}$$

and

$$\Omega_\beta := \{y \in C_l[0,\infty) : |y|_0 < \beta\}$$

respectively.

Conditions (2.3.2), (2.3.3) and (2.3.6) – (2.3.8) imply that

$$K : C_l[0,\infty) \to C_l[0,\infty) \text{ is continuous and completely continuous}$$

[137,Chpt.5]. The remainder of the proof now follows in a similar fashion to the proof of Theorem 2.2.1 with T replaced by ∞, $[0,T]$ by $[0,\infty)$, and K_1 replaced by K_3. We omit the detail. □

Remark 2.3.1. Condition (2.3.10) may be replaced with

$$(2.3.12) \qquad \begin{cases} \text{there exists } \beta > 0,\ \beta \neq \alpha, \text{ such that} \\ \dfrac{\beta}{K_4 f(M\beta)} < 1, \text{ where } K_4 := \sup_{t \in [0,\infty)} \displaystyle\int_a^b k(t,s)\, ds. \end{cases}$$

Example 2.3.1. Suppose

$$k(t,s) = k_1(t)k_2(s), \quad t \in [0,\infty), \text{ a.e. } s \in [0,\infty),$$

where

$$k_1 \in C_l[0,\infty) \text{ with } k_1(t) \geq 0,\ \ t \in [0,\infty) \text{ and } |k_1|_0 \neq 0,$$

and

$$k_2 \in L^1[0,\infty) \text{ with } k_2(s) \geq 0, \text{ a.e. } s \in [0,\infty).$$

Then k satisfies (2.3.2) – (2.3.6) of Theorem 2.3.1. To see this note that we can choose

$$\tilde{k}(s) = k_1(\infty)k_2(s), \text{ a.e. } s \in [0,\infty)$$

and

$$\kappa(s) = |k_1|_0 k_2(s), \text{ a.e. } s \in [0,\infty).$$

Also since $0 \neq |k_1|_0$, there exists an interval $[a,b] \subset [0,\infty)$ and $0 < M < 1$ such that $\min_{t\in[a,b]} k_1(t) \geq M|k_1|_0$.

Example 2.3.2. If there exists $\tilde{k} \in L^1[0,\infty)$ such that $\lim_{t\to\infty} k(t,s) = \tilde{k}(s)$, a.e. $s \in [0,\infty)$, and there exists $a \in L^1[0,\infty)$ such that $0 \leq k(t,s) \leq a(s)$, then by the Lebesgue dominated convergence theorem, k satisfies (2.3.6). Note that (2.3.5) is also satisfied. In addition, if a is such that there exists $[a,b] \subset [0,\infty)$ and $0 < M < 1$ such that $Ma(s) \leq k(t,s)$, $t \in [a,b]$, a.e. $s \in [0,\infty)$, then clearly (2.3.4) holds.

Analogues of Theorem 2.2.2 and Theorem 2.2.3 which ensure the existence of one or more solutions respectively of (2.3.1) in $C_l[0,\infty)$ also exist. We just state the analogue of Theorem 2.2.2 here.

Theorem 2.3.2. *Suppose* (2.3.2), (2.3.3), (2.3.6), (2.3.8), (2.3.9) *and*

$$h \in C_l[0,\infty) \text{ with } h(t) \geq 0, \quad t \in [0,\infty) \tag{2.3.13}$$

are true. Then (2.3.1) *has at least one solution* $y \in C_l[0,\infty)$ *such that* $0 \leq y(t) < \infty$, $t \in [0,\infty)$.

Finally we present a slightly more general result for (2.3.1) which uses Theorem 2.2.1, Theorem 2.2.2 and an inductive argument using the Arzéla-Ascoli theorem to ensure the existence of at least two nonnegative solutions y_1, $y_2 \in BC[0,\infty)$ of (2.3.1).

Theorem 2.3.3. *Suppose* (2.3.2) – (2.3.5), (2.3.8),

$$h \in BC[0,\infty) \text{ with } h(t) \geq 0, \; t \in [0,\infty) \text{ and } \min_{t\in[a,b]} h(t) \geq M|h|_0, \tag{2.3.14}$$

$$\begin{cases} \text{there exists } \alpha_2 > \alpha_1 > 0 \text{ with } M\alpha_2 > \alpha_1, \\ \text{and such that (2.3.9) holds with } \alpha = \alpha_1, \; \alpha_2 \end{cases} \tag{2.3.15}$$

and

$$\begin{cases} \text{there exists } \beta > \alpha_2 \text{ such that} \\ \dfrac{\beta}{K_5 f(M\beta)} < 1, \text{ where } K_5 := \sup_{t\in[a,b]} \displaystyle\int_a^b k(t,s)\,ds \end{cases} \tag{2.3.16}$$

hold. Then (2.3.1) *has at least two nonnegative solutions* y_1, $y_2 \in BC[0,\infty)$ *with*

$$0 \leq |y_1|_0 \leq \alpha_1 < M\alpha_2 \leq |y_2|_0 \leq \beta \text{ and } \min_{t\in[a,b]} y_2(t) \geq M\alpha_2.$$

Notation. For each $n \in \mathbf{N}$, we let $|y|_n := \sup_{t\in[0,n]} |y(t)|$, and recall that $|y|_0 = \sup_{t\in[0,\infty)} |y(t)|$.

PROOF: Without loss of generality assume that $[a,b] \subseteq [0,1]$, and consider the integral equation

$$(2.3.17) \qquad y(t) = h(t) + \int_0^1 k(t,s)f(y(s))\,ds, \quad t \in [0,1].$$

Note the hypotheses of Theorem 2.2.1 (with (2.2.9) replaced by (2.2.14)) hold with $T = 1$ and $\alpha = \alpha_2 < \beta$, while the hypotheses of Theorem 2.2.2 are satisfied with $T = 1$ and $\alpha = \alpha_1$. In particular, (2.2.8) holds with $\alpha = \alpha_1$, α_2 since from (2.3.15) with $\alpha = \alpha_1$, α_2, we have

$$\begin{aligned} \alpha &> |h|_0 + K_3 f(\alpha) = \sup_{t\in[0,\infty)} h(t) + \left(\sup_{t\in[0,\infty)} \int_0^\infty k(t,s)\,ds\right) f(\alpha) \\ &\geq \sup_{t\in[0,1]} h(t) + \left(\sup_{t\in[0,1]} \int_0^1 k(t,s)\,ds\right) f(\alpha) = |h|_1 + K_1 f(\alpha). \end{aligned}$$

Similarly (2.2.14) follows from (2.3.16) since

$$\begin{aligned} \beta &< K_5 f(M\beta) = \left(\sup_{t\in[a,b]} \int_a^b k(t,s)\,ds\right) f(M\beta) \\ &\leq \left(\sup_{t\in[0,1]} \int_a^b k(t,s)\,ds\right) f(M\beta) = K_2 f(M\beta). \end{aligned}$$

Hence by Theorem 2.2.1 and Theorem 2.2.2, (2.3.17) has at least two nonnegative solutions $y_{1,1}$, $y_{1,2} \in C[0,1]$ such that

$$0 \leq |y_{1,1}|_1 < \alpha_1 < \alpha_2 < |y_{1,2}|_1 < \beta \text{ with } \min_{t\in[a,b]} y_{1,2}(t) \geq M|y_{1,2}|_1 > M\alpha_2.$$

In a similar fashion, we have that for each $n \in \mathbf{N}$, the integral equation

$$(2.3.18) \qquad y(t) = h(t) + \int_0^n k(t,s)f(y(s))\,ds, \quad t \in [0,n]$$

has at least two nonnegative solutions $y_{n,1}$, $y_{n,2} \in C[0,n]$ such that

$$0 \leq |y_{n,1}|_n < \alpha_1 < \alpha_2 < |y_{n,2}|_n < \beta \text{ with } \min_{t\in[a,b]} y_{n,2}(t) \geq M|y_{n,2}|_n > M\alpha_2.$$

Since for each $k = 1, 2, \ldots$, the sequences $\{y_{n,1}\}_{n\geq k}$ and $\{y_{n,2}\}_{n\geq k}$ are uniformly bounded and equicontinuous on $[0, k]$ [94], we can use an inductive argument and the Arzéla-Ascoli theorem (as in [94]), to obtain two functions z_1, $z_2 \in BC[0, \infty)$ (solutions of (2.3.1)) such that

$$0 \leq |z_1|_0 \leq \alpha_1 < M\alpha_2 \leq |z_2|_0 \leq \beta \text{ with } \min_{t\in[a,b]} z_2(t) \geq M\alpha_2.$$

We omit the detail and instead we refer the reader to [94]. □

Remark 2.3.2. If there exists $\sigma \in [0, \infty)$ with

$$\int_a^b k(\sigma, s)\, ds = \sup_{t\in[0,\infty)} \int_a^b k(t, s)\, ds,$$

then K_5 could be replaced by $K_4 = \sup_{t\in[0,\infty)} \int_a^b k(t, s)\, ds$. If $\sigma \in [0, 1]$ then the proof follows exactly as in Theorem 2.3.3; we need only note that

$$K_4 f(M\beta) = \left(\sup_{t\in[0,1]} \int_a^b k(t, s)\, ds\right) f(M\beta).$$

(If $\sigma > 1$ then an obvious adjustment is necessary.)

2.4. Multiple nonnegative L^p solutions.

Up to now in this chapter we have been concerned with proving the existence of nonnegative (and positive) solutions of Hammerstein integral equations, defined on either a compact interval or the half-line. Most of our results have relied on Krasnoselskii's fixed point theorem, however we have also used the nonlinear alternative and indeed a combination of both Krasnoselskii's theorem and the nonlinear alternative to obtain additional results. In all cases however we sought solutions that were continuous on the relevant interval.

In the final section of this chapter we present some analogous results for certain classes of Hammerstein integral equations, except that now we request the solutions to lie in the space of L^p-integrable functions. In particular, the results in this chapter will follow using Krasnoselskii's fixed point theorem. As in Section 2.2, the existence of additional nonnegative L^p solutions of the integral equation in question may – in certain cases – be obtained using the nonlinear alternative. However to avoid repetition we omit these results here.

Consider the nonlinear integral equation

$$y(t) = \int_0^T k(t, s) f(s, y(s))\, ds, \quad \text{a.e. } t \in [0, T]. \tag{2.4.1}$$

We would like to know what conditions one requires on k and f in order for this equation to have a positive solution $y \in L^p[0,T]$ where $1 \le p < \infty$. Here by a positive function y we mean $y(t) > 0$ for a.e. $t \in [0,T]$.

Notation. As in Chapter 1, we will use $||.||_p$ to denote the norm on $L^p[0,T]$ with

$$||y||_p := \left(\int_0^T |y(t)|^p \, dt \right)^{\frac{1}{p}} \text{ if } 1 \le p < \infty,$$

while

$$||y||_p := \text{ess} - \sup_{t \in [0,T]} |y(t)| \text{ if } p = \infty.$$

Ultimately we will apply Krasnoselskii's fixed point theorem to obtain the desired result. This means that we must construct an appropriate cone in $L^p[0,T]$ and ensure that the relevant integral operator satisfies the conditions of this fixed point theorem on the cone.

We start by defining the operators $K_L : L^{p_2}[0,T] \to L^p[0,T]$ and $F : L^p[0,T] \to L^{p_2}[0,T]$ by

$$K_L y(t) := \int_0^T k(t,s) y(s) \, ds, \text{ a.e. } t \in [0,T] \tag{2.4.2}$$

and

$$Fy(t) := f(t, y(t)), \text{ a.e. } t \in [0,T] \tag{2.4.3}$$

respectively. Here $1 \le p_2 < \infty$ and p_1 is such that $1/p_1 + 1/p_2 = 1$.

Concentrating first on the operator $F : L^p[0,T] \to L^{p_2}[0,T]$ suppose that

$$\begin{cases} f : [0,T] \times \mathbf{R} \to \mathbf{R} \text{ is a Carathéodory function; that is} \\ (i) \text{ the map } t \mapsto f(t,y) \text{ is measurable for all } y \in \mathbf{R}, \\ (ii) \text{ the map } y \mapsto f(t,y) \text{ is continuous for almost all} \\ t \in [0,T], \end{cases} \tag{2.4.4}$$

and

$$\begin{cases} \text{there exists } a_1 \in L^{p_2}[0,T] \text{ and } a_2 > 0 \text{ such that} \\ |f(t,y)| \le a_1(t) + a_2 |y|^{\frac{p}{p_2}}, \text{ a.e. } t \in [0,T] \end{cases} \tag{2.4.5}$$

hold. Then we have that

$$f(t, y(t)) \in L^{p_2}[0,T] \text{ for } y \in L^p[0,T]$$

and from [89], we see that

(2.4.6) $$F : L^p[0,T] \to L^{p_2}[0,T] \text{ is continuous and bounded}$$

is true. From (2.4.5) we obtain

$$|f(t,y(t))|^{p_2} \le 2^{p_2-1}\left(a_1^{p_2}(t) + a_2^{p_2}|y(t)|^p\right), \text{ a.e. } t \in [0,T]$$

and hence

$$\int_0^T |f(t,y(t))|^{p_2}\, dt \le 2^{p_2-1}\left(\int_0^T a_1^{p_2}(t)\, dt + a_2^{p_2}\int_0^T |y(t)|^p\, dt\right).$$

Thus defining $\psi \in C[0,\infty)$ by

(2.4.7) $$\psi(t) := 2^{\frac{p_2-1}{p_2}}\left(||a_1||_{p_2}^{p_2} + a_2^{p_2}t^p\right)^{\frac{1}{p_2}}, \quad t \in [0,\infty)$$

we see that

(2.4.8) $$\begin{cases} \text{there exists } \psi \in C[0,\infty) \text{ such that} \\ ||Fy||_{p_2} \le \psi(||y||_p) \text{ for all } y \in L^p[0,T] \end{cases}$$

holds. Finally since we are interested in positive solutions of (2.4.1) we will assume that

(2.4.9) $$f(t,y) > 0 \text{ for all } y > 0 \text{ and a.e. } t \in [0,T]$$

is satisfied. Thus defining the cones

$$C_p := \{y \in L^p[0,T] : y(t) \ge 0 \text{ a.e. } t \in [0,T]\}$$

and

$$C_{p_2} := \{y \in L^{p_2}[0,T] : y(t) \ge 0 \text{ a.e. } t \in [0,T]\}$$

we see from (2.4.6) and (2.4.9) that

(2.4.10) $$F : C_p \to C_{p_2} \text{ is a bounded, continuous operator.}$$

We now turn our attention to the operator $K_L : L^{p_2}[0,T] \to L^p[0,T]$. Suppose that

(2.4.11) $k : [0,T] \times [0,T] \to \mathbf{R}$ is such that $(t,s) \mapsto k(t,s)$ is measurable

and

$$(2.4.12)\qquad \begin{cases} \text{there exists } 0 < M \le 1,\ k_1 \in L^p[0,T] \text{ and } k_2 \in L^{p_1}[0,T] \\ \text{such that } 0 < k_1(t),\ k_2(t) \text{ a.e. } t \in [0,T] \\ \text{and } Mk_1(t)k_2(s) \le k(t,s) \le k_1(t)k_2(s), \\ \text{for a.e. } t \in [0,T],\ \text{a.e. } s \in [0,T]. \end{cases}$$

hold. From (2.4.12) we have immediately that

$$\left(\int_0^T \left(\int_0^T |k(t,s)|^p\,dt\right)^{\frac{p_1}{p}} ds\right)^{\frac{1}{p_1}} \le ||k_1||_p||k_2||_{p_1} \equiv M_0 < \infty.$$

Thus from Chapter 1 or [137,P.47-49] we have that

$$K_L : L^{p_2}[0,T] \to L^p[0,T] \text{ is a completely continuous operator.}$$

In fact due to the positivity of the kernel k we have that

$$(2.4.13)\qquad K_L : C_{p_2} \to C_p \text{ is a completely continuous operator}$$

is true. However we can be more specific about the range of K_L.

If $y \in C_{p_2}$ then we know from (2.4.12) that

$$|K_L y(t)|^p \le k_1^p(t)\left(\int_0^T k_2(s)y(s)\,ds\right)^p, \text{ a.e. } t \in [0,T],$$

that is,

$$(2.4.14)\qquad ||K_L y||_p \le ||k_1||_p \int_0^T k_2(s)y(s)\,ds.$$

However we also have from (2.4.12) that for $y \in C_{p_2}$,

$$K_L y(t) \ge M\,k_1(t)\int_0^T k_2(s)y(s)\,ds,\ \text{a.e. } t \in [0,T].$$

Combining this inequality with (2.4.14) we have that for $y \in C_{p_2}$,

$$(2.4.15)\qquad K_L y(t) \ge M\frac{k_1(t)}{||k_1||_p}||K_L y||_p,\ \text{a.e. } t \in [0,T].$$

Thus defining

$$(2.4.16)\qquad a(t) := M\frac{k_1(t)}{||k_1||_p},\ \text{a.e. } t \in [0,T],$$

we see that $a \in L^p[0,T]$, $a(t) > 0$, a.e. $t \in [0,T]$ and $||a||_p = M \le 1$. Therefore if we define the cone $C_{p,a}$ by

$$C_{p,a} := \{y \in L^p[0,T] : y(t) \ge a(t)||y||_p, \text{ a.e. } t \in [0,T]\},$$

then by (2.4.15) and (2.4.13) we have

(2.4.17) $\quad K_L : C_{p_2} \to C_{p,a}$ is a completely continuous operator.

Finally since $C_{p,a} \subseteq C_p$, if we let

$$Ky(t) := K_L F y(t) = \int_0^T k(t,s) f(s,y(s))\, ds, \text{ a.e. } t \in [0,T],$$

then (2.4.10) and (2.4.17) combine to give

(2.4.18) $\quad K : C_{p,a} \to C_{p,a}$ is a completely continuous operator.

Now that it is clearer what operator and cone we intend to use in Krasnoselskii's fixed point theorem, we state the following existence result for (2.4.1).

Theorem 2.4.1. *Assume that p, p_1 and p_2 are such that $1 \le p_1 \le p < \infty$ and $1/p_1 + 1/p_2 = 1$. Suppose that $f : [0,T] \times \mathbf{R} \to \mathbf{R}$ satisfies* (2.4.4), (2.4.5), (2.4.9) *and*

(2.4.19) $\quad f(t,y)$ *is nondecreasing in y for a.e. $t \in [0,T]$,*

while $k : [0,T] \times [0,T] \to \mathbf{R}$ satisfies (2.4.11) *and* (2.4.12). *In addition assume that*

(2.4.20) $\quad$ *there exists $\alpha > 0$ such that* $1 < \dfrac{\alpha}{||k_1||_p ||k_2||_{p_1} \psi(\alpha)}$

and

(2.4.21) $$\begin{cases} \text{there exists } \beta > 0,\ \beta \ne \alpha, \text{ such that} \\ 1 > \dfrac{\beta}{M||k_1||_p \displaystyle\int_0^T k_2(s) f(s, a(s)\beta)\, ds} \end{cases}$$

(Here $\psi \in C[0,\infty)$ is as described in (2.4.7) *and $a \in L^p[0,T]$ is as described in* (2.4.16).*) Then* (2.4.1) *has at least one positive solution $y \in L^p[0,T]$ and either*

(A) $0 < \alpha < ||y||_p < \beta$ *and* $y(t) \ge a(t)\alpha$ *a.e.* $t \in [0,T]$ *if* $\alpha < \beta$

or

$$\textbf{(B)}\ 0 < \beta < ||y||_p < \alpha \text{ and } y(t) \geq a(t)\beta \text{ a.e. } t \in [0,T] \text{ if } \beta < \alpha$$

holds.

PROOF: Define

$$\Omega_\alpha := \{y \in L^p[0,T] : ||y||_p < \alpha\}$$

and

$$\Omega_\beta := \{y \in L^p[0,T] : ||y||_p < \beta\}.$$

Assume that $\beta < \alpha$. (A similar argument holds if $\alpha < \beta$.) It is clear from (2.4.18) that

$$K : C_{p,a} \cap (\overline{\Omega}_\alpha \backslash \Omega_\beta) \to C_{p,a} \text{ is a completely continuous operator.}$$

If we show that in addition

$$||Ky||_p \leq ||y||_p \text{ for } y \in C_{p,a} \cap \partial\Omega_\alpha \tag{2.4.22}$$

and

$$||Ky||_p \geq ||y||_p \text{ for } y \in C_{p,a} \cap \partial\Omega_\beta \tag{2.4.23}$$

are true, then by Krasnoselskii's fixed point theorem the operator K has a fixed point in $C_{p,a} \cap (\overline{\Omega}_\alpha \backslash \Omega_\beta)$. This in turn implies that (2.4.1) has at least one solution $y \in L^p[0,T]$ such that $\beta \leq ||y||_p \leq \alpha$ and $y(t) \geq a(t)\beta$ for a.e. $t \in [0,T]$.

Suppose then that $y \in C_{p,a} \cap \partial\Omega_\alpha$. In particular $||y||_p = \alpha$. Then (2.4.12), Hölder's inequality, (2.4.8) and (2.4.20) give

$$\begin{aligned} ||Ky||_p &\leq ||k_1||_p ||k_2||_{p_1} ||Fy||_{p_2} \leq ||k_1||_p ||k_2||_{p_1} \psi(||y||_p) \\ &= ||k_1||_p ||k_2||_{p_1} \psi(\alpha) < \alpha = ||y||_p \end{aligned}$$

and thus (2.4.22) is satisfied.

Next suppose that $y \in C_{p,a} \cap \partial\Omega_\beta$. Then $||y||_p = \beta$ and note also that $y(t) \geq a(t)\beta$ for a.e. $t \in [0,T]$. Conditions (2.4.12) and (2.4.19) give

$$\begin{aligned} \int_0^T |Ky(t)|^p\, dt &\geq M^p ||k_1||_p^p \left(\int_0^T k_2(s) f(s, y(s))\, ds \right)^p \\ &\geq M^p ||k_1||_p^p \left(\int_0^T k_2(s) f(s, a(s)\beta)\, ds \right)^p. \end{aligned}$$

This inequality along with (2.4.21) implies that

$$||Ky||_p \geq M||k_1||_p \left(\int_0^T k_2(s) f(s, a(s)\beta)\, ds \right) > \beta = ||y||_p$$

–thus (2.4.23) holds and the theorem is proved. □

Example 2.4.1. Let p, p_1 and p_2 be as in Theorem 2.4.1 and suppose that $f(s, y) = y^n$ where $0 \leq n < 1$. It is immediate that f satisfies (2.4.4), (2.4.9) and (2.4.19). To ensure that (2.4.5) holds (with $a_1 = 0$ and $a_2 = 1$), we must make the additional assumption that $np_2 \leq p$. This fact, along with Hölder's inequality gives

$$\int_0^T |f(s, y(s))|^{p_2}\, ds = \int_0^T |y(s)|^{np_2}\, ds \leq T^m ||y||_p^{np_2} \tag{2.4.24}$$

where $m := (p - np_2)/p$. In addition from (2.4.24) we can define $\psi(t) := Ct^n$ where $C := T^{m/p_2}$. It remains to show that there exist α and β that satisfy (2.4.20) and (2.4.21) respectively, with the above choice of f and ψ. Since $0 \leq n < 1$, it is possible to choose $0 < \alpha$ (large enough) and $0 < \beta$ (small enough) to satisfy

$$1 < \frac{\alpha^{1-n}}{C||k_1||_p||k_2||_{p_1}}$$

and

$$1 > \frac{\beta^{1-n}}{M||k_1||_p \int_0^T k_2(s) a^n(s)\, ds}$$

respectively. In addition we see that $0 < \beta < \alpha$ since

$$\begin{aligned} \beta^{1-n} &< MC||k_1||_p||k_2||_{p_1}||a||_p^n = M^{1+n}C||k_1||_p||k_2||_{p_1} \\ &\leq C||k_1||_p||k_2||_{p_1} < \alpha^{1-n}. \end{aligned}$$

For integral equations with similar nonlinearities, we refer the reader to the references. In particular [63,64,100,101,105,106,151] discuss integral equations where the nonlinearity f exhibits sublinear growth as above.

Example 2.4.2. Let p, p_1 and p_2 be as in Theorem 2.4.1 and suppose that $f(s, y) = y^n$ where now $n > 1$ and we assume again that $np_2 \leq p$. Arguing as in Example 2.4.1, we see that f satisfies (2.4.4), (2.4.5), (2.4.9) and (2.4.19) and ψ can once again be given by $\psi(t) := Ct^n$ where C is as defined in the above example. Now since $n > 1$, one can choose $0 < \alpha$ (small enough) to ensure that

$$1 < \frac{1}{C||k_1||_p||k_2||_{p_1}\alpha^{n-1}}.$$

In addition β can then be chosen (large enough) such that $0 < \alpha < \beta$ and

$$1 > \frac{1}{M||k_1||_p \left(\int_0^T k_2(s)a^n(s)\,ds\right)\beta^{n-1}}.$$

Integral equations whose nonlinear part exhibits superlinear growth are also discussed in [63,64,100,101,105,106].

While the results in [63,64] also rely on Krasnoselskii's fixed point theorem, the advantage of Theorem 2.4.1 for the two examples discussed here, is that we are only required to know how the nonlinearity f behaves at two points, α and β.

Remark 2.4.1. In Theorem 2.4.1 we sought the existence of a positive solution $y \in L^p[0,T]$ of (2.4.1). It is possible to modify the hypotheses of this theorem to obtain the existence of at least one *nonnegative* solution of (2.4.1), that is, a solution $y \in L^p[0,T]$ such that $y(t) \geq 0$, a.e. $t \in [0,T]$. Replacing condition (2.4.12) with

$$(2.4.12)^\star \quad \begin{cases} \text{there exists } 0 < M \leq 1,\ k_1 \in L^p[0,T] \text{ and } k_2 \in L^{p_1}[0,T] \\ \text{such that } 0 \leq k_1(t),\, k_2(t) \text{ a.e. } t \in [0,T],\ 0 < ||k_1||_p,\ ||k_2||_{p_1} \\ \text{and } Mk_1(t)k_2(s) \leq k(t,s) \leq k_1(t)k_2(s) \\ \text{a.e. } t \in [0,T],\ \text{a.e. } s \in [0,T] \end{cases}$$

and assuming that $\int_0^T k_2(s)f(s,a(s)\beta)\,ds > 0$, for all $\beta > 0$, will ensure the result.

For the remainder of this section we will concentrate on establishing results analogous to Theorem 2.4.1, for two variations of equation (2.4.1). Firstly we consider the Hammerstein integral equation

$$(2.4.25) \qquad y(t) = h(t) + \int_0^T k(t,s)f(s,y(s))\,ds, \text{ a.e. } t \in [0,T].$$

The hypotheses and proof of Theorem 2.4.1 can be modified easily to ensure that this nonlinear equation has at least one positive solution $y \in L^p[0,T]$.

We give some details of the proof here since it is necessary to consider a slightly different cone to the one used in the proof of Theorem 2.4.1.

Theorem 2.4.2. *Suppose that* $f : [0,T] \times \mathbf{R} \to \mathbf{R}$ *satisfies* (2.4.4), (2.4.5), (2.4.9) *and* (2.4.19), *while* $k : [0,T] \times [0,T] \to \mathbf{R}$ *satisfies* (2.4.11) *and* (2.4.12). *In addition assume that*

$$(2.4.26) \qquad \begin{cases} h \in L^p[0,T] \text{ and } h(t) \geq a(t)||h||_p, \text{ a.e. } t \in [0,T]; \\ \text{here } a \in L^p[0,T] \text{ is as described in (2.4.16)}, \end{cases}$$

$$(2.4.27) \quad \text{there exists } \alpha > 0 \text{ such that } 1 < \frac{\alpha}{4^{\frac{p-1}{p}}(||h||_p + ||k_1||_p||k_2||_{p_1}\psi(\alpha))}$$

and

$$(2.4.28) \qquad \begin{cases} \text{there exists } \beta > 0, \ \beta \neq \alpha, \text{ such that} \\ 1 > \dfrac{\beta}{2^{\frac{1-p}{p}}\left(||h||_p + M||k_1||_p \displaystyle\int_0^T k_2(s) f(s, \tilde{a}(s)\beta)\, ds\right)} \end{cases}$$

are true. (The function ψ *in* (2.4.27) *is such that* (2.4.8) *holds, while* $\tilde{a}$ *in condition* (2.4.28) *is as given in* (2.4.33).*) Then* (2.4.25) *has at least one positive solution* $y \in L^p[0,T]$ *and either*

(A) $0 < \alpha < ||y||_p < \beta$ *and* $y(t) \geq a(t)\alpha$ *a.e.* $t \in [0,T]$ *if* $\alpha < \beta$

or

(B) $0 < \beta < ||y||_p < \alpha$ *and* $y(t) \geq a(t)\beta$ *a.e.* $t \in [0,T]$ *if* $\beta < \alpha$

holds.

PROOF: From the proof of Theorem 2.4.1 we see that (2.4.10) is true, that is,

$$F : C_p \to C_{p_2} \text{ is a bounded, continuous operator,}$$

where $Fy(t) := f(t, y(t))$, a.e. $t \in [0,T]$. Defining the operator $\tilde{K} : L^{p_2}[0,T] \to L^p[0,T]$ by

$$\tilde{K}y(t) := h(t) + \int_0^T k(t,s)y(s)\, ds, \text{ a.e. } t \in [0,T],$$

we can show (once again using results from Chapter 1 or [137,P.47-49] in addition to (2.4.26)) that

$$\tilde{K} : L^{p_2}[0,T] \to L^p[0,T] \text{ is a completely continuous operator.}$$

Indeed, due to the positive nature of the functions involved, it is clear that

$$\tilde{K} : C_{p_2} \to C_p \text{ is a completely continuous operator.} \tag{2.4.29}$$

However, as in the proof of Theorem 2.4.1, we want to be more specific about the range of $\tilde{K}$ in this case.

If $y \in C_{p_2}$, then from (2.4.12), (2.4.26) and using the facts that

$$\begin{cases} (a+b)^p \leq 2^{p-1}(a^p + b^p) \text{ and} \\ (a+b)^{\frac{1}{p}} \leq 2^{\frac{p-1}{p}}(a^{\frac{1}{p}} + b^{\frac{1}{p}}) \text{ for } a, b > 0,\ p \geq 1, \end{cases} \tag{2.4.30}$$

we see that

$$|\tilde{K}y(t)|^p \leq 2^{p-1}\left(|h(t)|^p + k_1^p(t)\left(\int_0^T k_2(s)y(s)\,ds\right)^p\right), \text{ a.e. } t \in [0,T],$$

that is,

$$||\tilde{K}y||_p^p \leq 2^{p-1}\left(||h||_p^p + ||k_1||_p^p\left(\int_0^T k_2(s)y(s)\,ds\right)^p\right),$$

and therefore

$$||\tilde{K}y||_p \leq 4^{\frac{p-1}{p}}\left(||h||_p + ||k_1||_p\int_0^T k_2(s)y(s)\,ds\right). \tag{2.4.31}$$

However we also have from (2.4.12) and (2.4.26) that for $y \in C_{p2}$,

$$\begin{aligned} \tilde{K}y(t) &\geq a(t)||h||_p + Mk_1(t)\int_0^T k_2(s)y(s)\,ds \\ &= M\frac{k_1(t)}{||k_1||_p}\left(||h||_p + ||k_1||_p\int_0^T k_2(s)y(s)\,ds\right), \text{ a.e. } t \in [0,T]. \end{aligned}$$

Combining this with (2.4.31) yields

$$\tilde{K}y(t) \geq 4^{\frac{1-p}{p}}M\frac{k_1(t)}{||k_1||_p}||\tilde{K}y||_p, \text{ a.e. } t \in [0,T],. \tag{2.4.32}$$

Thus defining

$$\tilde{a}(t) := 4^{\frac{1-p}{p}}M\frac{k_1(t)}{||k_1||_p}, \text{ a.e. } t \in [0,T], \tag{2.4.33}$$

we see that $\tilde{a} \in L^p[0,T]$, $\tilde{a}(t) > 0$, a.e. $t \in [0,T]$ and $||\tilde{a}||_p = 4^{\frac{1-p}{p}} M \leq 1$. Therefore if we define the cone $C_{p,\tilde{a}}$ by

$$C_{p,\tilde{a}} := \{y \in L^p[0,T] : y(t) \geq \tilde{a}(t)||y||_p, \text{ a.e. } t \in [0,T]\},$$

then by (2.4.29) and (2.4.32) we have

$$K_L : C_{p_2} \to C_{p,\tilde{a}} \text{ is a completely continuous operator.}$$

Finally if we let

$$Ky(t) := \tilde{K}Fy(t) = h(t) + \int_0^T k(t,s)f(s,y(s))\,ds, \text{ a.e. } t \in [0,T],$$

we obtain

$$K : C_{p,\tilde{a}} \to C_{p,\tilde{a}} \text{ is a completely continuous operator.}$$

In order to apply Krasnoselskii's fixed point theorem to obtain the desired result it remains to show that

$$||Ky||_p \leq ||y||_p \text{ for } y \in C_{p,\tilde{a}} \cap \partial\Omega_\alpha \tag{2.4.34}$$

and

$$||Ky||_p \geq ||y||_p \text{ for } y \in C_{p,\tilde{a}} \cap \partial\Omega_\beta \tag{2.4.35}$$

are true. (Here Ω_α and Ω_β are as defined in the proof of Theorem 2.4.1.) Suppose first that $y \in C_{p,\tilde{a}} \cap \Omega_\alpha$, implying in particular that $||y||_p = \alpha$. This fact, (2.4.31), Hölder's inequality, (2.4.8) and (2.4.27) give

$$\begin{aligned} ||\tilde{K}y||_p &\leq 4^{\frac{p-1}{p}}(||h||_p + ||k_1||_p||k_2||_{p_1}\psi(||y||_p)) \\ &= 4^{\frac{p-1}{p}}(||h||_p + ||k_1||_p||k_2||_{p_1}\psi(\alpha)) < \alpha = ||y||_p \end{aligned}$$

– thus (2.4.34) holds.

For $y \in C_{p,\tilde{a}} \cap \Omega_\beta$ we have that $||y||_p = \beta$ and $y(t) \geq \tilde{a}(t)\beta$, a.e. $t \in [0,T]$. This information, along with (2.4.12) and (2.4.19), gives

$$\begin{aligned} |Ky(t)|^p &\geq |h(t)|^p + M^p k_1^p(t) \left(\int_0^T k_2(s)f(s,y(s))\,ds \right)^p \\ &\geq |h(t)|^p + M^p k_1^p(t) \left(\int_0^T k_2(s)f(s,\tilde{a}(s)\beta)\,ds \right)^p. \end{aligned}$$

If we recall that $(a+b)^{\frac{1}{p}} \geq 2^{\frac{1-p}{p}}(a^{\frac{1}{p}}+b^{\frac{1}{p}})$, for $a,b>0$ and $p \geq 1$, we now obtain from (2.4.28),

$$||Ky||_p \geq 2^{\frac{1-p}{p}} \left(||h||_p + M||k_1||_p \int_0^T k_2(s)f(s,\tilde{a}(s)\beta)\, ds \right) > \beta = ||y||_p,$$

– and (2.4.35) is proved. □

Finally we state a result for a second variation of (2.4.1), namely,

$$(2.4.36) \qquad y(t) = h(t) + \int_0^\infty k(t,s)f(s,y(s))\, ds, \text{ a.e. } t \in [0,\infty).$$

Theorem 2.4.3. *Suppose that* $f : [0,\infty) \times \mathbf{R} \to \mathbf{R}$ *satisfies* (2.4.4), (2.4.5), (2.4.9) *and* (2.4.19), $k : [0,\infty) \times [0,\infty) \to \mathbf{R}$ *satisfies* (2.4.11) *and* (2.4.12) *and* $h : [0,\infty) \to \mathbf{R}$ *satisfies* (2.4.26), *in addition to* (2.4.27) *and* (2.4.28) *being true, where in all of the above,* $T = \infty$ *and* $[0,T]$ *is replaced by* $[0,\infty)$. *Then* (2.4.36) *has at least one positive solution* $y \in L^p[0,\infty)$ *and either*

(A) $0 < \alpha < ||y||_p < \beta$ *and* $y(t) \geq a(t)\alpha$ *a.e.* $t \in [0,\infty)$ *if* $\alpha < \beta$

or

(B) $0 < \beta < ||y||_p < \alpha$ *and* $y(t) \geq a(t)\beta$ *a.e.* $t \in [0,\infty)$ *if* $\beta < \alpha$

holds.

Notation. Here $||.||_p$ denotes the norm on $L^p[0,\infty)$ with

$$||y||_p := \left(\int_0^\infty |y(t)|^p\, dt \right)^{\frac{1}{p}} \text{ if } 1 \leq p < \infty,$$

while

$$||y||_p := \text{ess} - \sup_{t\in[0,\infty)} |y(t)| \text{ if } t = \infty.$$

PROOF: Proving that equation (2.4.36) has at least one positive solution $y \in L^p[0,\infty)$ follows almost identically to the proof of Theorem 2.4.1 and Theorem 2.4.2. The only real difference arises when we want to show that

(2.4.37) $K_\infty : L^{p_2}[0,\infty) \to L^p[0,\infty)$ is a completely continuous operator,

where now

$$(2.4.38) \qquad K_\infty y(t) := h(t) + \int_0^\infty k(t,s)y(s)\, ds, \text{ a.e. } t \in [0,\infty).$$

In the proofs of Theorem 2.4.1 and Theorem 2.4.2 we refer the reader to Chapter 1 or [137,P.47-49], where the Riesz compactness criteria are used, to show that both $K_L : L^{p_2}[0,T] \to L^p[0,T]$ and $\tilde{K} : L^{p_2}[0,T] \to L^p[0,T]$ are completely continuous. Slightly different compactness criteria are required to show that (2.4.37) is true (see Theorem 1.2.14), however the reader can find the details of this argument in [137,P.66-68]. We omit the rest of the details since they are identical to those already outlined in the above proofs. $\square$

CHAPTER 3

POSITIVE SOLUTIONS OF SINGULAR INTEGRAL EQUATIONS

3.1. Introduction.

In this chapter we extend the results of the previous section to provide existence results which guarantee the existence of positive solutions of singular integral equations. Both Fredholm and Volterra integral equations are discussed and Krasnoselskii's fixed point theorem is the main fixed point result that is used.

In Section 3.2 we examine the singular integral equation

$$y(t) = \int_0^T k(t,s)[f(s,y(s)) + g(s,y(s))]\,ds, \ t \in [0,T]. \tag{3.1.1}$$

By singular we mean that for $y = 0$, $g(t,y)$ is undefined. We seek the existence of positive, continuous solutions of (3.1.1). Of course, due to the singular nature of this equation, it transpires that the desired solution will have to be strictly greater than zero almost everywhere on the interval $[0,T]$. We apply Krasnoselskii's fixed point theorem, and illustrate that if the cone is chosen appropriately, this fixed point result provides us with an easy, natural treatment of such singular integral equations.

We apply the results of Section 3.2, to show in Section 3.3, that under certain conditions on f and g, the boundary value problem

$$\begin{cases} -y'' = f(t,y) + g(t,y), \ t \in (0,1) \\ \alpha y(0) - \beta y'(0) = 0 \\ \gamma y(1) + \delta y'(1) = 0, \end{cases} \tag{3.1.2}$$

has at least one positive solution $y \in C[0,T]$. This equation has been discussed in [63,64,97].

One may observe from the results in Section 3.2, that the conditions placed on the kernel k are quite restrictive. On the other hand, it is these conditions that allow us to consider a rather large class of functions $f + g$. However, narrowing down this class of functions, allows us to lift some of these conditions on the kernel. We present a result in Section 3.4 which illustrates this point. Again, it is Krasnoselskii's fixed point theorem and an appropriate choice of cone, that provides us with our result.

In Section 3.5 we note that in certain cases, the existence of positive solutions of nonlinear integral equations of the form

$$y(t) = h(t) + \int_0^T k(t,s)[f(s,y(s)) + g(s,y(s))]\,ds, \ t \in [0,T]. \tag{3.1.3}$$

can be assured using Schauder's fixed point theorem. Using this fixed point result, some of the more restrictive conditions that were placed on the kernel in Section 3.2, may now be lifted. However the compromise is that we loose the ready availability of a multiple solutions result that was possible in certain cases using Krasnoselskii's fixed point theorem in Section 3.2.

Finally in Section 3.6 and Section 3.7, we discuss the existence of positive solutions of singular and nonsingular Volterra integral equations. This is a topic which has been discussed in some detail in the literature (see [36-38,73-75,161]), where many of the results are obtained using comparison theorems. In keeping with the spirit of this and the previous chapter, our results will follow from Krasnoselskii's fixed point theorem. The existence of positive solutions of certain types of nonlinear integral inequalities lies at the heart of the results in these two sections. We spend much time considering the inequalities that arise from the study of the nonsingular and singular Volterra integral equations,

$$y(t) = \int_0^t k(t,s)f(y(s))\,ds, \ t \in [0,T] \tag{3.1.4}$$

and

$$y(t) = \int_0^t k(t,s)g(y(s))\,ds, \ t \in [0,T] \tag{3.1.5}$$

respectively. Section 3.6 deals with the existence of continuous solutions of (3.1.5) and (3.1.6), while Section 3.7 is devoted to proving the existence of positive solutions $y \in L^p[0,T]$, $1 \le p < \infty$, of (3.1.4).

3.2. Continuous positive solutions of a singular integral equation.

The existence of positive, continuous solutions of the singular Fredholm integral equation

$$y(t) = \int_0^T k(t,s)[f(s,y(s)) + g(s,y(s))]\,ds, \ t \in [0,T] \tag{3.2.1}$$

is the focus of our attention in this section. We assume that $f : [0,T] \times [0,\infty) \to [0,\infty)$ is nondecreasing in y, while $g : [0,T] \times (0,\infty) \to [0,\infty)$ is nonincreasing in y. We are also allowing for the eventuality that the integral equation may be *singular*, that is, we consider functions g for which $g(t,0)$ may be undefined. Our main existence result will follow from Krasnoselskii's fixed point theorem which we also used in the examination of nonsingular integral equations in Chapter 2. Applying Krasnoselskii's fixed point theorem, we observe that the (integral) operator must map a relevant cone intersected with an annulus back into the cone. Careful selection of the cone and annulus provides us with an easy treatment of integral equations with singular nonlinearities. In addition, the solution will inherit properties from both the cone and annulus, providing us in many cases with a more detailed description of the solution than perhaps was initially anticipated.

Notation. For $y \in L^p[0,T]$ the norm is given by $||y||_p := (\int_0^T |y(t)|^p\,dt)^{\frac{1}{p}}$ for $1 \le p < \infty$ and $||y||_p := \operatorname{ess\,sup}_{t\in[0,T]} |y(t)|$ for $p = \infty$. However, to avoid repetition, throughout this section for $p = \infty$ we will use $(\int_0^T |y(t)|^p\,dt)^{1/p}$ to denote $||y||_\infty$.

Theorem 3.2.1. *Let $1 \le p \le \infty$ be a constant and q be such that $1/p + 1/q = 1$. Suppose that*

$$\left\{\begin{array}{l} \text{there exists } a \in C[0,T] \text{ and } t^\star \in [0,T] \text{ with } a(t) > 0 \text{ for} \\ \text{a.e. } t \in [0,T] \text{ and } a(t^\star) > 0, \text{ in addition to } \kappa \in L^p[0,T] \\ \text{with } \kappa(t) \ge 0 \text{ a.e. } t \in [0,T] \text{ and } ||\kappa||_p > 0, \text{ such that} \\ a(t)\kappa(s) \le k(t,s) \text{ for all } t \in [0,T], \text{ a.e. } s \in [0,T], \end{array}\right. \tag{3.2.2}$$

$$k_t(s) := k(t,s) \le \kappa(s) \text{ for all } t \in [0,T], \text{ a.e. } s \in [0,T], \tag{3.2.3}$$

$$\text{the map } t \mapsto k_t \text{ is continuous from } [0,T] \text{ to } L^p[0,T], \tag{3.2.4}$$

$$\left\{\begin{array}{l} f : [0,T] \times [0,\infty) \to [0,\infty) \text{ is Carathéodory,} \\ \text{with } f(t,y) \text{ nondecreasing in } y, \end{array}\right. \tag{3.2.5}$$

$$\left\{\begin{array}{l} g : [0,T] \times (0,\infty) \to [0,\infty) \text{ is Carathéodory, with} \\ g(t,0) \text{ undefined, and } g(t,y) \text{ nonincreasing in } y, \end{array}\right. \tag{3.2.6}$$

$$f(t,y) + g(t,y) > 0 \text{ for } y > 0,\ t \in [0,T], \tag{3.2.7}$$

(3.2.8) $\displaystyle\int_0^T [f(s,R)+g(s,\tilde{R}a(s))]^q\,ds < \infty$ *for any constants* $R, \tilde{R} > 0$,

(3.2.9) $$\begin{cases} \text{there exists } R_1 > 0 \text{ such that} \\ 1 < \dfrac{R_1}{\displaystyle\int_0^T \kappa(s)[f(s,R_1)+g(s,R_1a(s))]\,ds} \end{cases}$$

and

(3.2.10) $$\begin{cases} \text{there exists } R_2 > 0,\ R_2 \neq R_1,\ \text{such that} \\ 1 > \dfrac{R_2}{a(t^\star)\displaystyle\int_0^T \kappa(s)[f(s,R_2a(s))+g(s,R_2)]\,ds} \end{cases}$$

hold. Then (3.2.1) *has at least one* **positive** *solution* $y \in C[0,T]$ *and either*

(A) $0 < R_1 < |y|_0 < R_2$ *and* $y(t) \geq a(t)R_1 \geq 0,\ t \in [0,T]$ *if* $R_1 < R_2$

or

(B) $0 < R_2 < |y|_0 < R_1$ *and* $y(t) \geq a(t)R_2 \geq 0,\ t \in [0,T]$ *if* $R_2 < R_1$

holds.

PROOF: Define the operator K by

$$Ky(t) := \int_0^T k(t,s)[f(s,y(s))+g(s,y(s))]\,ds,\ t \in [0,T]$$

and let the cone C_a be given by

$$C_a := \{y \in C[0,T] : y(t) \geq a(t)|y|_0 \text{ for all } t \in [0,T]\}.$$

In addition define Ω_{R_1} and Ω_{R_2} by

$$\Omega_{R_1} := \{y \in C[0,T] : |y|_0 < R_1\}$$

and

$$\Omega_{R_2} := \{y \in C[0,T] : |y|_0 < R_2\}$$

respectively and suppose in what follows that $R_2 < R_1$. (A similar argument holds if $R_1 < R_2$.) Since we intend to apply Krasnoselskii's fixed point theorem (Theorem 2.1.1) we first show

(3.2.11) $$K : C_a \cap (\overline{\Omega}_{R_1} \backslash \Omega_{R_2}) \to C_a \text{ is a compact operator}$$

(and hence completely continuous).

Let $y \in C_a \cap (\overline{\Omega}_{R_1} \backslash \Omega_{R_2})$. Then

$$0 < R_2 \le |y|_0 \le R_1 \text{ and } 0 < R_2 a(t) \le y(t) \le R_1, \text{ a.e. } t \in [0,T]$$

and (3.2.5) and (3.2.6) yield

$$0 < f(t, y(t)) + g(t, y(t)) \le f(t, R_1) + g(t, R_2 a(t)) \text{ a.e. } t \in [0,T].$$

(Note that since $a(t) > 0$, a.e. $t \in [0,T]$ and hence $0 < R_2 a(t) \le y(t)$, a.e. $t \in [0,T]$ we have that $g(t, y(t))$ is measurable for a.e. $t \in [0,T]$.) From this inequality and (3.2.8) we obtain

$$\text{(3.2.12)} \quad \int_0^T [f(s, y(s)) + g(s, y(s))]^q \, ds$$
$$\le \int_0^T [f(s, R_1) + g(s, R_2 a(s))]^q \, ds < \infty,$$

that is, $f(t, y(t)) + g(t, y(t)) \in L^q[0,T]$ for $y \in C_a \cap (\overline{\Omega}_{R_1} \backslash \Omega_{R_2})$. Now for $t, t' \in [0,T]$ we have from (3.2.3) and (3.2.12) that

$$|Ky(t) - Ky(t')|$$
$$\le \left(\int_0^T |k_t(s) - k_{t'}(s)|^p \, ds \right)^{\frac{1}{p}} \left(\int_0^T [f(s, R_1) + g(s, R_2 a(s))]^q \, ds \right)^{\frac{1}{q}}$$

holds and hence from (3.2.4) we see that

$$\text{(3.2.13)} \quad |Ky(t) - Ky(t')| \to 0 \text{ as } t \to t' \text{ for } y \in C_a \cap (\overline{\Omega}_{R_1} \backslash \Omega_{R_2})$$

is true. Consequently

$$K : C_a \cap (\overline{\Omega}_{R_1} \backslash \Omega_{R_2}) \to C[0,T] \text{ is well-defined.}$$

In addition, for $y \in C_a \cap (\overline{\Omega}_{R_1} \backslash \Omega_{R_2})$, condition (3.2.2) yields

$$Ky(t) \ge a(t) \int_0^T \kappa(s)[f(s, y(s)) + g(s, y(s))] \, ds, \ t \in [0,T],$$

while (3.2.3) ensures that

$$|Ky|_0 \le \int_0^T \kappa(s)[f(s, y(s)) + g(s, y(s))] \, ds, \ t \in [0,T].$$

Combining both of these inequalities we see that for $y \in C_a \cap (\overline{\Omega}_{R_1} \backslash \Omega_{R_2})$,

$$Ky(t) \geq a(t)|Ky|_0$$

and therefore

$$K : C_a \cap (\overline{\Omega}_{R_1} \backslash \Omega_{R_2}) \to C_a \text{ is well-defined.}$$

We next use the Arzéla-Ascoli theorem to prove that the operator $K : C_a \cap (\overline{\Omega}_{R_1} \backslash \Omega_{R_2}) \to C_a$ is compact. For $y \in C_a \cap (\overline{\Omega}_{R_1} \backslash \Omega_{R_2})$, we obtain from (3.2.3) and (3.2.12)

$$|Ky|_0 \leq ||\kappa||_p \left(\int_0^T [f(s, R_1) + g(s, R_2 a(s))]^q \, ds \right)^{\frac{1}{q}} \equiv M < \infty$$

– thus $K(C_a \cap (\overline{\Omega}_{R_1} \backslash \Omega_{R_2}))$ is uniformly bounded. In addition (3.2.13) immediately guarantees the equicontinuity of $K(C_a \cap (\overline{\Omega}_{R_1} \backslash \Omega_{R_2}))$. Therefore the compactness of $K : C_a \cap (\overline{\Omega}_{R_1} \backslash \Omega_{R_2}) \to C_a$ now follows from the Arzéla-Ascoli theorem. In conclusion we have shown that (3.2.11) is true.

To apply Krasnoselskii's fixed point theorem it remains to show that

(3.2.14) $$|Ky|_0 < |y|_0 \text{ for } y \in C_a \cap \partial\Omega_{R_1}$$

and

(3.2.15) $$|Ky|_0 > |y|_0 \text{ for } y \in C_a \cap \partial\Omega_{R_2}$$

hold. Let $y \in C_a \cap \partial\Omega_{R_1}$. Then $|y|_0 = R_1$ and

$$0 \leq a(t)R_1 \leq y(t) \leq R_1, \ t \in [0, T],$$

therefore by (3.2.3), (3.2.5), (3.2.6) and (3.2.9),

$$\begin{aligned} |Ky|_0 &\leq \int_0^T \kappa(s)[f(s, y(s)) + g(s, y(s))] \, ds \\ &\leq \int_0^T \kappa(s)[f(s, R_1) + g(s, R_1 a(s))] \, ds < R_1 = |y|_0. \end{aligned}$$

Hence (3.2.14) is true.

Now let $y \in C_a \cap \partial\Omega_{R_2}$. Then $|y|_0 = R_2$ and in addition

$$0 \leq a(t)R_2 \leq y(t) \leq R_2, \ t \in [0, T].$$

Then from (3.2.2), (3.2.5), (3.2.6) and (3.2.10) we obtain

$$\begin{aligned} |Ky|_0 \geq Ky(t^\star) &\geq a(t^\star)\int_0^T \kappa(s)[f(s,y(s))+g(s,y(s))]\,ds \\ &\geq a(t^\star)\int_0^T \kappa(s)[f(s,R_2a(s))+g(s,R_2)]\,ds \\ &> R_2 = |y|_0. \end{aligned}$$

Thus (3.2.15) holds.

With the conditions of Krasnoselskii's fixed point theorem satisfied, we are guaranteed that the operator K has a fixed point in $C_a \cap (\overline{\Omega}_{R_1}\backslash\Omega_{R_2})$, that is, (3.2.1) has a positive solution $y \in C[0,T]$ with $0 < R_2 < |y|_0 < R_1$ and $y(t) \geq R_2a(t) \geq 0$, $t \in [0,T]$. (It is clear than an analogous result holds if $R_1 < R_2$.) □

Remark 3.2.1. Note that the solution $y \in C[0,T]$ of (3.2.1) guaranteed by Theorem 3.2.1 is positive whenever $a \in C[0,T]$ is, since either

$$y(t) \geq R_1a(t) \geq 0,\ \ t\in[0,T],\ \text{ or }\ y(t) \geq R_2a(t) \geq 0,\ \ t\in[0,T]$$

is true. Clearly then if a is strictly positive on $[0,T]$, that is, $a(t) > 0$ for $t \in [0,T]$, then the solution too must be strictly positive on $[0,T]$.

The assumption that $g : [0,T]\times(0,\infty) \to [0,\infty)$ is such that $g(t,0)$ is undefined in Theorem 3.2.1 forces $a \in C[0,T]$ in (3.2.2) to be strictly positive for a.e. $t \in [0,T]$. However this assumption on $a \in C[0,T]$ can be relaxed if $g(t,0)$ is defined for a.e. $t \in [0,T]$. We just state the result as the proof is almost identical to that of the proof of Theorem 3.2.1.

Theorem 3.2.2. *Let $1 \leq p \leq \infty$ be a constant and q be such that $1/p + 1/q = 1$. Suppose that* (3.2.3), (3.2.4), (3.2.9), (3.2.10),

$$(3.2.16)\quad \begin{cases} \text{there exists } a \in C[0,T] \text{ and } t^\star \in [0,T] \text{ with } a(t) \geq 0, \\ t \in [0,T] \text{ and } a(t^\star) > 0, \text{ in addition to } \kappa \in L^p[0,T] \\ \text{with } \kappa(t) \geq 0 \text{ a.e. } t \in [0,T] \text{ and } ||\kappa||_p > 0, \text{ such that} \\ a(t)\kappa(s) \leq k(t,s) \text{ for all } t \in [0,T],\ \text{a.e. } s \in [0,T], \end{cases}$$

$$(3.2.17)\quad \begin{cases} f : [0,T]\times[0,\infty) \to [0,\infty) \text{ is } L^q\text{-Carathéodory}, \\ \text{with } f(t,y) \text{ nondecreasing in } y, \end{cases}$$

$$(3.2.18)\quad \begin{cases} g : [0,T]\times[0,\infty) \to [0,\infty) \text{ is } L^q\text{-Carathéodory}, \\ \text{with } g(t,y) \text{ nonincreasing in } y \end{cases}$$

and

$$(3.2.19)\qquad \begin{cases} \displaystyle\int_0^T \kappa(s)[f(s,R)+g(s,Ra(s))]\,ds > 0 \text{ and} \\ \displaystyle\int_0^T \kappa(s)[f(s,Ra(s))+g(s,R)]\,ds > 0 \text{ for any } R > 0, \end{cases}$$

hold. Then (3.2.1) *has at least one* **positive** *solution* $y \in C[0,T]$ *and either*

(A) $0 < R_1 < |y|_0 < R_2$ *and* $y(t) \geq a(t)R_1 \geq 0,\ t \in [0,T]$ *if* $R_1 < R_2$

or

(B) $0 < R_2 < |y|_0 < R_1$ *and* $y(t) \geq a(t)R_2 \geq 0,\ t \in [0,T]$ *if* $R_2 < R_1$

holds.

Remark 3.2.2. It may not be immediately obvious what type of functions f and g satisfy conditions (3.2.9) and (3.2.10) of Theorem 3.2.1. In fact these conditions are quite general and are satisfied by a fairly large class of functions. We illustrate below with some examples. While we will discuss examples of functions f and g which satisfy (3.2.9) and (3.2.10), to simplify the examples and avoid getting lost in details, we will assume that k satisfies the hypotheses of a special case of Theorem 3.2.1.

The following result is a special case of Theorem 3.2.1 where $a \in C[0,T]$ is strictly positive and $f : [0,T] \times [0,\infty) \to [0,\infty)$ and $g : [0,T] \times (0,\infty) \to [0,\infty)$ are both continuous.

Theorem 3.2.3. *Suppose that* (3.2.3), (3.2.4) *with* $p = 1$, *and* (3.2.9) *hold and suppose also that there exists* $t^\star \in [0,T]$ *such that* (3.2.10) *is true. In addition assume that*

$$(3.2.20)\qquad \begin{cases} \text{there exists } a \in C[0,T] \text{ with } a(t) > 0 \text{ for } t \in [0,T], \\ \text{in addition to } \kappa \in L^1[0,T] \text{ with } \kappa(t) \geq 0 \text{ a.e. } t \in [0,T] \\ \text{and } ||\kappa||_1 > 0 \text{ such that } a(t)\kappa(s) \leq k(t,s) \\ \text{for all } t \in [0,T], \text{ a.e. } s \in [0,T] \end{cases}$$

and

$$(3.2.21)\qquad \begin{cases} f : [0,T] \times [0,\infty) \to [0,\infty) \text{ is continuous,} \\ \text{with } f(t,y) \text{ nondecreasing in } y, \\ g : [0,T] \times (0,\infty) \to [0,\infty) \text{ is continuous,} \\ \text{with } g(t,y) \text{ nonincreasing in } y \\ \text{and } f(t,y)+g(t,y) > 0 \text{ for } y > 0,\ t \in [0,T] \end{cases}$$

hold. Then (3.2.1) *has at least one* **strictly positive** *solution* $y \in C[0,T]$ *and either*

(A) $0 < R_1 < |y|_0 < R_2$ *and* $y(t) \geq a(t)R_1 > 0,\ t \in [0,T]$ *if* $R_1 < R_2$

or

(B) $0 < R_2 < |y|_0 < R_1$ *and* $y(t) \geq a(t)R_2 > 0,\ t \in [0,T]$ *if* $R_2 < R_1$

holds.

PROOF: Conditions (3.2.2) – (3.2.7) and (3.2.9) – (3.2.10) of Theorem 3.2.1 hold with $p = 1$. The strict positivity of $a \in C[0,T]$ along with the fact that f and g are continuous implies that (3.2.8) holds with $q = \infty$. The result now follows from Theorem 3.2.1. □

Remark 3.2.3. The above theorem is similar to a result established in [101] for (3.2.1). Note that since $a \in C[0,T]$ in (3.2.20) is such that $a(t) > 0$, $t \in [0,T]$, we see from (3.2.3) and (3.2.20) that the kernel k also satisfies

$$(3.2.22) \qquad \begin{cases} \text{there exists } 0 < M \leq 1 \text{ and } \kappa \in L^1[0,T] \text{ such that} \\ k(t,s) \geq M\kappa(s) \geq 0 \text{ for all } t \in [0,T], \text{ a.e. } s \in [0,T], \end{cases}$$

where $M := \min_{t \in [0,T]} a(t)$. It is such a condition that is placed on the kernel in the treatment of (3.2.1) in Theorem 2.1 in [101]. Here Krasnoselskii's fixed point theorem is used to prove the existence of at least one positive solution of (3.2.1) and the details of the proof are similar to those in the proof of Theorem 3.2.1. In this case the appropriate cone is given by

$$C_M := \{y \in C[0,T] : y(t) \geq M|y|_0 \text{ for all } t \in [0,T]\}.$$

In the following three examples we assume for simplicity that a and κ are as described in Theorem 3.2.3.

Example 3.2.1. Suppose that $f(t,y) = y^n$, $n \geq 0$ and $g \equiv 0$. Clearly (3.2.21) is satisfied. Fix $t^\star \in [0,T]$ and note that since $0 < a(t) \leq 1$, we have for all $n \geq 0$,

$$a(t^\star)\int_0^T \kappa(s)a^n(s)\,ds \leq ||\kappa||_1 = \int_0^T \kappa(s)\,ds.$$

Firstly assume that $0 \leq n < 1$, that is, f exhibits sublinear growth. Then (3.2.9) and (3.2.10) are satisfied with $R_2 < R_1$ such that

$$0 < R_2^{1-n} < a(t^\star)\int_0^T \kappa(s)a^n(s)\,ds \leq ||\kappa||_1 < R_1^{1-n}.$$

Alternatively, assume that f is superlinear, that is, that $f(y) = y^n$ where $n > 1$. Now (3.2.9) and (3.2.10) are satisfied with $R_1 < R_2$ such that

$$R_1^{n-1} < \frac{1}{||\kappa||_1} \leq \frac{1}{a(t^\star)\displaystyle\int_0^T \kappa(s)a^n(s)\,ds} < R_2^{n-1}.$$

Example 3.2.2. Suppose that $f \equiv 0$ and $g(t,y) = y^{-n}$, $n \geq 0$. It is easy to see that (3.2.21) holds. Fix $t^\star \in [0,T]$ and observe that

$$a(t^\star)||\kappa||_1 \leq ||\kappa||_1 = \int_0^T \kappa(s)\,ds \leq \int_0^T \kappa(s)a^{-n}(s)\,ds.$$

On this occasion (3.2.9) and (3.2.10) are satisfied with $R_2 < R_1$ such that

$$R_2^{1+n} < a(t^\star)||\kappa||_1 \leq \int_0^T \kappa(s)a^{-n}(s)\,ds < R_1^{1+n}.$$

Example 3.2.3. Let $f(t,y) = 1 + y^m + y^n$, $0 \leq m < 1 < n$ and $g \equiv 0$. Fix $t^\star \in [0,T]$. Since

$$\frac{R}{a(t^\star)\displaystyle\int_0^T \kappa(s)[1 + a^m(s)R^m + a^n(s)R^n]\,ds} \to 0 \text{ as } R \to 0^+ \text{ and } R \to \infty,$$

there exists $0 < r < \tilde{r}$ such that (3.2.10) is satisfied with both $R_2 = R_2'$ and $R_2 = R_2''$ where $R_2' \in (0,r)$ and $R_2'' \in (\tilde{r},\infty)$. In addition if

$$\sup_{R\in[0,\infty)} \frac{R}{||\kappa||_1[1 + R^m + R^n]} > 1$$

then there exists $R_1 > 0$ which satisfies (3.2.9). Here $0 < R_2' < R_1 < R_2''$.

Remark 3.2.4. One advantage of the theorems in this section is that in certain cases it may be possible to apply the theorem repeatedly to an integral equation to yield **multiple** positive solutions. For example, suppose that the kernel k satisfies the hypotheses of Theorem 3.2.3 and the functions f and g are as described above in Example 3.2.3. Then one application of Theorem 3.2.3 will yield a positive solution $y_1 \in C[0,T]$ of (3.2.1) where

$$0 < R_2' < |y_1|_0 < R_1 \text{ and } 0 < a(t)R_2' \leq y_1(t) < R_1,\ t \in [0,T],$$

while a second application will guarantee the existence of a solution $y_2 \in C[0,T]$ such that

$$0 < R_1 < |y_2|_0 < R_2'' \text{ and } 0 < a(t)R_1 \leq y_2(t) < R_2'', \ t \in [0,T].$$

This idea was illustrated in more detail for nonsingular integral equations in Chapter 2 (in particular see Theorem 2.2.3).

3.3. An application.

In this section we apply the main result of the previous section – Theorem 3.2.1 – to prove that under certain conditions on f and g, the boundary value problem

$$\begin{cases} -y'' = f(t,y) + g(t,y), \ t \in (0,1) \\ \alpha y(0) - \beta y'(0) = 0 \\ \gamma y(1) + \delta y'(1) = 0, \end{cases} \tag{3.3.1}$$

similar to that discussed in [63,64,97], has one (or more) continuous, positive solution(s). Here α, β, γ, $\delta \geq 0$ and

$$\rho := \gamma\beta + \alpha\gamma + \alpha\delta > 0. \tag{3.3.2}$$

Since the boundary value problem is equivalent to the integral equation

$$y(t) = \int_0^1 G(t,s)[f(s,y(s)) + g(s,y(s))]\, ds, \tag{3.3.3}$$

where the Green's function is given by

$$G(t,s) = \frac{1}{\rho}\begin{cases} (\gamma + \delta - \gamma t)(\beta + \alpha s), & 0 \leq s \leq t \leq 1 \\ (\beta + \alpha t)(\gamma + \delta - \gamma s), & 0 \leq t \leq s \leq 1, \end{cases} \tag{3.3.4}$$

we show that under certain hypotheses on f and g the conditions of Theorem 3.2.1 are satisfied with

$$k(t,s) := G(t,s),$$

thus giving the desired result for (3.3.1) and (3.3.3).

We concentrate on the kernel and show that $k(t,s) = G(t,s)$ satisfies (3.2.2) – (3.2.4) with $p = \infty$ and $T = 1$. As in [63], for convenience we let

$$\phi(t) := \gamma + \delta - \gamma t \quad \text{and} \quad \psi(t) := \beta + \alpha t, \ t \in [0,1].$$

Since ϕ is nonincreasing, while ψ is nondecreasing we obtain

$$k(t,s) \le k(s,s), \;\; 0 \le t,s \le 1.$$

Define $\kappa \in C[0,1]$ by

$$\kappa(s) := k(s,s) = G(s,s), \;\; s \in [0,1] \tag{3.3.5}$$

and observe that

$$\kappa(s) = k(s,s) > 0 \text{ for } s \in (0,1).$$

Moreover for $t \in (0,1)$ we see that

$$\frac{k(t,s)}{\kappa(s)} = \frac{k(t,s)}{k(s,s)} = \begin{cases} \dfrac{\gamma+\delta-\gamma t}{\gamma+\delta-\gamma s}, & 0 \le s \le t < 1 \\ \dfrac{\beta+\alpha t}{\beta+\alpha s}, & 0 < t \le s \le 1 \end{cases} \tag{3.3.6}$$

$$\ge \begin{cases} \dfrac{\gamma+\delta-\gamma t}{\gamma+\delta}, & 0 \le s \le t < 1 \\ \dfrac{\beta+\alpha t}{\beta+\alpha}, & 0 < t \le s \le 1. \end{cases}$$

Therefore defining $a \in C[0,1]$ by

$$\begin{cases} a(t) := \dfrac{(\gamma+\delta-\gamma t)(\beta+\alpha t)}{(\gamma+\delta)(\beta+\alpha)} = \dfrac{\phi(t)\psi(t)}{\eta}, \;\; t \in [0,1], \\ (\text{where } \eta := (\gamma+\delta)(\beta+\alpha)) \end{cases} \tag{3.3.7}$$

we see from (3.3.6) that

$$k(t,s) \ge a(t)\kappa(s), \;\; 0 \le t,s \le 1.$$

Furthermore one can verify that

$$a(t) > 0 \text{ for } t \in (0,1).$$

We have thus shown that

$$\begin{cases} \text{there exists } a \in C[0,1] \text{ with } a(t) > 0 \text{ for } t \in (0,1), \\ \text{in addition to } \kappa \in C[0,1] \text{ with } \kappa(t) > 0 \text{ for } t \in (0,1), \\ \text{such that } a(t)\kappa(s) \le k(t,s) \le \kappa(s) \text{ for all } t \in [0,1],\ s \in [0,1] \end{cases}$$

Thus $k(t,s) = G(t,s)$ satisfies (3.2.2) and (3.2.3) with $p = \infty$ and $T = 1$. Finally it is clear that (3.2.4) holds (again with $p = \infty$ and $T = 1$).

In addition if we assume that $f : [0,1] \times [0,\infty) \to [0,\infty)$ and $g : [0,1] \times (0,\infty) \to [0,\infty)$ are such that (3.2.5) – (3.2.10) with $q = 1$ and $T = 1$ are true (see previous section for examples) then we have the following result for (3.3.3) (and hence (3.3.1)).

Theorem 3.3.1. *Suppose that*

$$\begin{cases} G : [0,1] \times [0,1] \to [0,\infty) \text{ is as defined in (3.3.4)} \\ \text{where } \alpha,\ \beta,\ \gamma,\ \delta \geq 0 \text{ and } \rho := \gamma\beta + \alpha\gamma + \alpha\delta > 0 \end{cases} \tag{3.3.8}$$

holds and $f : [0,1] \times [0,\infty) \to [0,\infty)$ *and* $g : [0,1] \times (0,\infty) \to [0,\infty)$ *satisfy* (3.2.5) – (3.2.10) *with* $q = 1$ *and* $T = 1$, *where* $\kappa \in C[0,1]$ *and* $a \in C[0,1]$ *are as described in* (3.3.5) *and* (3.3.7) *respectively. Then* (3.3.3) *has at least one* **positive** *solution* $y \in C[0,T]$ *and either*

(A) $0 < R_1 < |y|_0 < R_2$ *and* $y(t) \geq a(t)R_1 \geq 0,\ t \in [0,T]$ *if* $R_1 < R_2$

or

(B) $0 < R_2 < |y|_0 < R_1$ *and* $y(t) \geq a(t)R_2 \geq 0,\ t \in [0,T]$ *if* $R_2 < R_1$

holds.

PROOF: The result follows from the above analysis and an application of Theorem 3.2.1. □

3.4. Existence results for a class of singular integral equations.

Recall the singular integral equation

$$y(t) = \int_0^T k(t,s)[f(s,y(s)) + g(s,y(s))]\,ds,\ t \in [0,T], \tag{3.4.1}$$

discussed in Section 3.2. One observation from the main result in Section 3.2, namely Theorem 3.2.1, is that the conditions on the kernel k are quite restrictive. On examination we see that one reason for this is that we are considering a rather large class of nonlinear functions $f + g$ and as a compromise, we transfer to the shoulders of the kernel the onus of ensuring that the image of the appropriate space under the integral operator K, lies in the chosen cone. Therefore it seems reasonable to ask how we might strengthen the conditions on the nonlinear part $f + g$, allowing it to better utilise the properties of the domain on which it operates and thus (hopefully) reducing the restrictions on the kernel.

To some extent we saw this happen in Chapter 2, where we discussed (3.4.1) but with $g \equiv 0$. In this case it was not essential that the solution $y \in C[0,T]$ be strictly greater than zero almost everywhere on the interval $[0,T]$. Therefore, in for example Theorem 2.2.1, the conditions on the kernel k are not as restrictive as those described in Theorem 3.2.1.

Suppose then that we have a kernel $k : [0,T] \times [0,T] \to \mathbf{R}$ that satisfies

$$0 \le k_t = k(t,s) \in L^1[0,T] \text{ for each } t \in [0,T], \tag{3.4.2}$$

and

$$\text{the map } t \mapsto k_t \text{ is continuous from } [0,T] \text{ to } L^1[0,T], \tag{3.4.3}$$

and suppose that K_1 and K_2 are such that

$$\begin{cases} K_1 := \sup_{t\in[0,T]} \int_0^T k(t,s)\,ds, \\ K_2 := \inf_{t\in[0,T]} \int_0^T k(t,s)\,ds > 0 \text{ and } K_1 \ne K_2 \end{cases} \tag{3.4.4}$$

holds. In addition let $K_L : C[0,T] \to C[0,T]$ be defined by

$$K_L y(t) := \int_0^T k(t,s)y(s)\,ds, \quad t \in [0,T].$$

Out of interest, let's restrict the linear operator K_L to the cone

$$C_M := \{y \in C[0,T] : y(t) \ge M|y|_0 \text{ for all } t \in [0,T]\}.$$

(Here $0 < M < 1$ is a predetermined constant.) Then for $y \in C_M$ we see that

$$|K_L y|_0 \le \sup_{t\in[0,T]} \int_0^T k(t,s)y(s)\,ds \le K_1|y|_0,$$

which in addition to (3.4.2) and (3.4.3) implies

$$K_L y(t) \ge \inf_{t\in[0,T]} \int_0^T k(t,s)y(s)\,ds \ge K_2 M|y|_0 \ge M\frac{K_2}{K_1}|K_L y|_0. \tag{3.4.5}$$

By (3.4.4) we have that $K_2/K_1 < 1$ and hence $1 > M > MK_2/K_1$. Now (3.4.5) implies that

$$K_L : C_M \to C_{M\frac{K_2}{K_1}}, \tag{3.4.6}$$

that is, K_L maps C_M to the larger cone C_{MK_2/K_1}. This suggests that we are looking for an operator F that takes C_M into a smaller cone.

Suppose then that $F : C[0,T] \to C[0,T]$ is given by

$$Fy(t) := y^m(t), \quad t \in [0,T] \text{ and } 0 \leq m < 1.$$

For $y \in C_M$ we see that $y(t) \geq M|y|_0 \geq 0$ for all $t \in [0,T]$, thus implying that

$$y^m(t) \geq M^m |y|_0^m = M^m |y^m|_0,$$

that is

$$y^m \in C_{M^m} := \{y \in C[0,T] : y(t) \geq M^m |y|_0 \text{ for all } t \in [0,T]\}.$$

Hence

$$F : C_M \to C_{M^m} \subseteq C_M. \tag{3.4.7}$$

The fact that C_{M^m} is a smaller cone than C_M (since $K_2/K_1 < 1$) is ideal for us, since if we define

$$M := \left(\frac{K_2}{K_1}\right)^{\frac{1}{1-m}} \tag{3.4.8}$$

then from (3.4.6) we see that

$$K_L : C_{M^m} \to C_{M^m \frac{K_2}{K_1}} = C_M. \tag{3.4.9}$$

Consequently (3.4.7) and (3.4.9) imply that $K = K_L F : C_M \to C_M$, and in particular (assuming that $R_2 < R_1$)

$$K : C_M \cap (\overline{\Omega}_{R_1} \backslash \Omega_{R_2}) \to C_M,$$

where as before we define

$$\Omega_{R_i} := \{y \in C[0,T] : |y|_0 < R_i\}, \; i = 1, 2.$$

Alternatively suppose that

$$Fy(t) := y^{-m}(t), \quad t \in [0,T] \text{ and } 0 \leq m < 1$$

and define M as in (3.4.8). If $y \in C_M \cap (\overline{\Omega}_{R_1} \backslash \Omega_{R_2})$ then $0 < MR_2 \leq M|y|_0 \leq y(t) \leq |y|_0 \leq R_1$. Consequently for $y \in C_M \cap (\overline{\Omega}_{R_1} \backslash \Omega_{R_2})$,

$$|Ky|_0 \leq \sup_{t \in [0,T]} \int_0^T k(t,s) y^{-m}(s)\, ds \leq K_1 M^{-m} |y|_0^{-m},$$

while

$$\begin{aligned} Ky(t) &\geq \inf_{t\in[0,T]} \int_0^T k(t,s)y^{-m}(s)\,ds \\ &\geq K_2|y|_0^{-m} \geq \frac{K_2}{K_1}M^m|Ky|_0 = M|Ky|_0,\ t\in[0,T]. \end{aligned}$$

Therefore in this case also

$$K : C_M \cap (\overline{\Omega}_{R_1}\backslash\Omega_{R_2}) \to C_M.$$

We can summarise the above in the following theorem:

Theorem 3.4.1. *Suppose that* (3.4.2) – (3.4.4) *and either*

(3.4.10) $$f(y) = y^m \text{ where } 0 \leq m < 1$$

or

(3.4.11) $$f(y) = y^{-m} \text{ where } 0 \leq m < 1$$

hold. Then

(3.4.12) $$y(t) = \int_0^T k(t,s)f(y(s))\,ds,\ \ t\in[0,T]$$

has a positive solution $y \in C[0,T]$ *with* $0 < R_2 < |y|_0 < R_1$ *and* $y(t) \geq MR_2$, $t\in[0,T]$. *Here* M *is as defined in* (3.4.8), *and* R_1 *and* R_2 *are as described in* (3.4.13) *(see below) if* (3.4.10) *holds, whereas if* (3.4.11) *holds,* R_1 *and* R_2 *satisfy* (3.4.16) *(see below).*

PROOF: From the details outlined above and using some of the ideas from the proof of Theorem 3.2.1, one can show that

$$\begin{cases} K = K_LF : C_M \cap (\overline{\Omega}_{R_1}\backslash\Omega_{R_2}) \to C_M \text{ is a} \\ \text{continuous and completely continuous operator.} \end{cases}$$

In addition if (3.4.10) holds, then there exists $0 < R_2 < R_1$ such that

(3.4.13) $$0 < R_2^{1-m} < K_2M^m \leq K_2 \leq K_1 < R_1^{1-m},$$

implying that

(3.4.14) $$|Ky|_0 < |y|_0 \text{ for } y \in C_M \cap \partial\Omega_{R_1}$$

and

$$|Ky|_0 > |y|_0 \text{ for } y \in C_M \cap \partial\Omega_{R_2} \tag{3.4.15}$$

are true. The result now follows from Krasnoselskii's fixed point theorem.

Alternatively, suppose that (3.4.11) is true. Then there exists $0 < R_2 < R_1$ such that

$$0 < R_2^{1+m} < K_2 \le K_1 \le K_1 M^{-m} < R_1^{1+m}, \tag{3.4.16}$$

implying that (3.4.14) and (3.4.15) are satisfied. Once again the desired result follows directly from Krasnoselskii's fixed point theorem. □

Remark 3.4.1. In (3.4.4) we choose k such that $K_1 \neq K_2$. If $K_1 = K_2$ then for the cone C_M we would have that $M = 1$, and any solution of (3.4.12) would be a constant solution.

We now present a general form of Theorem 3.4.1 which, as above allows us to keep minimal conditions on the kernel k, while enabling us to consider certain nonlinear functions $f+g$ with f nondecreasing and g nonincreasing and possibly undefined at zero.

Theorem 3.4.2. *Suppose that* (3.4.2) – (3.4.4),

$$\begin{cases} f : [0,\infty) \to [0,\infty) \textit{ is continuous and nondecreasing,} \\ g : (0,\infty) \to [0,\infty) \textit{ is continuous and nonincreasing} \\ \textit{and } f+g : (0,\infty) \to (0,\infty), \end{cases} \tag{3.4.17}$$

$$\begin{cases} \textit{there exists a continuous function } \psi : (0,1) \to (0,\infty) \\ \textit{such that for any } 0 < \tilde{M} < 1 \textit{ and } u > 0, \textit{ we have} \\ f(\tilde{M}u) + g(u) \ge \psi(\tilde{M})[f(u) + g(\tilde{M}u)], \end{cases} \tag{3.4.18}$$

$$\textit{there exists } 0 < M < 1 \textit{ such that } \frac{M}{\psi(M)} \le \frac{K_2}{K_1}, \tag{3.4.19}$$

$$\textit{there exists } R_1 > 0 \textit{ such that } 1 < \frac{R_1}{K_1[f(R_1) + g(MR_1)]} \tag{3.4.20}$$

and

$$\begin{cases} \textit{there exists } R_2 > 0,\ R_2 \neq R_1, \textit{ such that} \\ 1 > \dfrac{R_2}{K_2[f(MR_2) + g(R_2)]} \end{cases} \tag{3.4.21}$$

hold. Then

$$y(t) = \int_0^T k(t,s)[f(y(s)) + g(y(s))]\,ds, \quad t \in [0,T] \tag{3.4.22}$$

has at least one positive solution $y \in C[0,T]$ *and either*

(A) $0 < R_1 < |y|_0 < R_2$ *and* $y(t) \geq MR_1, t \in [0,T]$ *if* $R_1 < R_2$

or

(B) $0 < R_2 < |y|_0 < R_1$ *and* $y(t) \geq MR_2, t \in [0,T]$ *if* $R_2 < R_1$

holds.

PROOF: Define the operator

$$Ky(t) := \int_0^T k(t,s)[f(y(s)) + g(y(s))]\,ds, \quad t \in [0,T]$$

and the cone

$$C_M := \{y \in C[0,T] : y(t) \geq M|y|_0, \ t \in [0,T]\},$$

where M is as defined in (3.4.19). Suppose that $R_2 < R_1$ (a similar argument holds if $R_1 < R_2$) and let $y \in C_M \cap (\overline{\Omega}_{R_1} \backslash \Omega_{R_2})$. We have that $0 < M|y|_0 \leq y(t) \leq |y|_0$, $t \in [0,T]$ and therefore from (3.4.17) one obtains

$$|Ky|_0 \leq K_1[f(|y|_0) + g(M|y|_0)]. \tag{3.4.23}$$

In addition, (3.4.17) – (3.4.19) and (3.4.23) gives for $t \in [0,T]$,

$$\begin{aligned} Ky(t) &\geq K_2[f(M|y|_0) + g(|y|_0)] \\ &\geq K_2\psi(M)[f(|y|_0) + g(M|y|_0)] \\ &\geq \frac{K_2\psi(M)}{K_1}|Ky|_0 \geq M|Ky|_0. \end{aligned}$$

and hence $K : C_M \cap (\overline{\Omega}_{R_1} \backslash \Omega_{R_2}) \rightarrow C_M$. The continuity and complete continuity of $K : C_M \cap (\overline{\Omega}_{R_1} \backslash \Omega_{R_2}) \rightarrow C_M$ follows in a similar fashion to that detailed in Theorem 3.2.1.

It remains to show that (3.4.14) and (3.4.15) hold. Once again the technique (using (3.4.20) and (3.4.21)) is similar to that illustrated in Theorem 3.2.1, therefore we omit the detail. The desired result now follows from Krasnoselskii's fixed point theorem. □

Obviously Theorem 3.4.1 is a special case of the above, with $g \equiv 0$ and f as described in either (3.4.10) or (3.4.11). In both of these cases $\psi(M) = M^m$ and $M = (K_2/K_1)^{\frac{1}{1-m}}$ satisfies (3.4.18) and (3.4.19) respectively. We finish this section with an example.

Example 3.4.1. Suppose that

$$f(y) = y^m \text{ and } g(y) = y^{-n}, \text{ where } 0 \le n \le m \le 1.$$

Then for any $M \in (0,1)$,

$$\begin{aligned} f(Mu) + g(u) &= M^m u^m + u^{-n} = M^m[u^m + M^{-m}u^{-n}] \\ &\ge M^m[u^m + M^{-n}u^{-n}] = M^m[f(u) + g(Mu)] \end{aligned}$$

– thus f and g satisfy (3.4.18) with $\psi(M) = M^m$ and (3.4.19) is satisfied with $0 < M = (K_2/K_1)^{\frac{1}{1-m}}$.

Alternatively suppose that

$$f(y) = y^m \text{ and } g(y) = y^{-n}, \text{ where now } 0 \le m \le n \le 1.$$

Then for any $M \in (0,1)$,

$$\begin{aligned} f(Mu) + g(u) &= M^m u^m + u^{-n} \ge M^n u^m + u^{-n} \\ &= M^n[u^m + M^{-n}u^{-n}] = M^n[f(u) + g(Mu)] \end{aligned}$$

– thus f and g satisfy (3.4.18) with $\psi(M) = M^n$ and (3.4.19) is satisfied with $0 < M = (K_2/K_1)^{\frac{1}{1-n}}$.

3.5. Schauder's fixed point theorem and positive solutions.

Up to now in this chapter we have used Krasnoselskii's fixed point theorem to determine the existence of positive solutions of various types of singular integral equations. The advantage of this technique is that repeated application of Krasnoselskii's fixed point theorem (when the nonlinearities allow) can provide us with a multiple solutions result. However, as already mentioned in Section 3.4, in certain cases we are forced to place rather strict conditions on the kernel k, to ensure that the integral operator brings the appropriate cone intersected with the annulus, back into the cone.

In this section, we use Schauder's fixed point theorem to provide us with existence results for the singular integral equation

$$(3.5.1) \qquad y(t) = h(t) + \int_0^T k(t,s)[f(s,y(s)) + g(s,y(s))]\,ds, \ t \in [0,T].$$

The benefit in using this approach is that we are able to consider a large class of kernels k. The drawback is that obtaining multiple solutions of the integral equation becomes more difficult.

Notation. Throughout this section we let

$$K_1 := \sup_{t\in[0,T]} \int_0^T k(t,s)\,ds \quad \text{and} \quad K_2 := \inf_{t\in[0,T]} \int_0^T k(t,s)\,ds. \tag{3.5.2}$$

Theorem 3.5.1. *Suppose that*

$$h \in C[0,T] \text{ with } h(t) \geq 0,\ t \in [0,T], \tag{3.5.3}$$

$$0 < k_t = k(t,s) \in L^1[0,T] \text{ for each } t \in [0,T], \tag{3.5.4}$$

$$\text{the map } t \mapsto k_t \text{ is continuous from } [0,T] \text{ to } L^1[0,T], \tag{3.5.5}$$

$$\begin{cases} f : [0,\infty) \to [0,\infty) \text{ is continuous and nondecreasing,} \\ g : (0,\infty) \to [0,\infty) \text{ is continuous and nonincreasing} \\ \text{and } f+g : (0,\infty) \to (0,\infty) \end{cases} \tag{3.5.6}$$

and

$$\begin{cases} \text{there exists } 0 < \beta < \alpha \text{ such that} \\ \dfrac{\beta}{\min\limits_{t\in[0,T]} h(t) + K_2[f(\beta)+g(\alpha)]} < 1 < \dfrac{\alpha}{|h|_0 + K_1[f(\alpha)+g(\beta)]} \\ \text{(here } K_1 \text{ and } K_2 \text{ are as defined in (3.5.2))} \end{cases} \tag{3.5.7}$$

hold. Then (3.5.1) *has at least one positive solution* $y \in C[0,T]$ *such that* $0 < \beta < y(t) < \alpha$, $t \in [0,T]$.

PROOF: Define $f^\star : \mathbf{R} \to [f(\beta), f(\alpha)]$ and $g^\star : \mathbf{R} \to [g(\alpha), g(\beta)]$ by

$$f^\star(y) := \begin{cases} f(\alpha) & , y \geq \alpha \\ f(y) & , \beta \leq y \leq \alpha \\ f(\beta) & , y \leq \beta, \end{cases} \quad \text{and} \quad g^\star(y) := \begin{cases} g(\alpha) & , y \geq \alpha \\ g(y) & , \beta \leq y \leq \alpha \\ g(\beta) & , y \leq \beta \end{cases}$$

respectively. In addition let the operator $K^\star$ be given by

$$K^\star y(t) := h(t) + \int_0^T k(t,s)[f^\star(y(s)) + g^\star(y(s))]\,ds,\ t \in [0,T],$$

and define a convex subset $C \subset C[0,T]$ by

$$C := \{y \in C[0,T] : |y|_0 \leq \alpha\}.$$

We will use Schauder's fixed point theorem (Theorem 1.2.3) to show that $K^\star$ has a fixed point in C, and consequently we must ensure that

$$K^\star : C \to C \text{ is a continuous, compact operator} \tag{3.5.8}$$

holds.

Let $y \in C$. Then for $t, t' \in [0,T]$ we have that

$$|K^\star y(t') - K^\star y(t)| \leq |h(t') - h(t)| + \int_0^T |k_{t'}(s) - k_t(s)|[f^\star(y(s)) + g^\star(y(s))]\, ds.$$

Furthermore (3.5.6) yields

$$\begin{aligned} &|K^\star y(t') - K^\star y(t)| \\ &\qquad \leq |h(t') - h(t)| + \left(\int_0^T |k_{t'}(s) - k_t(s)|\, ds\right)[f(\alpha) + g(\beta)]. \end{aligned} \tag{3.5.9}$$

From this inequality and (3.5.5) we see that

$$|K^\star y(t') - K^\star y(t)| \to 0 \text{ as } t \to t',$$

and hence $K^\star : C \to C[0,T]$ is well-defined. In addition note that for $y \in C$, (3.5.6) and (3.5.7) imply

$$|K^\star y|_0 \leq |h|_0 + K_1[f(\alpha) + g(\beta)] < \alpha, \tag{3.5.10}$$

and consequently $K^\star : C \to C$ is well-defined.

We next show that $K^\star : C \to C$ is a continuous operator. Let $y_n \to y$ in $C[0,T]$. Then for any $t \in [0,T]$,

$$\begin{aligned} |K^\star y_n(t) - K^\star y(t)| \;\leq\; K_1 \Bigg[&\sup_{t \in [0,T]} |f^\star(y_n(t)) - f^\star(y(t))| \\ + &\sup_{t \in [0,T]} |g^\star(y_n(t)) - g^\star(y(t))| \Bigg] \end{aligned}$$

and hence by (3.5.6) we have that $K^\star : C \to C$ is a continuous operator.

Finally we use the Arzéla-Ascoli theorem to show that $K^\star : C \to C$ is compact. Immediately we obtain from (3.5.10) that $K^\star(C)$ is uniformly bounded, while the equicontinuity of $K^\star(C)$ follows from (3.5.9).

The Arzéla-Ascoli theorem thus guarantees that $K^\star : C \to C$ is a compact operator.

We have therefore shown that (3.5.8) is true and hence by Schauder's fixed point theorem, $K^\star : C \to C$ has a fixed point $y \in C$, that is,

$$y(t) = h(t) + \int_0^T k(t,s)[f^\star(y(s)) + g^\star(y(s))]\,ds, \ t \in [0,T],$$

has a solution $y \in C[0,T]$. From (3.5.7) however we see that

$$|y|_0 \le |h|_0 + K_1[f(\alpha) + g(\beta)] < \alpha$$

and

$$y(t) \ge \min_{t\in[0,T]} h(t) + K_2[f(\beta) + g(\alpha)] > \beta, \ t \in [0,T],$$

that is, $\beta < y(t) < \alpha$, $t \in [0,T]$. Consequently, by the definition of $f^\star$ and $g^\star$, we have that $y \in C[0,T]$ is a solution of (3.5.1). □

Example 3.5.1. Suppose for simplicity that $h \equiv 0$ in Theorem 3.5.1 and let $f(y) = y^p$, $0 \le p < 1$ and $g(y) = y^{-q}$, $0 \le q < 1$. Since it is possible to find (for any positive $K_2 \le K_1$) $0 < \beta < \alpha$ such that

$$\frac{\alpha^q \beta}{\alpha^q \beta^p + 1} < K_2 \le K_1 < \frac{\alpha \beta^q}{\alpha^p \beta^q + 1}$$

holds, then it is clear that this choice of f and g satisfies (3.5.6) and (3.5.7).

If $h \in C[0,T]$ in Theorem 3.5.1 is strictly positive, that is, $h(t) > 0$, $t \in [0,T]$, then defining $m > 0$ by

$$m := \min_{t\in[0,T]} h(t) > 0,$$

we note that provided

(3.5.11) there exists $\alpha > 0$ such that $1 < \dfrac{\alpha}{|h|_0 + K_1[f(\alpha) + g(m)]}$

holds, the hypotheses of Theorem 3.5.1 are satisfied with $\beta = m$. We thus state the following special case of Theorem 3.5.1 that occurs when $h \in C[0,T]$ is strictly positive.

Theorem 3.5.2. *Suppose that* (3.5.4) – (3.5.6) *and* (3.5.11) *hold along with*

(3.5.12) $h \in C[0,T]$ *with* $h(t) > 0$ *for* $t \in [0,T]$.

Then (3.5.1) *has at least one positive solution* $y \in C[0,T]$. *In addition* $0 < h(t) < y(t) < \alpha$, $t \in [0,T]$.

We now want to extend these ideas to obtain analogous results for singular integral equations of the form

$$y(t) = h(t) + \int_0^\infty k(t,s)[f(s,y(s)) + g(s,y(s))]\,ds, \ \ t \in [0,\infty). \tag{3.5.13}$$

By revising the conditions imposed on h, k, f and g in Theorem 3.5.1, one can once again apply Schauder's fixed point theorem, this time to obtain the existence of at least one positive solution $y \in C_l[0,\infty)$ of (3.5.13).

Theorem 3.5.3. *Suppose that* (3.5.6),

$$h \in C_l[0,\infty) \text{ with } h(t) \geq 0, \ \ t \in [0,\infty), \tag{3.5.14}$$

$$0 < k_t(s) = k(t,s) \in L^1[0,\infty) \text{ for each } t \in [0,\infty), \tag{3.5.15}$$

$$\text{the map } t \mapsto k_t \text{ is continuous from } [0,\infty) \text{ to } L^1[0,\infty), \tag{3.5.16}$$

$$\begin{cases} \text{there exists } \tilde{k} \in L^1[0,\infty) \text{ such that} \\ k_t \to \tilde{k} \text{ in } L^1[0,\infty) \text{ as } t \to \infty \end{cases} \tag{3.5.17}$$

and

$$\begin{cases} \text{there exists } 0 < \beta < \alpha \text{ such that} \\ \dfrac{\beta}{\inf_{t\in[0,\infty)} h(t) + K_4[f(\beta) + g(\alpha)]} < 1 < \dfrac{\alpha}{|h|_0 + K_3[f(\alpha) + g(\beta)]} \\ (\text{here } K_3 \text{ and } K_4 \text{ are as defined in (3.5.19) below}) \end{cases} \tag{3.5.18}$$

hold. Then (3.5.13) *has at least one positive solution* $y \in C_l[0,\infty)$ *such that* $0 < \beta < y(t) < \alpha$, $t \in [0,\infty)$.

Notation. Let

$$K_3 := \sup_{t\in[0,\infty)} \int_0^\infty k(t,s)\,ds, \text{ and } K_4 := \inf_{t\in[0,\infty)} \int_0^\infty k(t,s)\,ds. \tag{3.5.19}$$

(From (3.5.15) – (3.5.17) one can show that K_3 and K_4 are well defined.)

PROOF: Let $f^\star$ and $g^\star$ be as defined in the proof of Theorem 3.5.1 let the operator $K^\star$ be given by

$$K^\star y(t) := h(t) + \int_0^\infty k(t,s)[f^\star(y(s)) + g^\star(y(s))]\,ds, \ t \in [0,\infty).$$

Defining the convex subset C of $C_l[0,\infty)$ by

$$C := \{y \in C_l[0,\infty) : |y|_0 \le \alpha\},$$

one can show that

$$K^\star : C \to C \text{ is a continuous, compact operator.}$$

The analysis is similar to that in Theorem 3.5.1 therefore we omit the detail. The theorem now follows from Schauder's fixed point theorem. Once again we omit the detail since it is almost identical to that in the proof of Theorem 3.5.1. □

Rather than require the solution of (3.5.13) to lie in $C_l[0,\infty)$, suppose we seek solutions that are in the space $BC[0,\infty) \subseteq C[0,\infty)$. Here we denote the set of all continuous functions on $[0,\infty)$ by $C[0,\infty)$. Recall that this is not a normed space, rather it is a Fréchet space and for $y \in C[0,\infty)$ we define, for each $m \in \{1,2,\ldots\}$, the seminorm $\rho_{0,m}(y)$ by

$$\rho_{0,m}(y) := \sup_{t\in[0,m]} |y(t)|.$$

Recall also that a subset C of $C[0,\infty)$ is bounded if there exists a positive function $\alpha \in C[0,\infty)$ such that $|y(t)| \le \alpha(t)$, for all $t \in [0,\infty)$ and $y \in C$. In addition recall that a subset C of $C[0,\infty)$ is compact if it is compact on each compact subinterval of $[0,\infty)$.

We conclude the section by presenting a result which relies on the Schauder-Tychonoff theorem (see Theorem 3.5.4 below for the statement of this theorem) to ensure that (3.5.13) has a positive solution in the Fréchet space $C[0,\infty)$. Conditions on h, k, f and g will then further imply that in fact the solution belongs to $BC[0,\infty)$.

Theorem 3.5.4. Schauder-Tychonoff Theorem.
Let D be a closed, convex subset of a locally convex, Hausdorff space E. Assume that $N : D \to D$ is continuous and $N(D)$ is relatively compact in E. Then N has at least one fixed point in D.

Theorem 3.5.5. *Suppose that* (3.5.6), (3.5.15), (3.5.16), (3.5.18),

(3.5.20) $\qquad h \in BC[0,\infty)$ *with* $h(t) \ge 0, \ t \in [0,\infty)$

and

(3.5.21) *the map* $t \mapsto k_t$ *is bounded from* $[0,\infty)$ *to* $L^1[0,\infty)$

hold. Then (3.5.13) *has at least one positive solution* $y \in C_l[0,\infty)$ *such that* $0 < \beta < y(t) < \alpha$, $t \in [0,\infty)$.

Remark 3.5.1. It is easy to check that conditions (3.5.15), (3.5.16) and (3.5.21) ensure that K_3 and K_4 as given in (3.5.19) are well defined.

PROOF: The idea behind the proof is quite similar to the idea in the proof of Theorem 3.5.3 except that now we use the Schauder-Tychonoff fixed point theorem.

Define $f^\star$, $g^\star$ and $K^\star$ as in the proof of Theorem 3.5.3. Let the closed, convex subset C, required by the Schauder-Tychonoff theorem, be given by

$$C := \{y \in C[0,\infty) : y \in BC[0,\infty) \text{ and } |y|_0 < \alpha\}.$$

Firstly note that $K^\star : C \to C$ is well-defined. Let $y \in C$. Then $K^\star y \in C[0,\infty)$ since for any $t, t' \in [0,\infty)$, we have from (3.5.15), (3.5.16) and (3.5.20) that

$$(3.5.22)\quad |K^\star y(t) - K^\star y(t')| \le |h(t) - h(t')| + \left(\int_0^\infty |k(t,s) - k(t',s)|\, ds\right)[f(\alpha) + g(\beta)] \to 0 \text{ as } t \to t'$$

holds. In addition, it is clear from (3.5.18) that

$$(3.5.23)\quad |K^\star y|_0 \le |h|_0 + K_3[f(\alpha) + g(\beta)] < \alpha \text{ for all } y \in C[0,\infty)$$

is true, therefore we see that $K^\star : C \to C$ is well defined. (Note in fact that $K^\star : C[0,\infty) \to C$ is well defined.)

Secondly we show that $K^\star(C)$ is relatively compact in $C \subset C[0,\infty)$. To do this we must show that $K^\star(C)$ is uniformly bounded and equicontinuous on each compact subinterval of $[0,\infty)$. It is immediate from (3.5.23) that $K^\star(C)$ is in fact uniformly bounded on $[0,\infty)$, while the equicontinuity of $K^\star(C)$ on each compact subinterval of $[0,\infty)$ follows directly from (3.5.22). Hence $K^\star(C)$ (and also $K^\star(C[0,\infty))$) is relatively compact in $C[0,\infty)$.

Lastly we are required to show that $K^\star : C \to C$ is continuous. Suppose that $y_n \to y$ in $C \subseteq C[0,\infty)$, that is, $y_n \to y$ in $C[0,m]$ for each $m \in \{1,2,\ldots\}$. Clearly this implies the pointwise convergence of y_n to y on $[0,\infty)$. Coupling this fact with (3.5.6) we obtain for each $t \in [0,\infty)$ that

$$k_t(s)[f(y_n(s)) + g(y_n(s))] \to k_t(s)[f(y(s)) + g(y(s))] \text{ a.e. } s \in [0,\infty).$$

In addition for each $t \in [0, \infty)$

$$\begin{aligned} 0 &\le k_t(s)[f(y_n(s)) + g(y_n(s))] \\ &\le k_t(s)[f(\alpha) + g(\beta)], \text{ a.e. } s \in [0, \infty), \text{ for all } n \in \mathbf{N}. \end{aligned}$$

Consequently by the Lebesgue dominated convergence theorem we have

$$K^\star y_n(t) \to K^\star y(t), \text{ for each } t \in [0, \infty) \text{ as } n \to \infty. \tag{3.5.24}$$

Now fix $m \in \{1, 2, \ldots\}$. Since $[0, m]$ is compact, combining (3.5.22) and (3.5.24) yields

$$K^\star y_n \to K^\star y \text{ in } C[0, m] \text{ as } n \to \infty.$$

Obviously this is true for any $m \in \{1, 2, \ldots\}$ and therefore $K^\star y_n \to K^\star y$ in $C[0, \infty)$ as $y_n \to y$ in C.

In summary we have that

$$K^\star : C \to C \text{ is a continuous and compact operator}$$

and therefore by the Schauder-Tychonoff theorem, $K^\star$ has a fixed point $y \in C$. Clearly $0 < \beta < y(t) < \alpha$, $t \in [0, \infty)$ and consequently we have the desired result. □

3.6. Krasnoselskii's fixed point theorem and positive solutions of Volterra integral equations.

In Chapter 2 and up to now in Chapter 3, our discussion has centred on the existence of positive solutions of various types of nonlinear Fredholm integral equations. However, the existence of positive solutions of nonlinear Volterra integral equations is also deserving of attention and has been studied in some detail in the literature. In particular we refer the reader to [36-38,74,75,161] and their references. Comparison theorems have been used in a number of cases to obtain the desired results.

In this section we use Krasnoselskii's fixed point theorem to discuss the existence of positive solutions $y \in C[0, T]$ of the nonsingular Volterra integral equation

$$y(t) = \int_0^t k(t, s) f(y(s))\, ds, \ t \in [0, T], \tag{3.6.1}$$

and the (possibly) singular Volterra integral equation

$$y(t) = \int_0^t k(t, s) g(y(s))\, ds, \ t \in [0, T]. \tag{3.6.2}$$

Once again we take singular to mean that $g(0)$ is undefined. In each case we seek the existence of a solution $y \in C[0,T]$, such that $y(0) = 0$ and $y(t) > 0$ for $t \in (0,T]$. While it will be necessary later in the paper to consider (3.6.1) and (3.6.2) separately to obtain some results of interest, initially – to avoid repeating certain arguments – we consider the more general Volterra integral equation

$$y(t) = \int_0^t k(t,s)[f(y(s)) + g(y(s))]\,ds, \;\; t \in [0,T]. \tag{3.6.3}$$

Define the operator K by

$$Ky(t) := \int_0^t k(t,s)[f(y(s)) + g(y(s))]\,ds, \;\; t \in [0,T], \tag{3.6.4}$$

and Ω_{R_1} and Ω_{R_2} by

$$\Omega_{R_1} := \{y \in C[0,T] : |y|_0 < R_1\}$$

and

$$\Omega_{R_2} := \{y \in C[0,T] : |y|_0 < R_2\}$$

respectively. We assume in what follows that $0 < R_2 < R_1$. (A similar argument holds if $0 < R_1 < R_2$.)

As we have already seen, to apply Krasnoselskii's fixed point theorem, we must first define an appropriate cone C_a and ensure that

$$K : C_a \cap (\overline{\Omega}_{R_1} \backslash \Omega_{R_2}) \to C_a \text{ is completely continuous} \tag{3.6.5}$$

holds. Let the cone C_a be defined by

$$C_a := \{y \in C[0,T] : y \text{ is nondecreasing and } y(t) \geq a(t)|y|_0\}.$$

The function a which satisfies

$$a \in C[0,T], \;\; a(0) = 0 \text{ and } a(t) > 0 \text{ for } t \in (0,T], \tag{3.6.6}$$

is a solution of an integral inequality which will be described later. Note that if $y \in C_a$, then since y is nondecreasing and $y(t) \geq 0$, $t \in [0,T]$, we have that $|y|_0 = y(T) \geq 0$.

To ensure that (3.6.5) holds we place the following conditions on k, f and g :

Let $1 \le p \le \infty$ be a constant and q be such that $1/p + 1/q = 1$. Suppose that

$$\text{(3.6.7)} \qquad \begin{cases} 0 < k(t,s) \text{ for all } t \in (0,T], \text{ a.e. } s \in [0,t], \\ \text{and } k_t(s) := k(t,s) \in L^p[0,t] \text{ for each } t \in [0,T], \end{cases}$$

$$\text{(3.6.8)} \qquad \begin{cases} \text{for any } t, t' \in [0,T], \ \int_0^{t^\star} |k_t(s) - k_{t'}(s)|^p\, ds \to 0 \\ \text{as } t \to t', \text{ where } t^\star = \min\{t, t'\}, \end{cases}$$

$$\text{(3.6.9)} \quad k(t_1,s) - k(t_2,s) > 0, \text{ for a.e. } s \in [0,t_2], \text{ where } 0 < t_2 < t_1 \le T,$$

$$\text{(3.6.10)} \qquad \begin{cases} f : [0,\infty) \to [0,\infty) \text{ is continuous and nondecreasing,} \\ g : (0,\infty) \to [0,\infty) \text{ is continuous and nonincreasing,} \\ \text{with } f(y) + g(y) > 0 \text{ for } y > 0 \end{cases}$$

and

$$\text{(3.6.11)} \qquad \int_0^T g(Ra(s))^q\, ds < \infty \text{ for any constant } R > 0$$

hold.

Let $y \in C_a \cap (\overline{\Omega}_{R_1} \backslash \Omega_{R_2})$. Then

$$0 < R_2 \le |y|_0 \le R_1 \text{ and } 0 < R_2 a(t) \le y(t) \le R_1, \ t \in (0,T]$$

and (3.6.10) yields

$$0 < f(y(t)) + g(y(t)) \le f(R_1) + g(R_2 a(t)), \ t \in (0,T].$$

(Note that since $a(t) > 0$, $t \in (0,T]$ and hence $0 < R_2 a(t) \le y(t)$, $t \in (0,T]$ we have that $g(y(t))$ is measurable for a.e. $t \in [0,T]$.) From this inequality and (3.6.11) we obtain

$$\text{(3.6.12)} \qquad \begin{cases} \int_0^T [f(y(s)) + g(y(s))]^q\, ds \le \int_0^T [f(R_1) + g(R_2 a(s))]^q\, ds \\ \qquad \le 2^{q-1} \int_0^T [f(R_1)^q + g(R_2 a(s))^q]\, ds < \infty, \end{cases}$$

that is, $f(y(t)) + g(y(t)) \in L^q[0,T]$ for $y \in C_a \cap (\overline{\Omega}_{R_1} \backslash \Omega_{R_2})$. Now for $t, t' \in [0,T]$, $t' < t$, we have from (3.6.7) and (3.6.12) that

$$\begin{aligned} |Ky(t) - Ky(t')| &\leq \int_0^{t'} |k_t(s) - k_{t'}(s)||f(y(s)) + g(y(s))|\, ds \\ &+ \int_{t'}^{t} |k_t(s)||f(y(s)) + g(y(s))|\, ds \\ &\leq \left(\int_0^{t'} |k_t(s) - k_{t'}(s)|^p\, ds\right)^{\frac{1}{p}} \\ &\times \left(\int_0^{T} [f(R_1) + g(R_2 a(s))]^q\, ds\right)^{\frac{1}{q}} \\ &+ \left(\int_{t'}^{t} |k_t(s)|^p\, ds\right)^{\frac{1}{p}} \left(\int_0^{T} [f(R_1) + g(R_2 a(s))]^q\, ds\right)^{\frac{1}{q}} \end{aligned}$$

holds and hence from (3.6.7) and (3.6.8) we see that

$$|Ky(t) - Ky(t')| \to 0 \text{ as } t \to t' \text{ for } y \in C_a \cap (\overline{\Omega}_{R_1} \backslash \Omega_{R_2}) \tag{3.6.13}$$

is true. Consequently

$$K : C_a \cap (\overline{\Omega}_{R_1} \backslash \Omega_{R_2}) \to C[0,T] \text{ is well-defined.}$$

Also, for $y \in C_a \cap (\overline{\Omega}_{R_1} \backslash \Omega_{R_2})$, and $t_2, t_1 \in [0,T]$ with $0 < t_2 < t_1 \leq T$, we see from (3.6.7), (3.6.9) and (3.6.10) that

$$\begin{aligned} Ky(t_1) - Ky(t_2) &= \int_0^{t_2} [k(t_1,s) - k(t_2,s)][f(y(s)) + g(y(s))]\, ds \\ &+ \int_{t_2}^{t_1} k(t_1,s)[f(y(s)) + g(y(s))]\, ds > 0. \end{aligned}$$

Thus $Ky(t)$ is nondecreasing in t, for $y \in C_a \cap (\overline{\Omega}_{R_1} \backslash \Omega_{R_2})$.

In addition, if $y \in C_a \cap (\overline{\Omega}_{R_1} \backslash \Omega_{R_2})$, then there exists $R \in [R_2, R_1]$ such that $|y|_0 = y(T) = R$ and $0 < Ra(t) \leq y(t) \leq R$, $t \in (0,T]$. Now (3.6.7) and (3.6.10) yield

$$Ky(t) \geq \int_0^t k(t,s)[f(Ra(s)) + g(R)]\, ds, \ t \in [0,T].$$

Again from (3.6.7) and (3.6.10) we have that

$$|Ky|_0 = Ky(T) \leq \int_0^T k(T,s)[f(R) + g(Ra(s))]\, ds, \ t \in [0,T].$$

Combining both of these inequalities we see that

$$Ky(t) \geq \frac{\displaystyle\int_0^t k(t,s)[f(Ra(s)) + g(R)]\,ds}{\displaystyle\int_0^T k(T,s)[f(R) + g(Ra(s))]\,ds}|Ky|_0.$$

Thus if

$$(3.6.14)\qquad \begin{cases} \text{there exists } a \in C[0,T] \text{ such that } a(0) = 0,\ 0 < a(t) \leq 1, \\ t \in (0,T], \text{ and for any constant } R > 0,\ a \text{ satisfies} \\ \dfrac{\displaystyle\int_0^t k(t,s)[f(Ra(s)) + g(R)]\,ds}{\displaystyle\int_0^T k(T,s)[f(R) + g(Ra(s))]\,ds} \geq a(t),\ t \in [0,T] \end{cases}$$

is true, then from above we obtain

$$Ky(t) \geq a(t)|Ky|_0,\ t \in [0,T].$$

Therefore

$$K : C_a \cap (\overline{\Omega}_{R_1} \backslash \Omega_{R_2}) \to C_a \text{ is well-defined.}$$

(Condition (3.6.14) requires considerable discussion and we return to it later in the section.)

We next use the Arzéla-Ascoli theorem to prove that the operator $K : C_a \cap (\overline{\Omega}_{R_1} \backslash \Omega_{R_2}) \to C_a$ is compact. For $y \in C_a \cap (\overline{\Omega}_{R_1} \backslash \Omega_{R_2})$, we obtain from (3.6.7) and (3.6.12)

$$|Ky|_0 \leq ||k_T||_p \left(\int_0^T [f(R_1) + g(R_2 a(s))]^q\,ds \right)^{\frac{1}{q}} \equiv M < \infty$$

– thus $K(C_a \cap (\overline{\Omega}_{R_1} \backslash \Omega_{R_2}))$ is uniformly bounded. In addition (3.6.13) immediately guarantees the equicontinuity of $K(C_a \cap (\overline{\Omega}_{R_1} \backslash \Omega_{R_2}))$. Therefore the compactness of $K : C_a \cap (\overline{\Omega}_{R_1} \backslash \Omega_{R_2}) \to C_a$ now follows from the Arzéla-Ascoli theorem. In conclusion we have shown that (3.6.5) is true.

The existence of a positive solution of (3.6.3) will now follow from Krasnoselskii's fixed point theorem, provided either (*i*) or (*ii*) holds. Therefore suppose that

$$(3.6.15)\qquad \begin{cases} \text{there exists } R_1 > 0 \text{ such that} \\ 1 < \dfrac{R_1}{\displaystyle\int_0^T k(T,s)[f(R_1) + g(R_1 a(s))]\,ds} \end{cases}$$

and

$$\text{(3.6.16)} \qquad \begin{cases} \text{there exists } R_2 > 0, \ R_2 \neq R_1, \text{ such that} \\ 1 > \dfrac{R_2}{\displaystyle\int_0^T k(T,s)[f(R_2 a(s)) + g(R_2)]\,ds} \end{cases}$$

hold. We show that

$$\text{(3.6.17)} \qquad |Ky|_0 < |y|_0 \text{ for } y \in C_a \cap \partial\Omega_{R_1}$$

and

$$\text{(3.6.18)} \qquad |Ky|_0 > |y|_0 \text{ for } y \in C_a \cap \partial\Omega_{R_2}$$

hold. Let $y \in C_a \cap \partial\Omega_{R_1}$. Then $|y|_0 = R_1$ and

$$0 \le a(t)R_1 \le y(t) \le R_1, \ \ t \in [0,T],$$

therefore by (3.6.10) and (3.6.15),

$$|Ky|_0 = Ky(T) \le \int_0^T k(T,s)[f(R_1) + g(R_1 a(s))]\,ds < R_1 = |y|_0.$$

Hence (3.6.17) is true.

Now let $y \in C_a \cap \partial\Omega_{R_2}$. Then $|y|_0 = R_2$ and in addition

$$0 \le a(t)R_2 \le y(t) \le R_2, \ \ t \in [0,T].$$

Then from (3.6.10) and (3.6.16) we obtain

$$|Ky|_0 = Ky(T) \ge \int_0^T k(T,s)[f(R_2 a(s)) + g(R_2)]\,ds > R_2 = |y|_0.$$

Thus (3.6.18) holds.

With the conditions of Krasnoselskii's fixed point theorem satisfied, we are guaranteed that the operator K has a fixed point in $C_a \cap (\overline{\Omega}_{R_1} \backslash \Omega_{R_2})$, that is, (3.6.3) has a positive solution $y \in C[0,T]$ with $0 < R_2 < |y|_0 < R_1$ and $y(t) \ge R_2 a(t) \ge 0$, $t \in [0,T]$. (It is clear than an analogous result holds if $R_1 < R_2$.)

For completeness we state the theorem below.

Theorem 3.6.1. *Let $1 \le p \le \infty$ be a constant and q be such that $1/p + 1/q = 1$. Suppose that (3.6.7) – (3.6.11) and (3.6.14) – (3.6.16) hold. Then (3.6.3) has at least one positive solution $y \in C[0,T]$ and either*

(A) $0 < R_1 < |y|_0 < R_2$ *and* $y(t) \ge a(t)R_1 \ge 0, \ t \in [0,T]$ *if* $R_1 < R_2$

or

(B) $0 < R_2 < |y|_0 < R_1$ *and* $y(t) \geq a(t)R_2 \geq 0,\ t \in [0,T]$ *if* $R_2 < R_1$

holds.

Remark 3.6.1. Note that the solution $y \in C[0,T]$ of (3.6.3) guaranteed by Theorem 3.6.1 is positive whenever $a \in C[0,T]$ is. Consequently since $a(t) > 0,\ t \in (0,T]$, we have either

$$y(t) \geq R_1 a(t) > 0,\ t \in (0,T], \text{ or } y(t) \geq R_2 a(t) > 0,\ t \in (0,T].$$

Clearly condition (3.6.14) needs further discussion. However we are mainly interested in two special cases of (3.6.3) – firstly when $g \equiv 0$ and secondly when $f \equiv 0$ – namely we want to consider (3.6.1) and (3.6.2) respectively. In each of these cases, by placing integrability conditions on f and g, and considering certain classes of kernel k, it is possible to find an appropriate solution $a \in C[0,T]$ to the integral inequality in (3.6.14).

We first consider (3.6.1). This equation has received a lot of attention in the literature [36-38,74,75,161]. In particular the results we present below for (3.6.1) are variations of some of those found in [36-38,74,75,161]. Suppose that the hypotheses of Theorem 3.6.1 hold with $g \equiv 0$. For simplicity, further assume that

$$f(ab) = f(a)f(b) \text{ for } a, b \geq 0 \tag{3.6.19}$$

is true. Condition (3.6.14) now simplifies to

$$\begin{cases} \text{there exists } a \in C[0,T] \text{ such that } a(0) = 0, \\ 0 < a(t) \leq 1,\ t \in (0,T] \text{ and} \\ \dfrac{1}{K_T}\displaystyle\int_0^t k(t,s) f(a(s))\, ds \geq a(t),\ t \in [0,T] \\ \text{where } K_T := \displaystyle\int_0^T k(T,s)\, ds. \end{cases} \tag{3.6.20}$$

If we consider the class of kernels k that satisfy

$$\begin{cases} \text{there exists a constant } N > 0, \text{ such that} \\ k(t,s) \geq N > 0, \text{ for all } t \in [0,T], \text{ a.e. } s \in [0,t], \end{cases} \tag{3.6.21}$$

then we see that

$$\frac{1}{K_T}\int_0^t k(t,s) f(a(s))\, ds \geq \frac{N}{K_T}\int_0^t f(a(s))\, ds.$$

(See [36,161] for an alternative treatment of (3.6.1) where the kernel satisfies a similar condition.) By (3.6.21), the inequality in (3.6.20) will be satisfied if we can find $a \in C[0,T]$ for which

$$\frac{N}{K_T}\int_0^t f(a(s))\,ds = a(t), \;\; t \in [0,T] \tag{3.6.22}$$

holds. Differentiating both sides and integrating from 0 to t, where $t \in [0,T]$ we obtain

$$\frac{N}{K_T}t = \int_0^{a(t)} \frac{ds}{f(s)}, \;\; t \in [0,T].$$

Thus

$$a(t) = J^{-1}\left(\frac{N}{K_T}t\right), \;\; t \in [0,T] \tag{3.6.23}$$

provided

$$\begin{cases} J(1) < \infty \text{ where } J : [0,1] \to [0,\infty) \\ \text{is defined by } J(z) := \displaystyle\int_0^z \frac{ds}{f(s)}, \;\; z \in [0,1] \end{cases} \tag{3.6.24}$$

and

$$\frac{N}{K_T}T \in \text{dom } J^{-1} \tag{3.6.25}$$

are true. Thus (3.6.22) has a solution $a \in C[0,T]$, such that $a(0) = 0$ and it is easy to check that $0 < a(t) \le 1$, $t \in (0,T]$. Therefore (3.6.20) holds. We state the result below.

Theorem 3.6.2. *Suppose that f satisfies* (3.6.19), (3.6.24) *and* (3.6.25), *and assume that* (3.6.21) *and* (3.6.7) – (3.6.9) *hold with $p = 1$. In addition suppose that* (3.6.10) *and* (3.6.15) – (3.6.16) *are true with $g \equiv 0$ and $a \in C[0,T]$ as described in* (3.6.23). *Then* (3.6.1) *has at least one positive solution $y \in C[0,T]$ and either*

(A) $0 < R_1 < |y|_0 < R_2$ *and* $y(t) \ge a(t)R_1 \ge 0$, $t \in [0,T]$ *if* $R_1 < R_2$

or

(B) $0 < R_2 < |y|_0 < R_1$ *and* $y(t) \ge a(t)R_2 \ge 0$, $t \in [0,T]$ *if* $R_2 < R_1$

holds.

Example 3.6.1. Suppose that k satisfies the hypotheses of Theorem 3.6.2 and let $f(y) = y^n$, where $0 \leq n < 1$. Clearly f satisfies (3.6.10) and (3.6.19). Note that since

$$J(z) := \int_0^z \frac{ds}{s^n} = \frac{z^{1-n}}{1-n}, \ z \in [0,1]$$

we can readily see that (3.6.24) holds for this choice of f. In addition J^{-1} : $[0, 1/(1-n)] \to [0,1]$ is a nondecreasing function given by

$$J^{-1}(z) = [(1-n)z]^{\frac{1}{1-n}}, \ z \in [0, 1/(1-n)],$$

and hence (3.6.25) is true since $NT/K_T \leq 1 \leq 1/(1-n)$ by (3.6.21). We obtain

$$a(t) = \left[(1-n)\frac{N}{K_T}\right]^{\frac{1}{1-n}} t^{\frac{1}{1-n}}, \ t \in [0,T].$$

Finally note that if we choose $0 < R_2 < R_1$ such that

$$0 < R_2^{1-n} < \int_0^T k(T,s)a^n(s)\,ds \leq \int_0^T k(T,s)\,ds < R_1^{1-n},$$

then (3.6.15) and (3.6.16) are true. Thus by Theorem 3.6.2, (3.6.1) has a positive solution $y \in C[0,T]$ with $y(t) \geq R_2 a(t)$.

Equation (3.6.1) where k is a convolution kernel, namely

$$y(t) = \int_0^t k(t-s)f(y(s))\,ds, \ t \in [0,T], \tag{3.6.26}$$

arises in connection with nonlinear diffusion and percolation problems (see [38] and its references). In their treatment of (3.6.26) in [38], the authors require that a condition similar to (3.6.20) be satisfied. In particular, for a class of equations of the form

$$y(t) = \int_0^t (t-s)^{\alpha-1} f(y(s))\,ds, \ t \in [0,T], \ \alpha > 1, \tag{3.6.27}$$

they prove that a condition similar to (3.6.20) holds provided f satisfies what they refer to as the Osgood-Gripenberg condition. For completeness we present an existence result for (3.6.27) which relies on this condition and Theorem 3.6.1.

Theorem 3.6.3. *Suppose that f satisfies* (3.6.10) *with* $g \equiv 0$, (3.6.19) *and*

$$\begin{cases} F(1) < \infty \text{ where } F : [0,1] \to [0,\infty) \text{ is defined by} \\ F(z) := \displaystyle\int_0^z \left[\frac{s}{f(s)}\right]^{\frac{1}{\beta}} \frac{ds}{s}, \ z \in [0,1], \text{ and } \beta > \alpha > 1 \text{ is} \\ \text{such that } cT \in \text{dom } F^{-1} \text{ where } c := \dfrac{\alpha^{\frac{1}{\beta}}\beta}{T}\left(\dfrac{\beta-\alpha}{\beta-1}\right)^{\frac{\beta-1}{\beta}}. \end{cases} \tag{3.6.28}$$

Assume also that (3.6.15) *and* (3.6.16) *are true with* $g \equiv 0$, $k(T,s) = (T-s)^{\alpha-1}$, *a.e.* $s \in [0,T]$, $\alpha > 1$ *and* $a \in C[0,T]$ *given by*

$$a(t) := F^{-1}(ct), \ t \in [0,T] \text{ and } c \text{ as defined in (3.6.28).} \tag{3.6.29}$$

Then (3.6.27) *has at least one positive solution* $y \in C[0,T]$ *and either*

(A) $0 < R_1 < |y|_0 < R_2$ *and* $y(t) \geq a(t)R_1 \geq 0$, $t \in [0,T]$ *if* $R_1 < R_2$

or

(B) $0 < R_2 < |y|_0 < R_1$ *and* $y(t) \geq a(t)R_2 \geq 0$, $t \in [0,T]$ *if* $R_2 < R_1$

holds.

PROOF: It is clear that $k(t,s) = (t-s)^{\alpha-1}$, $\alpha > 1$ satisfies (3.6.7) – (3.6.9) with $p = 1$. In this case (3.6.20) (and hence (3.6.14)) can be rewritten as

$$\begin{cases} \text{there exists } a \in C[0,T] \text{ such that } a(0) = 0, \\ 0 < a(t) \leq 1, \ t \in (0,T] \text{ and} \\ \dfrac{1}{K_T}\displaystyle\int_0^t (t-s)^{\alpha-1} f(a(s))\, ds \geq a(t), \ t \in [0,T] \\ \text{where } K_T := \displaystyle\int_0^T s^{\alpha-1}\, ds = \dfrac{T^\alpha}{\alpha}. \end{cases} \tag{3.6.30}$$

For $c > 0$ as defined in (3.6.28), consider the initial value problem

$$\begin{cases} a'(t) = c a^{1-\frac{1}{\beta}} f(a)^{\frac{1}{\beta}}, \ t \in [0,T] \\ a(0) = 0. \end{cases} \tag{3.6.31}$$

Since f satisfies (3.6.28) one can show that (3.6.31) has a solution $a \in C[0,T]$ given by

$$a(t) = F^{-1}(ct), \ t \in [0,T].$$

Clearly $a(0) = 0$ and $0 < a(t) \leq 1$ for $t \in (0,T]$. Note also that

$$\beta^\beta a(t) = \left(\int_0^t c f(a(s))^{\frac{1}{\beta}}\, ds\right)^\beta, \ t \in [0,T],$$

which on using Hölder's inequality becomes

$$\beta^\beta a(t) \leq c^\beta \left(\int_0^t (t-s)^{\alpha-1} f(a(s))\, ds\right) \left(\int_0^t (t-s)^{-\frac{\alpha-1}{\beta-1}}\, ds\right)^{\beta-1}, \ t \in [0,T].$$

However by our choice of c, we have that

$$a(t) \leq \frac{1}{K_T} \int_0^t (t-s)^{\alpha-1} f(a(s))\, ds.$$

We have therefore shown (using an argument from [38]) that (3.6.28) ensures that (3.6.30) (and hence (3.6.14)) holds. The result now follows from Theorem 3.6.1. □

Example 3.6.2. Consider (3.6.27) with $f(y) = y^n$, $0 \leq n < 1$. Clearly f satisfies (3.6.10) and (3.6.19). For $\beta > \alpha > 1$, $F : [0,1] \to [0, \beta/(1-n)]$ is defined by

$$F(z) := \frac{\beta}{1-n} z^{\frac{1-n}{\beta}}, \ z \in [0,1] \text{ and } F(1) = \frac{\beta}{1-n} < \infty.$$

We want to verify that $cT \in \operatorname{dom} F^{-1}$, where $F^{-1} : [0, \beta/(1-n)] \to [0,1]$ is given by

$$F^{-1}(z) = \left[\frac{1-n}{\beta}\right]^{\frac{\beta}{1-n}} z^{\frac{\beta}{1-n}}, \ z \in \left[0, \frac{\beta}{1-n}\right].$$

Now

$$cT = \alpha^{\frac{1}{\beta}} \beta \left(\frac{\beta-\alpha}{\beta-1}\right)^{\frac{\beta-1}{\beta}} \leq \alpha^{\frac{1}{\beta}} \beta,$$

and therefore $\alpha^{\frac{1}{\beta}} \beta \leq \dfrac{\beta}{1-n}$ if β is chosen large enough such that $\alpha < \dfrac{1}{(1-n)^\beta}$. Hence for this choice of f and β, we have that (3.6.28) is true.

Observe that

$$a(t) = F^{-1}(ct) = \left[\frac{(1-n)\alpha^{\frac{1}{\beta}}}{T} \left(\frac{\beta-\alpha}{\beta-1}\right)^{\frac{\beta-1}{\beta}}\right]^{\frac{\beta}{1-n}} t^{\frac{\beta}{1-n}}, \ t \in [0,T]$$

and in a similar fashion to Example 3.6.1, one can show that there exists $0 < R_2 < R_1$ that satisfy

$$0 < R_2^{1-n} < \int_0^T (T-s)^{\alpha-1} a^n(s)\, ds < \int_0^T (T-s)^{\alpha-1}\, ds < R_1^{1-n}.$$

Thus (3.6.15) and (3.6.16) are satisfied and by Theorem 3.6.3, (3.6.27) with $f(y) = y^n$, $0 \leq n < 1$, has a positive solution $y \in C[0,T]$.

Finally in this section we consider the singular Volterra equation (3.6.2). Suppose that the conditions of Theorem 3.6.1 hold with $f \equiv 0$ and once again for simplicity we assume

$$g(ab) = g(a)g(b), \text{ and } g\left(\frac{1}{b}\right) = \frac{1}{g(b)}, \quad a, b > 0. \tag{3.6.32}$$

We are interested in finding a function $a \in C[0,T]$ which satisfies (3.6.14) with $f \equiv 0$, that is,

$$\begin{cases} \text{there exists } a \in C[0,T] \text{ such that } a(0) = 0, \\ 0 < a(t) \le 1, \ t \in (0,T] \text{ and} \\ \dfrac{K_t}{\displaystyle\int_0^T k(T,s)g(a(s))\,ds} \ge a(t), \ t \in [0,T] \\ \text{where } K_t := \displaystyle\int_0^t k(t,s)\,ds, \ t \in [0,T]. \end{cases} \tag{3.6.33}$$

In addition (3.6.11) must hold, which using (3.6.32) we can rewrite as

$$\int_0^T g(a(s))^q\,ds < \infty. \tag{3.6.34}$$

Assume that

$$G(y) = \frac{1}{yg(y)} \text{ is a decreasing function in } y \text{ for } y > 0, \tag{3.6.35}$$

holds and consider the equation

$$\frac{K_t}{\displaystyle\int_0^T k(T,s)g(a(s))\,ds} = a(t), \ t \in [0,T]. \tag{3.6.36}$$

Provided the kernel k satisfies

$$\int_0^T \left[g\left(\int_0^s k(s,z)\,dz\right)\right]^q ds < \infty, \tag{3.6.37}$$

then a solution of (3.6.36) may be given by

$$\begin{cases} a(t) = CK_t = C\displaystyle\int_0^t k(t,s)\,ds, \ t \in [0,T], \\ \text{where } 0 < C = G^{-1}\left(\displaystyle\int_0^T k(T,s)g\left(\int_0^s k(s,z)\,dz\right) ds\right). \end{cases} \tag{3.6.38}$$

Observe that $a \in C[0,T]$, $a(0) = 0$ and $a(t) > 0$ for $t \in (0,T]$. Also note from (3.6.32) that since $g(1/b) = 1/g(b)$, $b > 0$ we have

$$\frac{1}{G(y)} = G\left(\frac{1}{y}\right), \quad y > 0.$$

Hence for $t \in [0,T]$,

$$\begin{aligned} a(t) = CK_t \ &\leq \ G^{-1}\left(\int_0^T k(T,s)g\left(\int_0^s k(s,z)\,dz\right)\,ds\right)K_T \\ &\leq \ G^{-1}\left(K_T g(K_T)\right)K_T = G^{-1}\left(\frac{1}{G(K_T)}\right)K_T \\ &= \ G^{-1}\left(G\left(\frac{1}{K_T}\right)\right) = 1, \end{aligned}$$

Consequently $a \in C[0,T]$ satisfies (3.6.33) and (3.6.34). We thus have the following theorem for (3.6.2).

Theorem 3.6.4. *Let* $1 \leq p \leq \infty$ *be a constant and* q *be such that* $1/p + 1/q = 1$. *Suppose that* k *satisfies* (3.6.7) – (3.6.9) *and* (3.6.10) *is true with* $f \equiv 0$. *In addition assume that* (3.6.32), (3.6.35) *and* (3.6.37) *hold along with* (3.6.15) *and* (3.6.16) *with* $f \equiv 0$ *and* $a \in C[0,T]$ *as described in* (3.6.38). *Then* (3.6.2) *has at least one positive solution* $y \in C[0,T]$ *and either*

(A) $0 < R_1 < |y|_0 < R_2$ *and* $y(t) \geq a(t)R_1 \geq 0,\ t \in [0,T]$ *if* $R_1 < R_2$

or

(B) $0 < R_2 < |y|_0 < R_1$ *and* $y(t) \geq a(t)R_2 \geq 0,\ t \in [0,T]$ *if* $R_2 < R_1$

holds.

Example 3.6.3. Consider the singular integral equation

$$(3.6.39) \qquad y(t) = \int_0^t (t-s)^{\alpha-1} y^{-n}(s)\,ds, \ t \in [0,1], \ \alpha > 1, \ 0 \leq n < 1.$$

Clearly $k(t,s) := (t-s)^{\alpha-1}$, $\alpha > 1$, satisfies (3.6.7) – (3.6.9) with $p = \infty$, while $g(y) = y^{-n}$, $0 \leq n < 1$ ensures that (3.6.10), (3.6.32) and (3.6.35) hold. Notice that

$$\int_0^1 g\left(\int_0^s k(s,z)\,dz\right)\,ds = \alpha^n \int_0^1 s^{-\alpha n}\,ds = \frac{\alpha^n}{1-\alpha n} < \infty,$$

provided

$$1 < \alpha < 1/n \tag{3.6.40}$$

is true. Thus if the restriction (3.6.40) is placed on α and n, then (3.6.37) is true with $q = 1$. In addition we see that

$$a(t) = C\frac{t^\alpha}{\alpha}, \ t \in [0,1],$$

with $C = [\alpha^n B(1-\alpha n, \alpha)]^{-\frac{1}{1-n}}$.

Choosing $0 < R_2 < R_1$ such that

$$R_2^{1+n} < \int_0^1 (1-s)^{\alpha-1}\,ds \le \int_0^1 (1-s)^{\alpha-1} a^{-n}(s)\,ds < R_1^{1+n}$$

we have from Theorem 3.6.4 that (3.6.39) has a positive solution $y \in C[0,1]$ with $y(t) \ge R_2 a(t)$, $t \in [0,T]$ provided (3.6.40) is true.

3.7. Positive L^p solutions of Volterra integral equations.

In Section 3.6 of this chapter we used Krasnoselskii's fixed point theorem to obtain a general existence result for singular and nonsingular Volterra integral equations. To complete the discussion, we present in this section an existence principle which guarantees the existence of a positive solution $y \in L^p[0,T]$, $1 \le p < \infty$, of the nonsingular Volterra equation

$$y(t) = \int_0^t k(t,s) f(s, y(s))\,ds, \ \text{a.e. } t \in [0,T]. \tag{3.7.1}$$

Here by a *positive* solution we mean $y(t) > 0$, a.e. $t \in [0,T]$. The argument we use once again relies on Krasnoselskii's fixed point theorem and is analogous to those presented in Section 2.4 of the previous chapter.

Define the operator $K : L^p[0,T] \to L^p[0,T]$ by

$$Ky(t) := K_L F y(t) := \int_0^t k(t,s) f(s,y(s))\,ds, \ \text{a.e. } t \in [0,T],$$

where in turn $F : L^p[0,T] \to L^{p_2}[0,T]$ and $K_L : L^{p_2}[0,T] \to L^p[0,T]$ are given by

$$Fy(t) := f(t,y(t)) \text{ and } K_L y(t) := \int_0^t k(t,s) y(s)\,ds, \ \text{a.e. } t \in [0,T].$$

Here $1 \le p_2 < \infty$ and p_1 is such that $1/p_1 + 1/p_2 = 1$.

Examining first the nonlinearity f, we assume that

(3.7.2) $$f : [0,T] \times \mathbf{R} \to \mathbf{R} \text{ is a Carathéodory function,}$$

(3.7.3) $$\begin{cases} \text{there exists } a_1 \in L^{p_2}[0,T] \text{ and } a_2 > 0 \text{ such that} \\ |f(t,y)| \le a_1(t) + a_2|y|^{\frac{p}{p_2}}, \text{ a.e. } t \in [0,T] \end{cases}$$

and

(3.7.4) $$\begin{cases} f(t,y) \text{ is nondecreasing in } y \text{ for a.e. } t \in [0,T] \\ \text{and } f(t,y) > 0 \text{ for all } y > 0 \text{ and a.e. } t \in [0,T] \end{cases}$$

hold. Then by (3.7.2) and (3.7.3) we have that

$$f(t,y(t)) \in L^{p_2}[0,T] \text{ for } y \in L^p[0,T]$$

and from [89], we see that

(3.7.5) $$F : L^p[0,T] \to L^{p_2}[0,T] \text{ is continuous and bounded}$$

is true. Also from (3.7.3) we obtain

$$\int_0^T |f(t,y(t))|^{p_2}\,dt \le 2^{p_2-1}\left(\int_0^T a_1^{p_2}(t)\,dt + a_2^{p_2}\int_0^T |y(t)|^p\,dt\right)$$

and therefore defining $\psi \in C[0,\infty)$ by

(3.7.6) $$\psi(t) := 2^{\frac{p_2-1}{p_2}}\left(||a_1||_{p_2}^{p_2} + a_2^{p_2}t^p\right)^{\frac{1}{p_2}}, \quad t \in [0,\infty)$$

we see that

(3.7.7) $$\begin{cases} \text{there exists } \psi \in C[0,\infty) \text{ such that} \\ ||Fy||_{p_2} \le \psi(||y||_p) \text{ for all } y \in L^p[0,T] \end{cases}$$

holds.

Next concentrating on the operator $K_L : L^{p_2}[0,T] \to L^p[0,T]$, assume that

(3.7.8) $$\begin{cases} k : [0,T] \times [0,t] \to \mathbf{R} \text{ is such that } (t,s) \mapsto k(t,s) \text{ is} \\ \text{measurable and } 0 < k(t,s) \text{ a.e. } t \in [0,T], \text{ a.e. } s \in [0,t], \end{cases}$$

(3.7.9) $$\left(\int_0^T \left(\int_s^T [k(t,s)]^p\,dt\right)^{\frac{p_1}{p}}\right)^{\frac{1}{p_1}} \equiv M < \infty$$

and

(3.7.10) $k(t_1, s) - k(t_2, s) > 0$, for a.e. $s \in [0, t_2]$, a.e. $0 < t_2 < t_1 \le T$

are true. We have that (see Section 2.4)

$$K_L : L^{p_2}[0,T] \to L^p[0,T] \text{ is a completely continuous operator.}$$

This, combined with (3.7.5) implies that

$$(3.7.11) \qquad \begin{cases} K = K_L F : L^p[0,T] \to L^p[0,T] \\ \text{is a completely continuous operator} \end{cases}$$

holds. We now want to construct a cone in which the solution of (3.7.1) will reside.

Define the cone $C_{p,a}$ by

$$\begin{aligned} C_{p,a} \quad := \quad & \{y \in L^p[0,T] : y \text{ is nondecreasing and} \\ & y(t) \ge a(t)||y||_p \text{ for a.e. } t \in [0,T]\}. \end{aligned}$$

Here $a \in L^p[0,T]$ is such that $a(t) > 0$, a.e. $t \in [0,T]$ and is the solution of an integral inequality which will be described below. Also, to be specific, by *y nondecreasing* we mean that $y(t)$ is nondecreasing in t for a.e. $t \in [0,T]$. In addition let Ω_{R_3} and Ω_{R_4} be given by

$$\Omega_{R_3} := \{y \in L^p[0,T] : ||y||_p < R_3\}$$

and

$$\Omega_{R_4} := \{y \in L^p[0,T] : ||y||_p < R_4\}$$

respectively. We assume in what follows that $0 < R_4 < R_3$. (A similar argument holds if $0 < R_3 < R_4$.)

Let $y \in C_{p,a} \cap (\overline{\Omega}_{R_3} \backslash \Omega_{R_4})$. One can easily show using (3.7.4), (3.7.8) and (3.7.10), that $Ky(t)$ is nondecreasing in t, for a.e. $t \in [0,T]$. Also since $y \in C_{p,a} \cap (\overline{\Omega}_{R_3} \backslash \Omega_{R_4})$, there exists $R \in [R_4, R_3]$ such that $||y||_p = R$ and $0 < Ra(t) \le y(t)$, a.e. $t \in [0,T]$. Now (3.7.4) and (3.7.8) yield

$$Ky(t) \ge \int_0^t k(t,s) f(s, Ra(s))\, ds, \text{ a.e. } t \in [0,T].$$

In addition using Hölder's inequality we have that

$$\begin{aligned} ||Ky||_p^p &= \int_0^T |Ky(t)|^p\, dt = \int_0^T \left(\int_0^t k(t,s) f(s,y(s))\, ds \right)^p dt \\ &\le \left(\int_0^T \left(\int_0^t k(t,s)^{p_1}\, ds \right)^{\frac{p}{p_1}} dt \right) \left(\int_0^T f(s,y(s))^{p_2}\, ds \right)^{\frac{p}{p_2}}, \end{aligned}$$

and thus from (3.7.7) and (3.7.9) we obtain

$$||Ky||_p \le \tilde{K}||Fy||_{p_2} \le \tilde{K}\psi(||y||_p) = \tilde{K}\psi(R),$$

where

$$\tilde{K} = \left(\int_0^T \left(\int_0^t k(t,s)^{p_1}\,ds\right)^{\frac{p}{p_1}} dt\right)^{\frac{1}{p}}. \tag{3.7.12}$$

Therefore for $y \in C_{p,a} \cap (\overline{\Omega}_{R_3} \backslash \Omega_{R_4})$, with $||y||_p = R$, $R \in [R_4, R_3]$ we see that

$$Ky(t) \ge \frac{\int_0^t k(t,s) f(s, Ra(s))\,ds}{\tilde{K}\psi(R)} ||Ky||_p, \text{ a.e. } t \in [0,T].$$

Thus if

$$\left\{\begin{array}{l} \text{there exists } a \in L^p[0,T] \text{ such that } 0 < a(t) \text{ a.e. } t \in [0,T], \\ \text{with } ||a||_p \le 1, \text{ and for any constant } R > 0, \ a \text{ satisfies} \\ \dfrac{\int_0^t k(t,s) f(s, Ra(s))\,ds}{\tilde{K}\psi(R)} \ge a(t), \text{ a.e. } t \in [0,T] \end{array}\right. \tag{3.7.13}$$

is true, then from above we obtain

$$Ky(t) \ge a(t)||Ky||_p, \text{ a.e. } t \in [0,T].$$

This inequality, the fact that $Ky(t)$ is nondecreasing for a.e. $t \in [0,T]$, and (3.7.11) combine to ensure

$$\left\{\begin{array}{l} K : C_{p,a} \cap (\overline{\Omega}_{R_3} \backslash \Omega_{R_4}) \to C_{p,a} \\ \text{is a completely continuous operator.} \end{array}\right. \tag{3.7.14}$$

Finally suppose that

$$\text{there exists } R_3 > 0 \text{ such that } 1 < \frac{R_3}{\tilde{K}\psi(R_3)} \tag{3.7.15}$$

and

$$\left\{\begin{array}{l} \text{there exists } R_4 > 0, \ R_4 \ne R_3, \text{ such that} \\ 1 > \dfrac{R_4}{\left(\int_0^T \left(\int_0^t k(t,s) f(s, R_4 a(s))\,ds\right)^p dt\right)^{\frac{1}{p}}} \end{array}\right. \tag{3.7.16}$$

are true. We will show that these conditions in turn guarantee that

$$\|Ky\|_p \le \|y\|_p \text{ for } y \in C_{p,a} \cap \partial\Omega_{R_3} \tag{3.7.17}$$

and

$$\|Ky\|_p \ge \|y\|_p \text{ for } y \in C_{p,a} \cap \partial\Omega_{R_4} \tag{3.7.18}$$

hold. Therefore by Krasnoselskii's fixed point theorem the operator K has a fixed point in $C_{p,a} \cap (\overline{\Omega}_{R_3} \backslash \Omega_{R_4})$. This in turn implies that (3.7.1) has at least one solution $y \in L^p[0,T]$ such that $R_4 \le \|y\|_p \le R_3$ and $y(t) \ge R_4 a(t)$ for a.e. $t \in [0,T]$.

Suppose then that $y \in C_{p,a} \cap \partial\Omega_{R_3}$. In particular $\|y\|_p = R_3$. Then Hölder's inequality and (3.7.15) imply

$$\|Ky\|_p \le \tilde{K}\psi(\|y\|_p) = \tilde{K}\psi(R_3) < R_3 = \|y\|_p$$

and thus (3.7.17) is satisfied.

Next suppose that $y \in C_{p,a} \cap \partial\Omega_{R_4}$. Then $\|y\|_p = R_4$ and note also that $y(t) \ge R_4 a(t)$ for a.e. $t \in [0,T]$. Conditions (3.7.4), (3.7.8) and (3.7.16) give

$$\|Ky\|_p \ge \left(\int_0^T \left(\int_0^t k(t,s) f(s, R_4 a(s))\, ds \right)^p dt \right)^{\frac{1}{p}} > R_4 = \|y\|_p$$

– and (3.7.18) holds.

We now formally state the result.

Theorem 3.7.1. *Assume that p, p_1 and p_2 are such that $1 \le p_1 \le p < \infty$ and $1/p_1 + 1/p_2 = 1$. Suppose that $f : [0,T] \times \mathbf{R} \to \mathbf{R}$ satisfies* (3.7.2) – (3.7.4) *and $k : [0,T] \times [0,t] \to \mathbf{R}$ satisfies* (3.7.8) – (3.7.10). *Let $\psi : [0,\infty) \to [0,\infty)$ and $\tilde{K}$ be as defined in* (3.7.7) *and* (3.7.12) *respectively and assume that* (3.7.13), (3.7.15) *and* (3.7.16) *hold. Then* (3.7.1) *has at least one positive solution $y \in L^p[0,T]$ and either*

(A) $0 < R_3 < \|y\|_p < R_4$ *and* $y(t) \ge R_3 a(t)$ *a.e.* $t \in [0,T]$ *if* $R_3 < R_4$

or

(B) $0 < R_4 < \|y\|_p < R_3$ *and* $y(t) \ge R_4 a(t)$ *a.e.* $t \in [0,T]$ *if* $R_4 < R_3$

holds.

Example 3.7.1. Suppose that $f(t,y) = y^n$, $0 \leq n < 1$. Then assuming that $np_2 \leq p$, one can easily see that f satisfies (3.7.2) – (3.7.4). In addition

$$\int_0^T |f(s,y(s))|^{p_2}\,ds = \int_0^T |y(s)|^{np_2}\,ds \leq T^m ||y||_p^{np_2}$$

where $m := (p - np_2)/p$. Thus we can find $\psi \in C[0,\infty)$ which satisfies (3.7.7), by letting $\psi(t) := Ct^n$ with $C = T^{m/p_2}$.

Note in this case that (3.7.13) reduces to

$$(3.7.19)\qquad \begin{cases} \text{there exists } a \in L^p[0,T] \text{ with } 0 < a(t) \text{ a.e. } t \in [0,T] \\ \text{and } ||a||_p \leq 1, \text{ such that} \\ \dfrac{\int_0^t k(t,s)a^n(s)\,ds}{\tilde{K}T^{m/p_2}} \geq a(t), \text{ a.e. } t \in [0,T]. \end{cases}$$

Finally it is possible to find $0 < R_4 < R_3$ such that

$$R_4^{1-n} < \left(\int_0^T \left(\int_0^t k(t,s)a^n(s)\,ds\right)^p dt\right)^{\frac{1}{p}} \leq \tilde{K} < R_3^{1-n}.$$

This choice of R_3 and R_4 satisfies (3.7.15) and (3.7.16) respectively.

CHAPTER 4

PERIODIC AND ALMOST PERIODIC SOLUTIONS OF NONLINEAR INTEGRAL EQUATIONS

4.1. Introduction.

In this chapter, using various fixed point methods, we present results which guarantee the existence of periodic and almost periodic solutions of nonlinear integral equations.

In Section 4.2 we consider the nonlinear integral equation

$$y(t) = h(t) + \int_I k(t,s) f(s, y(s))\, ds, \quad t \in I \tag{4.1.1}$$

with a view to proving the existence of a continuous periodic or continuous almost periodic solution y of (4.1.1). Several admissibility results and existence principles are presented for this equation. The nonlinear alternative is the fixed point method used in this section.

Section 4.3 establishes the existence of one or more periodic or almost periodic solutions of the nonlinear integral equation

$$y(t) = \int_{\mathbf{R}} k(t,s) f(s, y(s))\, ds, \quad \text{for } t \in \mathbf{R}. \tag{4.1.2}$$

Here existence will be established via various fixed point methods. In particular we will use either a nonlinear alternative of Leray-Schauder type, Krasnoselskii's fixed point theorem in a cone, or the Leggett Williams fixed point theory. The results in this section improve and extend those in the literature [18,49,50,103]

In [45] the nonlinear integral equation

$$x(t) = \int_{t-\tau}^{t} f(s, x(s))\, ds, \quad \text{for } t \in \mathbf{R} \tag{4.1.3}$$

was interpreted as a model for the spread of certain infectious diseases with periodic contact rate that varies seasonally. In (4.1.3), $x(t)$ represents the proportion of infectives in the population at time t, $f(t, x(t))$ (which is periodic in t, that is, there exists $w > 0$ with $f(t+w, x) = f(t, x)$ for all $t \in \mathbf{R}$ and $x \geq 0$) is the proportion of new infectives per unit time and $\tau > 0$ is the length of time an individual remains infectious. Cooke and Kaplan in [45] were interested in obtaining a nontrivial periodic nonnegative solution to (4.1.3). They used Krasnoselskii's fixed point theorem in a cone to achieve this. Four years later Leggett and Williams [95] using their own fixed point theorem (which was a generalization of Krasnoselskii's) generalized the results in [45]. In Section 4.4, we establish using a variety of fixed point theorems, existence results for the nonlinear integral equation

$$x(t) = \int_{t-\tau}^{t} k(t,s) f(s, x(s))\, ds, \quad \text{for } t \in \mathbf{R}. \tag{4.1.4}$$

We discuss the existence of one and multiple periodic solutions of (4.1.4). Moreover our results improve and extend those in [45,78,95].

Before proceeding, we recall some facts about periodic and almost periodic functions. Let $0 < \omega < \infty$ and suppose I is an interval of $\mathbf{R}$ that contains at least one compact subinterval of length ω, which we denote by I_ω. We define $A_\omega(I)$ to be the subspace of $BC(I)$ (bounded continuous functions on I with values in $\mathbf{R}$) consisting of all ω-periodic mappings, that is, if $y \in A_\omega(I)$, then y is continuous on I and

$$y(t+\omega) = y(t), \text{ for all } t \text{ such that } t + \omega \in I.$$

The norm on $A_\omega(I)$ is denoted by $|.|_\omega$ and for $y \in A_\omega(I)$

$$|y|_\omega = \sup_{t\in I} |y(t)| = \sup_{t \in I_\omega} |y(t)|.$$

For I an infinite interval of $\mathbf{R}$, we denote by $AP(I)$, the space of continuous almost periodic functions on I with values in $\mathbf{R}$. A continuous function is said to be almost periodic (in the Bohr sense) if for any $\epsilon > 0$, there exists $l(\epsilon) > 0$ such that any interval of I of length $l(\epsilon)$ contains an element τ such that

$$|x(t+\tau) - x(t)| < \epsilon \text{ for all } t \text{ such that } t + \tau \in I.$$

Continuous almost periodic functions are bounded [50,57], with the norm on $AP(I)$ given by $|.|_0$. Note also that $A_\omega(I) \subset AP(I)$.

Finally we have the following compactness criteria for the space $AP(I)$, the proof of which is given in [50,P.143]:

Theorem 4.1.1. Compactness Criteria for $AP(I)$

The necessary and sufficient condition that $M \subseteq AP(I)$ be relatively compact is that the following properties hold true:

(*i*) *for any $t \in I$, the set of values of functions from M is relatively compact in* $\mathbf{R}$,

(*ii*) *M is equi-continuous,*

(*iii*) *M is equi-almost periodic.*

4.2. Admissibility results and existence principles.

Consider the nonlinear, Fredholm integral equation

$$y(t) = h(t) + \int_I k(t,s) f(s, y(s))\, ds, \ \ t \in I \tag{4.2.1}$$

where I is an interval of $\mathbf{R}$ (finite or infinite), $h : I \to \mathbf{R}$, $k : I \times I \to \mathbf{R}$ and $f : I \times \mathbf{R} \to \mathbf{R}$. In this section we assume that either $h \in AP(I)$ or $h \in A_\omega(I)$, and discuss conditions that k and f are required to satisfy in order for (4.2.1) to have a solution $y \in AP(I)$ or $y \in A_\omega(I)$ respectively.

Notation. Throughout this section when discussing the normed space $AP(I)$ we assume that I is an infinite interval of $\mathbf{R}$, and when considering $A_\omega(I)$ we assume that $0 < \omega < \infty$ and I is an interval of $\mathbf{R}$ that contains at least one compact subinterval of length ω, which we denote by I_ω.

Obviously by defining

$$Ky(t) := \int_I k(t,s) y(s)\, ds \tag{4.2.2}$$

and

$$Fy(t) := f(t, y(t)) \tag{4.2.3}$$

on the interval I, one can write (4.2.1) in the form of the operator equation

$$y(t) = h(t) + KFy(t), \ \ t \in I. \tag{4.2.4}$$

Throughout this section we will make repeated use of the following existence principle for (4.2.4), the proof of which follows easily from the nonlinear alternative (Theorem 1.2.1).

Theorem 4.2.1. Existence Principle.
Let E_1 and E_2 be normed spaces with norms denoted by $||.||_{E_1}$ and $||.||_{E_2}$ respectively. Suppose that

$$h \in E_1, \tag{4.2.5}$$

(4.2.6) $\qquad F : E_1 \to E_2$ *is a bounded and continuous operator*

and

(4.2.7) $K : E_2 \to E_1$ *is a continuous and completely continuous operator*

hold. In addition, suppose there exists a constant $M > 0$, *independent of* λ, *with* $||y||_{E_1} \neq M$ *for any solution* $y \in E_1$ *of*

$$y(t) = \lambda\,(h(t) + KFy(t))$$

for each $\lambda \in (0,1)$. *Then* (4.2.4) *has a solution* $y \in E_1$.

PROOF: It is easy to check using (4.2.5) – (4.2.7) that the operator $N : E_1 \to E_1$ defined by

$$Ny(t) := h(t) + KFy(t)$$

is continuous and completely continuous. Therefore the hypotheses of the nonlinear alternative are satisfied with $C = E = E_1$, $U = \{y \in E_1 : ||y||_{E_1} < M\}$ and $p^\star = 0$. Note that condition (ii) cannot occur. Consequently N has a fixed-point y in E_1, which is equivalent to saying that (4.2.4) has a solution $y \in E_1$. $\square$

For $h \in A_\omega(I)$ or $h \in AP(I)$, our aim is to impose conditions on k and f so that K and F as defined above satisfy the hypotheses of Theorem 4.2.1 with $E_1 = A_\omega(I)$ or $E_1 = AP(I)$ respectively. We thus ensure that (4.2.4), or equivalently (4.2.1) has a solution $y \in A_\omega(I)$ or $y \in AP(I)$ respectively.

In particular most of this section will be concerned with illustrating conditions for $k : I \times I \to \mathbf{R}$ such that the linear integral operator K satisfies

(4.2.8) $\qquad K : E_2 \to E_1$ is continuous and completely continuous

where E_2 and E_1 are certain normed function spaces. In [48,49] the pair of function spaces (E_2, E_1) is said to be admissible with respect to the operator K if for any $y \in E_2$, we have $Ky \in E_1$. Note that no requirement of continuity or complete continuity of K is made in this definition, however for convenience we extend the definition by saying that (E_2, E_1) is admissible with respect to K if (4.2.8) is satisfied.

Condition (4.2.8) is discussed in the literature [137,Chpt.4] where K is as defined in (4.2.2), $E_2 = L^p[0,T]$, $1 \le p \le \infty$ and $E_1 = C[0,T]$. Our first two admissibility results extend the conditions given on k in [137], so that (4.2.8) is now true with $E_2 = L^p(I)$, $1 \le p \le \infty$ and $E_1 = A_\omega(I)$ or $AP(I)$. (Note that I may now be an infinite interval.)

Theorem 4.2.2. Admissibility Result.
Let $1 \le p \le \infty$ be a constant, q be such that $\frac{1}{p} + \frac{1}{q} = 1$, $0 < \omega < \infty$ and I be an interval of $\mathbf{R}$ *that contains at least one compact subinterval I_ω of length ω. Assume that*

$$k_t(s) := k(t,s) \in L^q(I) \text{ for each } t \in I, \tag{4.2.9}$$

$$\text{the map } t \mapsto k_t \text{ is continuous from } I \text{ to } L^q(I) \tag{4.2.10}$$

and

$$\int_I |k_{t+\omega}(s) - k_t(s)|^q \, ds = 0 \text{ for all } t \text{ such that } t + \omega \in I \tag{4.2.11}$$

hold. Then

$$K : L^p(I) \to A_\omega(I) \text{ is continuous and completely continuous} \tag{4.2.12}$$

is true.

Remark 4.2.1. From (4.2.10) and the triangle inequality we have that $||k_t||_q$ is continuous on I. In addition, (4.2.11) and the triangle inequality implies that $||k_{t+\omega}||_q = ||k_t||_q$ for all t such that $t + \omega \in I$. Hence we see that

$$||k_t||_q \in A_\omega(I), \tag{4.2.13}$$

and therefore

$$\sup_{t \in I} ||k_t||_q = \sup_{t \in I_\omega} ||k_t||_q = M_0 < \infty. \tag{4.2.14}$$

Also notice that (4.2.11) implies for almost every $s \in I$ that

$$k(t+\omega, s) = k(t,s) \text{ for all } t \text{ such that } t + \omega \in I. \tag{4.2.15}$$

This latter condition may be easier to verify than (4.2.11).

PROOF OF THEOREM 4.2.2: Immediately from (4.2.9), (4.2.10) and Hölder's inequality we have that

$$K : L^p(I) \to C(I) \text{ is well-defined,}$$

since for t_1, $t_2 \in I$ and $y \in L^p(I)$,

$$|Ky(t_1) - Ky(t_2)| \le \left(\int_I |k_{t_1}(s) - k_{t_2}(s)|^q \, ds\right)^{\frac{1}{q}} \left(\int_I |y(s)|^p \, ds\right)^{\frac{1}{p}}. \tag{4.2.16}$$

Replacing t_1 and t_2 in (4.2.16) with $t+\omega \in I$ and $t \in I$ respectively, yields from (4.2.11)

$$|Ky(t+\omega) - Ky(t)| \leq ||k_{t+\omega} - k_t||_q ||y||_p = 0 \tag{4.2.17}$$

and therefore

$$K : L^p(I) \to A_\omega(I) \text{ is well-defined.}$$

From (4.2.14) and Hölder's inequality it follows easily that $K : L^p(I) \to A_\omega(I)$ is continuous, since if $y_n \to y$ in $L^p(I)$, then

$$|Ky_n - Ky|_\omega \leq \sup_{t\in I} ||k_t||_q ||y_n - y||_p \to 0. \tag{4.2.18}$$

We finally show that $K : L^p(I) \to A_\omega(I)$ is completely continuous. Let Ω be a bounded set in $L^p(I)$, that is there exists $M_1 > 0$ such that $||y||_p < M_1$ for all $y \in \Omega$ and consider $K\Omega$. We want to show that $K\Omega$ is relatively compact in $A_\omega(I)$. Let $(y_n)_{n=1}^\infty$ be a sequence in Ω and hence $(Ky_n)_{n=1}^\infty$ is a sequence in $K\Omega$. Now for all $t \in I_\omega$

$$|Ky_n(t)| \leq \sup_{t\in I_\omega} ||k_t||_q ||y_n||_p \leq \sup_{t\in I_\omega} ||k_t||_q M_1 < \infty,$$

showing that $(Ky_n)_{n=1}^\infty$ is uniformly bounded on I_ω. From (4.2.16) with t_1, $t_2 \in I_\omega$ we have that $(Ky_n)_{n=1}^\infty$ is also equicontinuous. Therefore the Arzéla-Ascoli theorem implies that there exists $Ky \in C(I_\omega)$ and a subsequence $(Ky_{n_k})_{k=1}^\infty$ of $(Ky_n)_{n=1}^\infty$ which converges to Ky. Let $Ky(t) = Ky(t+\omega)$, then since $Ky_{n_k}(t) = Ky_{n_k}(t+\omega)$ we have

$$Ky_{n_k} \to Ky \text{ in } A_\omega(I).$$

Consequently $K : L^p(I) \to A_\omega(I)$ is completely continuous and the result is proved. □

Remark 4.2.2. If $q = 1$ in (4.2.9) – (4.2.11), Theorem 4.2.2 implies that

$$K : L^\infty(I) \to A_\omega(I) \text{ is continuous and completely continuous.} \tag{4.2.19}$$

Consequently, since $A_\omega(I) \subset AP(I) \subset BC(I) \subset L^\infty(I)$, $K : X \to A_\omega(I)$ is well-defined, continuous and completely continuous if $X = A_\omega(I)$, $AP(I)$ or $BC(I)$.

Our second admissibility result gives conditions on k such that (4.2.8) is true with $E_2 = L^q(I)$, $1 \leq q \leq \infty$ as before, but now $E_1 = AP(I)$.

Theorem 4.2.3. Admissibility Result.
Let $1 \le p \le \infty$ be a constant, q be such that $\frac{1}{p} + \frac{1}{q} = 1$ and I be an interval of $\mathbf{R}$. *Assume that* (4.2.9) (4.2.10) *and*

$$(4.2.20) \qquad \begin{cases} \text{for any } \epsilon > 0, \text{ there exists } l(\epsilon) > 0 \text{ such that any} \\ \text{interval of length } l(\epsilon) \text{ in } I \text{ contains a } \tau \text{ such that} \\ ||k_{t+\tau} - k_t||_q = \left(\int_I |k_{t+\tau}(s) - k_t(s)|^q \, ds \right)^{\frac{1}{q}} < \epsilon \\ \text{for all } t \text{ such that } t + \tau \in I \end{cases}$$

hold. Then

$$(4.2.21) \qquad K : L^p(I) \to AP(I) \text{ is continuous and completely continuous}$$

is true.

Remark 4.2.3. As noted in Remark 4.2.1, (4.2.10) and the triangle inequality imply that $||k_t||_q \in C(I)$. Fix $\epsilon > 0$. Then the triangle inequality and (4.2.20) imply that there exists an $l(\epsilon) > 0$ such that any interval of length $l(\epsilon)$ contains an element τ such that for all t such that $t + \tau \in I$,

$$|\ ||k_{t+\tau}||_q - ||k_t||_q\ | \le ||k_{t+\tau} - k_t||_q < \epsilon.$$

Consequently,

$$(4.2.22) \qquad ||k_t||_q \in AP(I).$$

Therefore since almost periodic functions are bounded [50,57]

$$\sup_{t \in I} ||k_t||_q < \infty.$$

PROOF OF THEOREM 4.2.3: As seen in the proof of Theorem 4.2.2, $K : L^p(I) \to C(I)$ is well-defined by (4.2.9) and (4.2.10). We now show that

$$K : L^p(I) \to AP(I) \text{ is well-defined.}$$

Fix $\epsilon > 0$ and let $y \in L^p(I)$ be such that $||y||_p < M_2$. Then since k satisfies (4.2.20), there exists $l(\epsilon/M_2) > 0$ such that any interval of length $l(\epsilon/M_2)$ contains a τ such that

$$||k_{t+\tau} - k_t||_q < \frac{\epsilon}{M_2} \text{ for all } t \text{ such that } t + \tau \in I.$$

Hence

$$(4.2.23)\qquad \begin{cases} \text{for } \epsilon > 0, \text{ any interval of length } l(\epsilon/M_2) \\ \text{contains an element } \tau \text{ such that} \\ |Ky(t+\tau) - Ky(t)| \le ||k_{t+\tau} - k_t||_q ||y||_p < \dfrac{\epsilon}{M_2} M_2 = \epsilon. \end{cases}$$

Therefore for all t such that $t + \tau \in I$,

$$|Ky(t+\tau) - Ky(t)| < \epsilon$$

and $Ky \in AP(I)$ for $y \in L^p(I)$.

The continuity of K follows easily from (4.2.22) and Hölder's inequality, since if $y_n \to y$ in $L^p(I)$, then

$$|Ky_n - Ky|_0 \le \sup_{t \in I} ||k_t||_q ||y_n - y||_p \to 0.$$

To see that $K : L^p(I) \to AP(I)$ is completely continuous, let Ω be a bounded set in $L^p(I)$, that is there exists $M_3 > 0$ such that $||y||_p < M_3$ for all $y \in \Omega$, and consider $K\Omega$. If $K\Omega$ satisfies $(i) - (iii)$ of Theorem 4.1.1, the result follows. Using arguments similar to those used in the proof of Theorem 4.2.2, conditions (i) and (ii) are easily verified. Finally, replacing M_2 with M_3 in (4.2.23) shows that (iii) is also satisfied. □

Example 4.2.1. If for some $0 < \omega < \infty$ and I an interval of $\mathbf{R}$ that contains at least one compact subinterval of length ω,

$$k(t,s) = a(t)b(s), \;\; t \in I, \;\; a.e. \;\; s \in I,$$

where $a \in A_\omega(I)$ and $b \in L^q(I)$, then trivially k satisfies the hypotheses of Theorem 4.2.2. Alternatively, if $a \in AP(I)$, then k satisfies the properties of Theorem 4.2.3.

We now present two existence principles which follow from Theorem 4.2.1 with the help of Theorem 4.2.2 and Theorem 4.2.3 respectively.

Theorem 4.2.4. Existence Principle.
Let $1 \le p \le \infty$ be a constant, q be such that $\frac{1}{p} + \frac{1}{q} = 1$, $0 < \omega < \infty$ and I be an interval of $\mathbf{R}$ that contains at least one compact subinterval I_ω of length ω. Assume that (4.2.9) – (4.2.11) *hold along with*

$$(4.2.24)\qquad h \in A_\omega(I)$$

and

$$(4.2.25)\qquad f : I \times \mathbf{R} \to \mathbf{R} \textit{ is an } L^p - \textit{Carathéodory function.}$$

In addition, suppose that there exists a constant $M > 0$, *independent of* λ, *with* $|y|_\omega \neq M$ *for any solution* $y \in A_\omega(I)$ *of*

$$y(t) = \lambda\left(h(t) + \int_I k(t,s)f(s,y(s))\,ds\right), \quad t \in I \tag{4.2.26}$$

for each $\lambda \in (0,1)$. *Then* (4.2.1) *has a solution* $y \in A_\omega(I)$.

PROOF: Let $y \in BC(I)$ be such that $|y|_0 < r$. Then (4.2.25) ensures the existence of $\mu_r \in L^p(I)$ such that $|f(t,y(t))| \leq \mu_r(t)$, for *a.e.* $t \in I$. Consequently $F : BC(I) \to L^p(I)$ as defined in (4.2.3) is bounded since

$$\int_I |f(t,y(t))|^p\,dt \leq \int_I \mu_r^p(t)\,dt < \infty.$$

Also (4.2.25) and the Lebesgue dominated convergence theorem imply that $F : BC(I) \to L^p(I)$ is continuous. To see this suppose that $y_n \to y$ in $BC(I)$ and $|y_n|_0$, $|y|_0 < r$. Then from (4.2.25)

$$f(t,y_n(t)) \to f(t,y(t)) \text{ pointwise for a.e. } t \in I$$

and

$$|f(t,y_n(t)) - f(t,y(t))| \leq 2\mu_r(t), \quad a.e.\ t \in I.$$

Hence from the Lebesgue dominated convergence theorem we obtain

$$||Fy_n - Fy||_p^p = \int_I |f(t,y_n(t)) - f(t,y(t))|^p\,dt \to 0 \text{ as } n \to \infty.$$

Now since $A_\omega(I) \subset BC(I)$

$$F : A_\omega(I) \to L^p(I) \text{ is bounded and continuous.} \tag{4.2.27}$$

By Theorem 4.2.2, (4.2.12) is true where K is given by (4.2.2). It is therefore clear that the hypotheses of Theorem 4.2.1 are satisfied with $E_1 = A_\omega(I)$ and $E_2 = L^p(I)$ and hence (4.2.1) has a solution $y \in A_\omega(I)$. □

Theorem 4.2.5. Existence Principle.
Let $1 \leq p \leq \infty$ *be a constant,* q *be such that* $\frac{1}{p} + \frac{1}{q} = 1$ *and* I *be an interval of* **R**. *Assume that* (4.2.9), (4.2.10), (4.2.20), (4.2.25) *and*

$$h \in AP(I) \tag{4.2.28}$$

hold. In addition, suppose that there exists a constant $M > 0$, *independent of* λ, *with* $|y|_0 \neq M$ *for any solution* $y \in AP(I)$ *of* (4.2.26) *for each* $\lambda \in (0,1)$. *Then* (4.2.1) *has a solution* $y \in AP(I)$.

PROOF: Since $AP(I) \subset BC(I)$, we obtain from Theorem 4.2.4 that

$$F : AP(I) \to L^p(I) \text{ is bounded and continuous}$$

where F is as defined in (4.2.3). Theorem 4.2.3 implies that (4.2.21) is true. Hence the proof follows from Theorem 4.2.1 with $E_1 = AP(I)$ and $E_2 = L^p(I)$. □

Recall from Theorem 4.2.2 that if $k : I \times I \to \mathbf{R}$ satisfies (4.2.9) – (4.2.11), or equivalently, (4.2.9), (4.2.10) and (4.2.15), then the linear integral operator as given by (4.2.2) satisfies

$$K : L^p(I) \to A_\omega(I) \text{ is continuous and completely continuous.}$$

It seems reasonable that if one were to restrict the domain of definition of K to $A_\omega(I)$, then the periodicity properties of this space could be exploited to find alternative admissibility conditions for K. We find that this is in fact the case. However, not only will the domain of K play a more significant role in determining the conditions we place on $k : I \times I \to \mathbf{R}$, our admissibility results will also be sensitive to our choice of interval I. We therefore present two further admissibility results for K as defined by (4.2.2). In the first we let $I = \mathbf{R}$, while in the second $I = [0, T]$ where $0 < T < \infty$ is dependent on the period ω.

Consider the linear, integral operator

$$K_1 y(t) := \int_{\mathbf{R}} k(t,s) y(s)\, ds, \ t \in \mathbf{R}. \tag{4.2.29}$$

Theorem 4.2.6. Admissibility Result.
Let $0 < \omega < \infty$ and assume that (4.2.9) and (4.2.10) hold with $I = \mathbf{R}$ and $q = 1$, along with

$$k(t+\omega, s+\omega) = k(t,s), \text{ for all } t \in \mathbf{R}, \text{ a.e. } s \in \mathbf{R}. \tag{4.2.30}$$

Then

$$K_1 : A_\omega(I) \to A_\omega(I) \text{ is continuous and completely continuous} \tag{4.2.31}$$

is true.

PROOF: As in Theorem 4.2.2, (4.2.9) and (4.2.10) with $I = \mathbf{R}$ and $q = 1$ give that

$$K_1 : L^\infty(\mathbf{R}) \to C(\mathbf{R}) \text{ is well-defined.}$$

However for $y \in A_\omega(\mathbf{R}) \subset L^\infty(\mathbf{R})$, (4.2.30) further implies that $K_1 y \in A_\omega(\mathbf{R})$. To see this note that by making a change of variable in the first integral, using (4.2.30) and the fact that $y \in A_\omega(\mathbf{R})$, we obtain for $t \in \mathbf{R}$

$$\begin{aligned} K_1 y(t+\omega) - K_1 y(t) &= \int_{\mathbf{R}} k(t+\omega, s) y(s)\, ds - \int_{\mathbf{R}} k(t,s) y(s)\, ds \\ &= \int_{\mathbf{R}} k(t+\omega, s+\omega) y(s+\omega)\, ds - \int_{\mathbf{R}} k(t,s) y(s)\, ds \\ &= \int_{\mathbf{R}} k(t,s)\, [y(s+\omega) - y(s)]\, ds = 0. \end{aligned}$$

Now from (4.2.30) we see that by making a change of variable

$$\begin{aligned} \|k_{t+\omega}\|_1 &= \int_{\mathbf{R}} |k(t+\omega, s)|\, ds = \int_{\mathbf{R}} |k(t+\omega, s+\omega)|\, ds \\ &= \int_{\mathbf{R}} |k(t,s)|\, ds = \|k_t\|_1, \quad t \in \mathbf{R}. \end{aligned}$$

This fact along with (4.2.10) (with $I = \mathbf{R}$ and $q = 1$) imply that $\|k_t\|_1 \in A_\omega(\mathbf{R})$ and hence $\sup_{t \in \mathbf{R}} \|k_t\|_1 < \infty$ (see Remark 4.2.1). The remainder of the proof now follows in a similar fashion to the proof of Theorem 4.2.2. □

Example 4.2.2. Consider the convolution operator

$$\tilde{K} y(t) := \int_{\mathbf{R}} \tilde{k}(t-s) y(s)\, ds, \quad t \in \mathbf{R}. \tag{4.2.32}$$

Then for $\tilde{k} \in L^1(\mathbf{R})$, (4.2.9) and (4.2.10) with $I = \mathbf{R}$ and $q = 1$ are satisfied with

$$k(t,s) = \tilde{k}(t-s), \text{ a.e. } t, s \in \mathbf{R}. \tag{4.2.33}$$

In addition, $k(t,s) = \tilde{k}(t-s)$ satisfies (4.2.30) for **any** $\omega > 0$. Hence Theorem 4.2.6 implies that

$$\left\{ \begin{array}{l} \text{for all } \omega > 0, \ \ \tilde{K} : A_\omega(\mathbf{R}) \to A_\omega(\mathbf{R}) \\ \text{is continuous and completely continuous} \end{array} \right. \tag{4.2.34}$$

is true. Convolution kernels are discussed in detail in [48,49].

Remark 4.2.4. If $\tilde{k} = e^{-|t|}$, then by Example 4.2.2

$$k(t,s) = \tilde{k}(t-s) = e^{-|t-s|}$$

satisfies the hypotheses of Theorem 4.2.6. Note however that

$$k(t+\omega, s) = e^{-|t+\omega-s|} \neq e^{-|t-s|} = k(t,s)$$

and hence this particular kernel does not satisfy the hypotheses of Theorem 4.2.2 (with $I = \mathbf{R}$ and $q = 1$).

On the other hand, if we recall the kernel

$$k(t,s) = a(t)b(s), \quad t \in \mathbf{R}, \ \text{a.e. } s \in \mathbf{R}$$

discussed in Example 4.2.1, and if we assume that $b \in L^1(\mathbf{R})$ is not ω-periodic, then we have an example of a kernel that satisfies the hypotheses of Theorem 4.2.2 (with $I = \mathbf{R}$ and $q = 1$) but not those of Theorem 4.2.6.

In the proof of Theorem 4.2.6 the fact that K_1 is defined on $\mathbf{R}$ was significant when making the various changes of variable. Therefore one might ask whether for some $0 < T < \infty$ and kernel $k : [0,T] \times [0,T] \to \mathbf{R}$ satisfying an analogous condition to (4.2.30), an admissibility result for the linear, integral operator

$$K_2 y(t) := \int_0^T k(t,s)y(s)\,ds, \quad t \in [0,T] \tag{4.2.35}$$

could be established. The following is one answer to this question.

Theorem 4.2.7. Admissibility Result.
Let $0 < \omega < \infty$, $T = n\omega$ for some $n \in \mathbf{N}^+$ and assume that (4.2.9) and (4.2.10) hold with $I = [0,T]$ and $q = 1$, along with

$$k(t+\omega, s+\omega) = k(t,s) \ \text{for all } t \in [0, T-\omega], \ \text{a.e. } s \in [0, T-\omega] \tag{4.2.36}$$

and

$$\begin{cases} k(t+m\omega, s) = k(t, s+(n-m)\omega) \ \text{for all } t \in [0, T-m\omega], \\ \text{a.e. } s \in [0, T-(n-m)\omega] \ \text{where } m \in \{0,1,\dots,n\}. \end{cases} \tag{4.2.37}$$

Then

$$\begin{cases} K_2 : A_\omega[0,T] \to A_\omega[0,T] \ \text{is continuous} \\ \text{and completely continuous} \end{cases} \tag{4.2.38}$$

is true.

PROOF: Conditions (4.2.9) and (4.2.10) with $I = [0,T]$ and $q = 1$ imply that $K_2 : C[0,T] \to C[0,T]$ is well-defined. For $t \in [0, T-\omega]$ we see that

$$K_2 y(t+\omega) - K_2 y(t) = \int_0^T k(t+\omega, s)y(s)\,ds - \int_0^T k(t,s)y(s)\,ds$$

$$= \left(\int_0^{\omega} + \int_{\omega}^{T}\right) k(t+\omega,s)y(s)\,ds$$

$$- \left(\int_0^{T-\omega} + \int_{T-\omega}^{T}\right) k(t,s)y(s)\,ds. \tag{4.2.39}$$

Using (4.2.36) and by making a change of variable, we obtain for $y \in A_\omega[0,T]$ and $t \in [0, T-\omega]$ that

$$\begin{aligned} &\int_{\omega}^{T} k(t+\omega,s)y(s)\,ds - \int_0^{T-\omega} k(t,s)y(s)\,ds \\ = &\int_{\omega}^{T} k(t+\omega,s)y(s)\,ds - \int_0^{T-\omega} k(t+\omega,s+\omega)y(s)\,ds \\ = &\int_{\omega}^{T} k(t+\omega,s)[y(s)-y(s-\omega)]\,ds = 0, \end{aligned}$$

thus reducing (4.2.39) to

$$K_2y(t+\omega) - K_2y(t)$$

$$= \int_0^{\omega} k(t+\omega,s)y(s)\,ds - \int_{T-\omega}^{T} k(t,s)y(s)\,ds, \quad t \in [0, T-\omega]. \tag{4.2.40}$$

Since $T = n\omega$, some $n \in \mathbf{N}^+$, then once again by making a change of variable, recalling that $y \in A_\omega[0,T] = A_\omega[0,n\omega]$ and (4.2.37), we have from (4.2.40) for $t \in [0,(n-1)\omega]$,

$$\begin{aligned} K_2y(t+\omega) \;-\; &K_2y(t) \\ = &\int_0^{\omega} k(t+\omega,s)y(s)\,ds - \int_{(n-1)\omega}^{n\omega} k(t,s)y(s)\,ds \\ = &\int_0^{\omega} k(t+\omega,s)y(s)\,ds - \int_{(n-1)\omega}^{n\omega} k(t,s)y(s-(n-1)\omega)\,ds \\ = &\int_0^{\omega} [k(t+\omega,s) - k(t,s+(n-1)\omega)]y(s)\,ds = 0. \end{aligned}$$

Thus $K_2 : A_\omega[0,T] \to A_\omega[0,T]$ where $T = n\omega$, some $n \in \mathbf{N}^+$, is well-defined.

The continuity and complete continuity of K_2 follows using an argument similar to that used in the proof of Theorem 4.2.2. □

Example 4.2.3. Consider

$$Ky(t) = \int_0^{2\pi} \sin(t-s)y(s)\,ds, \quad t \in [0, 2\pi]. \tag{4.2.41}$$

Now $k : [0, 2\pi] \times [0, 2\pi] \to \mathbf{R}$ defined by $k(t, s) = \sin(t - s)$ satisfies (4.2.9) and (4.2.10) with $I = [0, 2\pi]$ and $q = 1$. Clearly for any $0 < \omega \le 2\pi$

$$k(t+\omega, s+\omega) = \sin(t+\omega-s-\omega) = \sin(t-s) = k(t, s), \quad 0 \le t, s \le 2\pi - \omega,$$

and therefore k also satisfies (4.2.36) with $T = 2\pi$. Finally for any $0 < \omega \le 2\pi$ that divides 2π, that is, $2\pi = n\omega$, some $n \in \mathbf{N}^+$, we have for $m \in \{0, 1, \ldots, n\}$ that

$$\begin{aligned} k(t+m\omega, s) &= \sin(t+m\omega-s) = \sin(t+m\omega-s-n\omega) \\ &= \sin(t-(s+(n-m)\omega)) = k(t, s+(n-m)\omega) \end{aligned}$$

since $\sin t$ has period $2\pi = n\omega$. Hence k satisfies (4.2.37) with $T = 2\pi$. Therefore Theorem 4.2.7 implies that $K : A_\omega[0, T] \to A_\omega[0, T]$ where K is as defined in (4.2.41), is continuous and completely continuous for any $\omega > 0$ that divides 2π.

In general $k(t, s) = \tilde{k}(t - s)$ where $\tilde{k} \in C[-T, T]$ satisfies (4.2.9) and (4.2.10) with $I = [0, T]$ and $q = 1$, and (4.2.36) for any $0 < \omega \le T$. In addition, if $\tilde{k}$ is T-periodic and $0 < \omega \le T$ is such that $T = n\omega$, some $n \in \mathbf{N}^+$, then k satisfies (4.2.37) since

$$\begin{aligned} k(t+m\omega, s) &= \tilde{k}(t+m\omega-s) = \tilde{k}(t+m\omega-n\omega-s) \\ &= \tilde{k}(t-(s+(n-m)\omega)), \end{aligned}$$

where $m \in \{0, 1, \ldots, n\}$, $t \in [0, (n-m)\omega]$, $s \in [0, m\omega]$.

To complete the section we state two existence principles, the first for

$$y(t) = h(t) + \int_{\mathbf{R}} k(t, s)f(s, y(s))\,ds, \quad t \in \mathbf{R}, \tag{4.2.42}$$

and the second for

$$y(t) = h(t) + \int_0^T k(t, s)f(s, y(s))\,ds, \quad t \in [0, T]. \tag{4.2.43}$$

Theorem 4.2.8. Existence Principle.

Let $0 < \omega < \infty$ and assume that (4.2.9) and (4.2.10) with $q = 1$, and (4.2.24) hold with $I = \mathbf{R}$. Suppose also that (4.2.30) and

$$\text{(4.2.44)} \qquad \begin{cases} f : \mathbf{R} \times \mathbf{R} \to \mathbf{R} \text{ is continuous and} \\ f(t+\omega, y) = f(t, y) \text{ for all } t, y \in \mathbf{R} \end{cases}$$

are true. In addition, suppose that there exists a constant $M > 0$, independent of λ, with $|y|_\omega \neq M$ for any solution $y \in A_\omega(\mathbf{R})$ of (4.2.26) with $I = \mathbf{R}$, for each $\lambda \in (0, 1)$. Then (4.2.42) has a solution $y \in A_\omega(\mathbf{R})$.

PROOF: From Theorem 4.2.6, (4.2.31) is true, hence if we show that

$$\text{(4.2.45)} \qquad F_1 y(t) = f(t, y(t)), \quad t \in \mathbf{R}$$

is such that

$$\text{(4.2.46)} \qquad F_1 : A_\omega(\mathbf{R}) \to A_\omega(\mathbf{R}) \text{ is bounded and continuous,}$$

it is clear that the result follows from Theorem 4.2.1 with $I = \mathbf{R}$, $E_1 = E_2 = A_\omega(\mathbf{R})$, $K = K_1$ and $F = F_1$ where K_1 and F_1 are as defined in (4.2.29) and (4.2.45) respectively. Clearly (4.2.44) gives that

$$F : BC(\mathbf{R}) \to BC(\mathbf{R}) \text{ is bounded and continuous}$$

and for $y \in A_\omega(\mathbf{R})$

$$\begin{aligned} Fy(t+\omega) &= f(t+\omega, y(t+\omega)) \\ &= f(t+\omega, y(t)) = f(t, y(t)) = Fy(t), \quad t \in \mathbf{R}. \end{aligned}$$

Hence (4.2.46) is true. □

Theorem 4.2.9. Admissibility Result.
Let $0 < \omega < \infty$, $T = n\omega$ for some $n \in \mathbf{N}^+$ and assume that (4.2.9) and (4.2.10) with $q = 1$, and (4.2.24) hold with $I = \mathbf{R}$. Suppose also that (4.2.36), (4.2.37) and

$$\text{(4.2.47)} \qquad \begin{cases} f : [0, T] \times \mathbf{R} \to \mathbf{R} \text{ is continuous and} \\ f(t+\omega, y) = f(t, y) \text{ for all } t \in [0, T-\omega], \ \ y \in \mathbf{R} \end{cases}$$

are true. In addition, suppose that there exists a constant $M > 0$, independent of λ, with $|y|_\omega \neq M$ for any solution $y \in A_\omega[0, T]$ of (4.2.26) with $I = \mathbf{R}$ for each $\lambda \in (0, 1)$. Then (4.2.43) has a solution $y \in A_\omega[0, T]$.

PROOF: Using Theorem 4.2.7 and an argument similar to the one used in the proof of Theorem 4.2.8, the result follows from Theorem 4.2.1 with

$I = [0, T]$, $E_1 = E_2 = A_\omega[0, T]$, $K = K_2$ where K_2 is as defined in (4.2.35) and

$$Fy(t) = F_2y(t) = f(t, y(t)), \quad t \in [0, T].$$

We omit the detail. □

Remark 4.2.5. Existence results which ensure the existence of a continuous solution of (4.2.1) are plentiful in the literature [137]. The techniques used in obtaining these results can be adapted to provide existence results for (4.2.1) when a solution $y \in A_\omega(I)$ or $y \in AP(I)$ is required.

Remark 4.2.6. One could also consider the equation

$$y(t) = h(t) + \int_I k(t, s) f(s, y(s))\, ds, \ \text{a.e. } t \in I \tag{4.2.48}$$

where I is a finite subinterval of $\mathbf{R}$. Results analogous to those presented for (4.2.1) in this section with I finite can be obtained which prove the existence of a solution $y \in L^p(I)$, $1 \le p < \infty$ of (4.2.48) such that

$$y(t) = y(t + \omega) \ \text{a.e. } t \text{ such that } t + \omega \in I.$$

We omit the detail as the results are readily established using the arguments presented in this paper along with those in [137,Chapter 4].

In [49] Corduneanu discusses (4.2.48) with $I = \mathbf{R}$ and convolution kernel, that is,

$$y(t) = h(t) + \int_{\mathbf{R}} \tilde{k}(t - s) f(s, y(s))\, ds, \ \text{a.e. } t \in \mathbf{R}. \tag{4.2.49}$$

A result is presented which proves the existence of a solution $y \in P_\omega(\mathbf{R})$ of (4.2.49). Here $P_\omega(\mathbf{R})$ is defined to be the space of locally integrable functions with period $\omega > 0$. We refer the reader to [49] for further details.

4.3. Existence theory.

Existence results which establish the existence of one or more periodic or almost periodic solutions of the nonlinear integral equation

$$y(t) = \int_{\mathbf{R}} k(t, s) f(s, y(s))\, ds, \ t \in \mathbf{R}, \tag{4.3.1}$$

are presented in this section. These results will be established via fixed point methods. To be specific, we will use either the nonlinear alternative (Theorem 1.2.1), Krasnoselskii's fixed point theorem (Theorem 2.1.1) or the Leggett Williams fixed point theorem which we now state.

Theorem 4.3.1. Leggett Williams Fixed Point Theorem.
Let $E := (E, ||.||)$ be a Banach space, $C \subset E$ a cone in E, $r_1 > 0$, $r_2 > 0$, $r_1 \neq r_2$ with $R := \max\{r_1, r_2\}$ and $r := \min\{r_1, r_2\}$. Define $\overline{C}_\eta := \{y \in C : ||y|| \leq \eta\}$ and $S_\eta := \{y \in C : ||y|| = \eta\}$. Let $N : \overline{C}_R \to C$ be a continuous, compact map with

(*i*) *there exists $u_0 \in C\backslash\{0\}$ with $Nu \not\leq u$ for $u \in S_{r_2} \cap C(u_0)$; here $C(u_0) := \{u \in C :$ there exists $\lambda > 0$ with $u \geq \lambda u_0\}$; and*

(*ii*) *$||Nu|| \leq ||u||$ for $u \in S_{r_1}$*

holding. Then N has at least one fixed point $y \in C$ with $r \leq ||y|| \leq R$.

We begin by using Krasnoselskii's fixed point theorem to establish an existence result for (4.3.1).

Theorem 4.3.2. *Let $1 \leq p \leq \infty$ be a constant and $q \geq 1$ be such that $\frac{1}{p} + \frac{1}{q} = 1$. Suppose the following conditions are satisfied:*

$$k_t(s) = k(t, s) \in L^q(\mathbf{R}) \text{ for each } t \in \mathbf{R}, \tag{4.3.2}$$

$$\text{the map } t \mapsto k_t \text{ is continuous from } \mathbf{R} \text{ to } L^q(\mathbf{R}), \tag{4.3.3}$$

$$\text{for each } t \in \mathbf{R},\ k(t, s) \geq 0 \text{ for a.e. } s \in \mathbf{R}, \tag{4.3.4}$$

$$\int_{\mathbf{R}} |k_{t+w}(s) - k_t(s)|^q \, ds = 0 \text{ for all } t \in \mathbf{R}, \tag{4.3.5}$$

$$f : \mathbf{R} \times [0, \infty) \to [0, \infty) \text{ is a } L^p - \text{Carathéodory function}, \tag{4.3.6}$$

$$\begin{cases} \text{there exists a function } \psi : [0, \infty) \to [0, \infty) \text{ continuous and} \\ \text{nondecreasing, a constant } A_0,\ 0 < A_0 \leq 1, \text{ and a function} \\ \theta \in L^p(\mathbf{R}) \text{ with } A_0\theta(s)\psi(y) \leq f(s, y) \leq \theta(s)\psi(y) \\ \text{for all } y \in [0, \infty) \text{ and a.e. } s \in \mathbf{R}, \end{cases} \tag{4.3.7}$$

$$\begin{cases} \text{there exists a continuous function } \phi : (0, 1) \to (0, \infty) \\ \text{such that for any } 0 < m < 1 \text{ and } u \geq 0, \text{ we have} \\ \psi(mu) \geq \phi(m)\psi(u), \end{cases} \tag{4.3.8}$$

$$K_2 := \inf_{t \in [0, w]} \int_{\mathbf{R}} \theta(s) k(t, s) \, ds > 0, \tag{4.3.9}$$

$$(4.3.10) \qquad \begin{cases} \textit{there exists } 0 < M < 1 \textit{ with } \dfrac{M}{\phi(M)} \le A_0 \dfrac{K_2}{K_1}, \\ \textit{where } K_1 := \sup\limits_{t\in[0,w]} \displaystyle\int_{\mathbf{R}} \theta(s)k(t,s)\,ds, \end{cases}$$

$$(4.3.11) \qquad \textit{there exists } \alpha > 0 \textit{ with } \alpha > K_1 \psi(\alpha)$$

and

$$(4.3.12) \qquad \textit{there exists } \beta > 0,\ \beta \ne \alpha,\ \textit{with } \beta < A_0 K_2 \psi(M\beta).$$

Then (4.3.1) *has at least one positive solution* $y \in A_w(\mathbf{R})$ *and either*

$$(A)\ 0 < \alpha < |y|_w < \beta \textit{ and } y(t) \ge M\alpha \textit{ for } t \in \mathbf{R} \textit{ if } \alpha < \beta$$

or

$$(B)\ 0 < \beta < |y|_w < \alpha \textit{ and } y(t) \ge M\beta \textit{ for } t \in \mathbf{R} \textit{ if } \beta < \alpha$$

holds.

Remark 4.3.1. Recall from Section 4.2 that (4.3.5) implies that $\|k_t\|_q \in A_w(\mathbf{R})$ and therefore $\sup_{t\in\mathbf{R}} \|k_t\|_q = \sup_{t\in[0,w]} \|k_t\|_q < \infty$. Recall also that (4.3.5) together with Chebychev's inequality implies for a.e. $s \in \mathbf{R}$ that $k(t+w,s) = k(t,s)$ for all $t \in \mathbf{R}$.

PROOF OF THEOREM 4.3.2: Let $E := (A_w(\mathbf{R}), |.|_w)$ and

$$\begin{aligned} C &:= \{u \in A_w(\mathbf{R}) : u(t) \ge M|u|_w \text{ for } t \in [0,w]\} \\ &= \{u \in A_w(\mathbf{R}) : u(t) \ge M|u|_w \text{ for } t \in \mathbf{R}\} \end{aligned}$$

where M is as defined in (4.3.10). Let $K : A_w(\mathbf{R}) \to E$ be defined by

$$Ky(t) := \int_{\mathbf{R}} k(t,s) f(s,y(s))\,ds \text{ for } t \in \mathbf{R};$$

here $y \in A_w(\mathbf{R})$. The results in Section 4.2 guarantee

$$(4.3.13) \qquad K : C \to A_w(\mathbf{R}) \text{ is continuous and completely continuous.}$$

Let

$$\Omega_\alpha := \{y \in A_w(\mathbf{R}) : |y|_w < \alpha\}$$

and

$$\Omega_\beta := \{y \in A_w(\mathbf{R}) : |y|_w < \beta\}.$$

We wish to apply Krasnoselskii's fixed point theorem. To do this we need to show that

$$K : C \to C, \tag{4.3.14}$$

$$|Ky|_w < |y|_w \text{ for } y \in C \cap \partial\Omega_\alpha = S_\alpha \tag{4.3.15}$$

and

$$|Ky|_w > |y|_w \text{ for } y \in C \cap \partial\Omega_\beta = S_\beta \tag{4.3.16}$$

hold.

Let $y \in C$. Then (4.3.7) implies for $t \in [0, w]$, that

$$|Ky(t)| \le \psi\left(|y|_w\right) \int_{\mathbf{R}} \theta(s)k(t,s)\,ds \le \psi\left(|y|_w\right) K_1,$$

therefore

$$|Ky|_w \le \psi\left(|y|_w\right) K_1. \tag{4.3.17}$$

On the other hand, since $y \in C$, we have that $y(t) \ge M|y|_w$ for $t \in \mathbf{R}$ and therefore (4.3.7), (4.3.8), (4.3.17) and (4.3.10) give for $t \in [0, w]$,

$$\begin{aligned} Ky(t) &\ge A_0 \int_{\mathbf{R}} \theta(s)k(t,s)\psi(y(s))\,ds \ge A_0\psi\left(M|y|_w\right) \int_{\mathbf{R}} \theta(s)k(t,s)\,ds \\ &\ge A_0\phi(M)\psi\left(|y|_w\right) \int_{\mathbf{R}} \theta(s)k(t,s)\,ds \ge K_2 A_0 \phi(M)\psi\left(|y|_w\right) \\ &\ge \frac{K_2}{K_1} A_0\phi(M)|Ky|_w \ge M|Ky|_w. \end{aligned}$$

Consequently $Ky \in C$ and (4.3.14) holds.

To show (4.3.15), let $y \in C \cap \partial\Omega_\alpha$. Then $|y|_w = \alpha$ and $y(t) \ge M\alpha$ for $t \in \mathbf{R}$. Now for $t \in [0, w]$, we have

$$|Ky(t)| \le \psi\left(|y|_w\right) \int_{\mathbf{R}} \theta(s)k(t,s)\,ds \le \psi(\alpha)K_1,$$

and this together with (4.3.11) yields

$$|Ky|_w \le \psi(\alpha)K_1 < \alpha = |y|_w. \tag{4.3.18}$$

Thus (4.3.15) is true.

Now let $y \in C \cap \partial\Omega_\beta$. Then $|y|_w = \beta$ and $M\beta \le y(t) \le \beta$ for $t \in \mathbf{R}$. Also for $t \in [0, w]$, we have using (4.3.7) and (4.3.12) that

$$\begin{aligned} Ky(t) &\ge A_0 \int_{\mathbf{R}} \theta(s)k(t,s)\psi(y(s))\,ds \\ &\ge A_0\psi(M\beta)\int_{\mathbf{R}} \theta(s)k(t,s)\,ds \\ &\ge A_0 K_2 \psi(M\beta) > \beta = |y|_w. \end{aligned}$$

Thus $|Ky|_w > |y|_w$ and (4.3.16) is true.

Krasnoselskii's fixed point theorem guarantees that there is a solution $y \in A_w(\mathbf{R})$ of (4.3.1) with $y \in C \cap \left(\overline{\Omega}_\alpha \backslash \Omega_\beta\right)$ if $\beta < \alpha$, whereas $y \in C \cap \left(\overline{\Omega}_\beta \backslash \Omega_\alpha\right)$ if $\alpha < \beta$. Finally note that $|y|_w \ne \alpha$ and $|y|_w \ne \beta$. To see this suppose $|y|_w = \alpha$. Then since $y = Ky$ we have (following the ideas used to prove (4.3.18)),

$$\alpha = |y|_w = |Ky|_w \le \psi(\alpha)K_1 < \alpha = |y|_w,$$

– a contradiction. A similar argument shows that $|y|_w \ne \beta$. □

Remark 4.3.2. It is possible to replace (4.3.2), (4.3.3), (4.3.5) and (4.3.6) with

(4.3.19) $$k_t(s) = k(t,s) \in L^1(\mathbf{R}) \text{ for each } t \in \mathbf{R},$$

(4.3.20) $$\text{the map } t \mapsto k_t \text{ is continuous from } \mathbf{R} \text{ to } L^1(\mathbf{R}),$$

(4.3.21) $$f : \mathbf{R} \times [0,\infty) \to [0,\infty) \text{ is continuous}$$

and

(4.3.22) $$f(t+w, y) = f(t,y) \text{ for all } t \in \mathbf{R} \text{ and } y \ge 0,$$

and the result in Theorem 4.3.2 is again true. We need only notice (see Section 4.2) that (4.3.19) – (4.3.22) imply that $K : C \to C$ is continuous and completely continuous.

With additional conditions on the nonlinearity and kernel, repeated application of Theorem 4.3.2 will yield additional positive, periodic solutions of (4.3.1). For completeness we provide one *multiple* solution result (the proof follows immediately from two applications of Theorem 4.3.2).

Theorem 4.3.3. *Let $1 \le p \le \infty$ be a constant and $q \ge 1$ be such that $\frac{1}{p} + \frac{1}{q} = 1$. Suppose that (4.3.2) – (4.3.10) hold. In addition assume there are constants $\gamma_0 < \gamma_1 < \gamma_2$ with the following satisfied:*

$$\text{there exists } \gamma_0 > 0 \text{ with } \gamma_0 < A_0 K_2 \psi(M\gamma_0), \tag{4.3.23}$$

$$\text{there exists } \gamma_1 > 0 \text{ with } \gamma_1 > K_1 \psi(\gamma_1) \tag{4.3.24}$$

and

$$\text{there exists } \gamma_2 > 0 \text{ with } \gamma_2 < A_0 K_2 \psi(M\gamma_2). \tag{4.3.25}$$

Then (4.3.1) has at least two positive solutions $y_1,\ y_2 \in A_w(\mathbf{R})$, with $0 < \gamma_0 < |y_1|_w < \gamma_1 < |y_2|_w < \gamma_2$ and $y_1(t) \ge M\gamma_0$ for $t \in \mathbf{R}$, $y_2(t) \ge M\gamma_1$ for $t \in \mathbf{R}$.

We next present a result for the case when (4.3.9) may not hold.

Theorem 4.3.4. *Let $1 \le p \le \infty$ be a constant and $q \ge 1$ be such that $\frac{1}{p} + \frac{1}{q} = 1$. Suppose that (4.3.2) – (4.3.6) and (4.3.11) hold. In addition assume the following conditions are satisfied:*

$$\left\{\begin{array}{l} \text{there exists } 0 < M < 1,\ \kappa \in L^q(\mathbf{R}), \text{ and an interval} \\ [a,b] \subseteq [0,w],\ a < b,\ \text{with } k(t,s) \ge M\kappa(s) \\ \text{for } t \in [a,b], \text{ and a.e. } s \in \mathbf{R}, \end{array}\right. \tag{4.3.26}$$

$$k(t,s) \le \kappa(s) \text{ for } t \in [0,w] \text{ and a.e. } s \in \mathbf{R}, \tag{4.3.27}$$

$$\left\{\begin{array}{l} \text{there exists a function } \psi : [0,\infty) \to [0,\infty) \text{ continuous} \\ \text{and nondecreasing, and } \theta \in L^p(\mathbf{R}), \text{ with} \\ f(s,y) \le \theta(s)\psi(y) \text{ for all } y \in [0,\infty), \text{ and a.e. } s \in \mathbf{R}, \end{array}\right. \tag{4.3.28}$$

$$\left\{\begin{array}{l} \text{there exists } \tau \in L^p[a,b] \text{ with } f(s,y) \ge \tau(s)\psi(y) \\ \text{for all } y \in [0,\infty), \text{ and a.e. } s \in [a,b] \end{array}\right. \tag{4.3.29}$$

and

$$\left\{\begin{array}{l} \text{there exists } \beta > 0,\ \beta \ne \alpha, \text{ with } \beta < K_3 \psi(M\beta) \\ \text{where } K_3 := \displaystyle\sup_{t\in[0,w]} \int_a^b k(t,s)\tau(s)\,ds. \end{array}\right. \tag{4.3.30}$$

Then (4.3.1) *has at least one positive solution* $y \in A_w(\mathbf{R})$, *and either*

$$(A)\ 0 < \alpha < |y|_w < \beta \text{ and } y(t) \geq M\alpha \text{ for } t \in [a,b] \text{ if } \alpha < \beta$$

or

$$(B)\ 0 < \beta < |y|_w < \alpha \text{ and } y(t) \geq M\beta \text{ for } t \in [a,b] \text{ if } \beta < \alpha$$

holds.

PROOF: Let E, K, Ω_α and Ω_β be as described in the proof of Theorem 4.3.2 and let

$$C := \{y \in A_w(\mathbf{R}) : y(t) \geq 0 \text{ for } t \in [0,w] \text{ and } \min_{t\in[a,b]} y(t) \geq M|y|_w\};$$

here M is as defined in (4.3.26). Again the results in Section 4.2 guarantee that $K : C \to E$ is continuous and completely continuous. We can apply Krasnoselskii's fixed point theorem if we show that (4.3.14) – (4.3.16) hold.

To see (4.3.14) let $y \in C$. Then

$$|Ky|_w = \sup_{t\in[0,w]} \int_{\mathbf{R}} k(t,s) f(s,y(s))\, ds$$

and this together with (4.3.27) gives

$$|Ky|_w \leq \int_{\mathbf{R}} \kappa(s) f(s,y(s))\, ds.$$

This inequality and (4.3.26) imply that

$$\begin{aligned} \min_{t\in[a,b]} Ky(t) &= \min_{t\in[a,b]} \int_{\mathbf{R}} k(t,s) f(s,y(s))\, ds \\ &\geq M \int_{\mathbf{R}} \kappa(s) f(s,y(s))\, ds \geq M|Ky|_w, \end{aligned}$$

and consequently (4.3.14) is true.

To verify (4.3.15), let $y \in C \cap \partial\Omega_\alpha$. Then $|y|_w = \alpha$. Conditions (4.2.11) and (4.3.28) yield

$$\begin{aligned} |Ky|_w &= \sup_{t\in[0,w]} \int_{\mathbf{R}} k(t,s) f(s,y(s))\, ds \leq \sup_{t\in[0,w]} \int_{\mathbf{R}} k(t,s)\theta(s)\psi(y(s))\, ds \\ &\leq \psi\left(|y|_w\right) \sup_{t\in[0,w]} \int_{\mathbf{R}} k(t,s)\theta(s)\, ds \leq \psi(\alpha) K_1 < \alpha = |y|_w, \end{aligned}$$

and therefore (4.3.15) holds.

To see (4.3.16) let $y \in C \cap \partial\Omega_\beta$ and consequently $|y|_w = \beta$ and $M\beta \leq y(t) \leq \beta$ for $t \in [a,b]$. Now (4.3.29) and (4.3.30) imply that

$$\begin{aligned} |Ky|_w &= \sup_{t\in[0,w]} \int_{\mathbf{R}} k(t,s)f(s,y(s))\,ds \geq \sup_{t\in[0,w]} \int_a^b k(t,s)f(s,y(s))\,ds \\ &\geq \sup_{t\in[0,w]} \int_a^b k(t,s)\tau(s)\psi(y(s))\,ds \geq \psi(M\beta) \sup_{t\in[0,w]} \int_a^b k(t,s)\tau(s)\,ds \\ &= K_3\psi(M\beta) > \beta = |y|_w, \end{aligned}$$

and (4.3.16) holds. We may now apply Krasnoselskii's fixed point theorem to obtain the desired result. □

Remark 4.3.3. It is possible to replace (4.3.2), (4.3.3), (4.3.5) and (4.3.6) with (4.3.19) – (4.3.22) and the result in Theorem 4.3.4 is again true.

Remark 4.3.4. As in Theorem 4.3.3, it is possible to combine Theorem 4.3.4 (A) and (B) to obtain a result which guarantees the existence of two (or more) nonnegative periodic solutions of (4.3.1).

It is also possible to use the nonlinear alternative (Theorem 1.2.1) to obtain an existence result for (4.3.1).

Theorem 4.3.5. *Let* $1 \leq p \leq \infty$ *be a constant and* $q \geq 1$ *be such that* $\frac{1}{p} + \frac{1}{q} = 1$. *Suppose that* (4.3.2) – (4.3.5) *and* (4.3.11) *hold. In addition assume the following conditions are satisfied:*

(4.3.31) $\quad f : \mathbf{R} \times \mathbf{R} \to [0,\infty)$ *is an* L^p – *Carathéodory function*

and

$$(4.3.32)\qquad \begin{cases} \text{there exists a function } \psi : [0,\infty) \to [0,\infty) \text{ continuous} \\ \text{and nondecreasing, and } \theta \in L^p(\mathbf{R}) \text{ with} \\ f(s,y) \leq \theta(s)\psi(y) \text{ for all } y \in \mathbf{R}, \text{ and a.e. } s \in \mathbf{R}. \end{cases}$$

Then (4.3.1) *has a nonnegative solution* $y \in A_w(\mathbf{R})$ *with* $|y|_w < \alpha$.

Remark 4.3.5. In the above theorem it is possible for $|y|_w$ to be zero in some applications.

Remark 4.3.6. It is possible to replace (4.3.2), (4.3.3), (4.3.5) and (4.3.31) with (4.3.19), (4.3.20) and

(4.3.33) $\quad f : \mathbf{R} \times \mathbf{R} \to [0,\infty)$ is continuous

and

$$f(t+w, y) = f(t, y) \text{ for all } t \in \mathbf{R} \text{ and } y \in \mathbf{R}, \tag{4.3.34}$$

and the result in Theorem 4.3.5 is again true.

PROOF OF THEOREM 4.3.5: We will apply the nonlinear alternative. Let

$$E := (A_w(\mathbf{R}), |.|_w), \ U := \{u \in A_w(\mathbf{R}) : |u|_w < \alpha\},$$

and $K : A_w(\mathbf{R}) \to A_w(\mathbf{R})$ be defined by

$$Ky(t) := \int_{\mathbf{R}} k(t,s) f(s, y(s))\, ds \text{ for } t \in \mathbf{R};$$

here $y \in A_w(\mathbf{R})$. Again the results in Section 4.2 guarantee that $K : A_w(\mathbf{R}) \to A_w(\mathbf{R})$ is continuous and completely continuous. Let $y \in A_w(\mathbf{R})$ be any solution of

$$y(t) = \lambda \left(\int_{\mathbf{R}} k(t,s) f(s, y(s))\, ds \right), \quad t \in \mathbf{R},$$

for $0 < \lambda < 1$. For $t \in [0, w]$ we have

$$\begin{aligned} |y(t)| &= y(t) \leq \int_{\mathbf{R}} k(t,s)\theta(s)\psi(y(s))\, ds \\ &\leq \psi(|y|_w) \int_{\mathbf{R}} k(t,s)\theta(s)\, ds \leq K_1 \psi(|y|_w) \end{aligned}$$

and therefore

$$|y|_w \leq K_1 \psi(|y|_w). \tag{4.3.35}$$

Now (4.3.11) and (4.3.35) imply that $|y|_w \neq \alpha$. Apply the nonlinear alternative (notice (ii) cannot occur). Consequently (4.3.1) has a nonnegative solution $y \in A_w(\mathbf{R})$ with $|y|_w < \alpha$ (note $|y|_w \leq \alpha$ from the nonlinear alternative and $|y|_w \neq \alpha$ by an argument similar to the one above). □

Remark 4.3.7. It is possible to combine Theorem 4.3.2 (A) (or Theorem 4.3.4 (A)) and Theorem 4.3.5 to obtain a result which guarantees the existence of two nonnegative periodic solutions of (4.3.1). (We saw a similar idea in Theorem 2.2.3 in Chapter 2.) We leave the details to the reader.

Next we use Theorem 4.3.1 to obtain an existence result for (4.3.1).

Theorem 4.3.6. *Let $1 \le p \le \infty$ be a constant and $q \ge 1$ be such that $\frac{1}{p} + \frac{1}{q} = 1$. Suppose that (4.3.2) – (4.3.6) and (4.3.9) hold. In addition assume the following conditions are satisfied:*

$$(4.3.36) \qquad \begin{cases} \text{there exists a nondecreasing continuous map} \\ \psi : [0,\infty) \to [0,\infty), \ \text{a continuous map} \\ \phi : [0,\infty) \to [0,\infty), \ \text{and } \theta \in L^p(\mathbf{R}) \\ \text{with } \phi(y)\theta(t) \le f(t,y) \le \psi(y)\theta(t) \\ \text{for all } y \in [0,\infty), \ \text{and a.e. } t \in \mathbf{R}, \end{cases}$$

$$(4.3.37) \qquad \text{there exists } r > 0 \text{ with } r < K_2\phi(r),$$

$$(4.3.38) \qquad \frac{\phi(x)}{x} \ \text{is nonincreasing on } (0,r)$$

and

$$(4.3.39) \qquad \begin{cases} \text{there exists } R > r \text{ with } R > \psi(R)K_1, \ \text{where} \\ K_1 := \sup_{t\in[0,w]} \int_{\mathbf{R}} \theta(s)k(t,s)\,ds. \end{cases}$$

Then (4.3.1) has a nonnegative solution $y \in A_w(\mathbf{R})$ with $r \le |y|_w < R$.

Remark 4.3.8. If $R \ge \psi(R)K_1$ in (4.3.39) then the result in Theorem 4.3.6 is again true if $r \le |y|_w < R$ is replaced by $r \le |y|_w \le R$.

PROOF OF THEOREM 4.3.6: Let $E = (A_w(\mathbf{R}), |.|_w)$ and

$$\begin{aligned} C &:= \{u \in A_w(\mathbf{R}) : u(t) \ge 0 \text{ for } t \in [0,w]\} \\ &= \{u \in A_w(\mathbf{R}) : u(t) \ge 0 \text{ for } t \in \mathbf{R}\}. \end{aligned}$$

Also let $u_0 \equiv 1$ and note that

$$\begin{aligned} C(u_0) &:= \{u \in C : \text{there exists } \lambda > 0 \text{ with } u(t) \ge \lambda \text{ for } t \in [0,w]\} \\ &= \{u \in C : u(t) > 0 \text{ for } t \in [0,w]\}. \end{aligned}$$

Define $K : C \to C$ by

$$Ky(t) := \int_{\mathbf{R}} k(t,s)f(s,y(s))\,ds \text{ for } t \in \mathbf{R};$$

here $y \in C$. The results in Section 4.2 guarantee that $K : C \to C$ is continuous and completely continuous. We want to apply Theorem 4.3.1, therefore we first show

$$(4.3.40) \qquad |Ky|_w \le |y|_w \text{ for } y \in S_R.$$

If $y \in S_R$ then $|y|_w = R$ and consequently for $t \in [0, w]$ we have that

$$Ky(t) \le \int_{\mathbf{R}} k(t,s)\theta(s)\psi(y(s))\,ds \le \psi\left(|y|_w\right) \int_{\mathbf{R}} k(t,s)\theta(s)\,ds \le \psi(R)K_1.$$

This together with (4.3.39) gives

$$|Ky|_w \le \psi(R)K_1 < R = |y|_w, \tag{4.3.41}$$

therefore (4.3.40) is true.

Next we prove

$$Ky \not\le y \text{ for } y \in S_r \cap C(u_0). \tag{4.3.42}$$

To see this let $y \in S_r \cap C(u_0)$, and hence

$$|y|_w = r \text{ and } r \ge y(t) > 0 \text{ for } t \in [0, w].$$

For $t \in [0, w]$ we have

$$\begin{aligned} Ky(t) &\ge \int_{\mathbf{R}} k(t,s)\theta(s)\phi(y(s))\,ds = \int_{\mathbf{R}} k(t,s)\theta(s)\frac{\phi(y(s))}{y(s)}y(s)\,ds \\ &\ge \frac{\phi(r)}{r}\int_{\mathbf{R}} k(t,s)\theta(s)y(s)\,ds. \end{aligned}$$

Let $t_0 \in [0, w]$ be such that $\min_{t\in[0,w]} y(t) = y(t_0)$, and this together with the previous inequality yields for $t \in [0, w]$ that

$$Ky(t) \ge \frac{\phi(r)}{r}y(t_0)\int_{\mathbf{R}} k(t,s)\theta(s)\,ds \ge \left(\frac{\phi(r)}{r}K_2\right)y(t_0).$$

Combining this inequality with (4.3.37) gives $Ky(t) > y(t_0)$ for $t \in [0, w]$, therefore in particular $Ky(t_0) > y(t_0)$. Thus (4.3.42) is true.

We apply Theorem 4.3.1 to deduce that (4.3.1) has a nonnegative solution $y \in A_w(\mathbf{R})$ with $r \le |y|_w \le R$. Note that if $|y|_w = R$, then from $y = Ky$ we have (follow the ideas from (4.3.40) to (4.3.41)),

$$R = |y|_w = |Ky|_w \le \psi(R)K_1 < R = |y|_w,$$

– a contradiction. □

For the remainder of this section, we turn our attention to existence results which establish the existence of a solution $y \in AP(\mathbf{R})$ of (4.3.1).

We first use Krasnoselskii's fixed point theorem to establish the existence of an $AP(\mathbf{R})$ solution of (4.3.1).

Theorem 4.3.7. *Let $1 \le p \le \infty$ be a constant and $q \ge 1$ be such that $\frac{1}{p} + \frac{1}{q} = 1$. Suppose that (4.3.2) – (4.3.4) and (4.3.6) – (4.3.8) hold. In addition assume the following conditions are satisfied:*

$$(4.3.43)\qquad \begin{cases} \text{for any } \epsilon > 0 \text{ there exists } l(\epsilon) > 0 \text{ such that any} \\ \text{interval of length } l(\epsilon) \text{ in } \mathbf{R} \text{ contains a } \tau \text{ such that} \\ \|k_{t+\tau} - k_t\|_q = (\int_{\mathbf{R}} |k_{t+\tau}(s) - k_t(s)|^q\, ds)^{\frac{1}{q}} < \epsilon \text{ for all } t \in \mathbf{R}, \end{cases}$$

$$(4.3.44)\qquad K_4 := \inf_{t\in\mathbf{R}} \int_{\mathbf{R}} \theta(s)k(t,s)\, ds > 0,$$

$$(4.3.45)\qquad \begin{cases} \text{there exists } 0 < M < 1 \text{ with } \dfrac{M}{\phi(M)} \le A_0 \dfrac{K_4}{K_5} \\ \text{where } K_5 := \sup_{t\in\mathbf{R}} \int_{\mathbf{R}} \theta(s)k(t,s)\, ds, \end{cases}$$

$$(4.3.46)\qquad \text{there exists } \alpha > 0 \text{ with } \alpha > K_5\psi(\alpha)$$

and

$$(4.3.47)\qquad \text{there exists } \beta > 0,\ \beta \ne \alpha, \text{ with } \beta < A_0K_4\psi(M\beta).$$

Then (4.3.1) has at least one positive solution $y \in AP(\mathbf{R})$ and either

$$(A)\ 0 < \alpha < |y|_0 < \beta \text{ and } y(t) \ge M\alpha \text{ for } t \in \mathbf{R} \text{ if } \alpha < \beta$$

or

$$(B)\ 0 < \beta < |y|_0 < \alpha \text{ and } y(t) \ge M\beta \text{ for } t \in \mathbf{R} \text{ if } \beta < \alpha$$

holds.

Remark 4.3.9. We know from Section 4.2 that $\|k_t\|_{L^q} \in AP(\mathbf{R})$ and since almost periodic functions are bounded [50] we have that $\sup_{t\in\mathbf{R}} \|k_t\|_{L^q} < \infty$.

PROOF OF THEOREM 4.3.7: Let $E := (AP(\mathbf{R}), |.|_0)$ and

$$C := \{u \in AP(\mathbf{R}) : u(t) \ge M|u|_0 \text{ for } t \in \mathbf{R}\}$$

where M is as defined in (4.3.45). Let $K : AP(\mathbf{R}) \to E$ be defined by

$$Ky(t) := \int_{\mathbf{R}} k(t,s)f(s,y(s))\,ds \text{ for } t \in \mathbf{R};$$

here $y \in AP(\mathbf{R})$. The results in Section 4.2 guarantee that

$$K : C \to AP(\mathbf{R}) \text{ is continuous and completely continuous.}$$

Let

$$\Omega_\alpha := \{y \in AP(\mathbf{R}) : |y|_0 < \alpha\}$$

and

$$\Omega_\beta := \{y \in AP(\mathbf{R}) : |y|_0 < \beta\}.$$

Essentially the same reasoning as in Theorem 4.3.2 (except that here we take $t \in \mathbf{R}$ instead of $t \in [0,w]$) guarantees that

$$K : C \to C,$$

$$|Ky|_0 < |y|_0 \text{ for } y \in C \cap \partial\Omega_\alpha = S_\alpha$$

and

$$|Ky|_0 > |y|_0 \text{ for } y \in C \cap \partial\Omega_\beta = S_\beta$$

hold. Now apply Krasnoselskii's fixed point theorem. $\square$

There are also analogues of Theorem 4.3.3–Theorem 4.3.6. We state these results here for the convenience of the reader.

Theorem 4.3.8. *Let* $1 \le p \le \infty$ *be a constant and* $q \ge 1$ *be such that* $\frac{1}{p} + \frac{1}{q} = 1$. *Suppose that* (4.3.2) – (4.3.4), (4.3.6) – (4.3.8), *and* (4.3.43) – (4.3.45) *hold. In addition assume there are constants* $\gamma_0 < \gamma_1 < \gamma_2$ *with the following satisfied:*

(4.3.48) *there exists* $\gamma_0 > 0$ *with* $\gamma_0 < A_0 K_4 \psi(M\gamma_0)$,

(4.3.49) *there exists* $\gamma_1 > 0$ *with* $\gamma_1 > K_5 \psi(\gamma_1)$

and

(4.3.50) *there exists* $\gamma_2 > 0$ *with* $\gamma_2 < A_0 K_4 \psi(M\gamma_2)$.

Then (4.3.1) *has at least two positive solutions* $y_1,\ y_2 \in AP(\mathbf{R})$ *with* $0 < \gamma_0 < |y_1|_0 < \gamma_1 < |y_2|_0 < \gamma_2$ *and* $y_1(t) \ge M\gamma_0$ *for* $t \in \mathbf{R}$, $y_2(t) \ge M\gamma_1$ *for* $t \in \mathbf{R}$.

Theorem 4.3.9. *Let $1 \le p \le \infty$ be a constant and $q \ge 1$ be such that $\frac{1}{p} + \frac{1}{q} = 1$. Suppose that (4.3.2) – (4.3.4) (4.3.6), (4.3.28), (4.3.43) and (4.3.46) hold. In addition assume the following conditions are satisfied:*

$$\text{(4.3.51)} \qquad \begin{cases} \text{there exists } 0 < M < 1,\ \kappa \in L^p(\mathbf{R}),\ \text{and an interval} \\ [a,b] \subseteq \mathbf{R},\ a < b,\ \text{with } k(t,s) \ge M\kappa(s) \\ \text{for } t \in [a,b],\ \text{and a.e. } s \in \mathbf{R}, \end{cases}$$

$$\text{(4.3.52)} \qquad k(t,s) \le \kappa(s) \text{ for } t \in \mathbf{R},\ \text{and a.e. } s \in \mathbf{R},$$

$$\text{(4.3.53)} \qquad \begin{cases} \text{there exists } \eta \in L^p[a,b] \text{ with } f(s,y) \ge \eta(s)\psi(y) \\ \text{for all } y \in [0,\infty),\ \text{and a.e. } s \in [a,b] \end{cases}$$

and

$$\text{(4.3.54)} \qquad \begin{cases} \text{there exists } \beta > 0,\ \beta \ne \alpha,\ \text{with} \\ \beta < K_6\psi(M\,\beta) \\ \text{where } K_6 := \sup_{t\in\mathbf{R}} \int_a^b k(t,s)\eta(s)\,ds. \end{cases}$$

Then (4.3.1) has at least one positive solution $y \in AP(\mathbf{R})$ and either

$$(A)\ 0 < \alpha < |y|_0 < \beta \text{ and } y(t) \ge M\alpha \text{ for } t \in [a,b] \text{ if } \alpha < \beta$$

or

$$(B)\ 0 < \beta < |y|_0 < \alpha \text{ and } y(t) \ge M\beta \text{ for } t \in [a,b] \text{ if } \beta < \alpha$$

hold.

Theorem 4.3.10. *Let $1 \le p \le \infty$ be a constant and $q \ge 1$ be such that $\frac{1}{p} + \frac{1}{q} = 1$. Suppose that (4.3.2) – (4.3.4), (4.3.6), (4.3.32), (4.3.43) and (4.3.46) hold. is satisfied. Then (4.3.1) has a nonnegative solution $y \in AP(\mathbf{R})$ with $|y|_0 < \alpha$.*

Theorem 4.3.11. *Let $1 \le p \le \infty$ be a constant and $q \ge 1$ be such that $\frac{1}{p} + \frac{1}{q} = 1$. Suppose that (4.3.2) – (4.3.4), (4.3.6), (4.3.36), (4.3.38), (4.3.43) and (4.3.44) hold. In addition assume the following conditions are satisfied:*

$$\text{(4.3.55)} \qquad \text{there exists } r > 0 \text{ with } r < K_4\phi(r)$$

and

$$\text{(4.3.56)} \qquad \begin{cases} \text{there exists } R > r,\ \text{with } R > \psi(R)K_5 \text{ where} \\ K_5 := \sup_{t\in\mathbf{R}} \int_{\mathbf{R}} \theta(s)k(t,s)\,ds. \end{cases}$$

Then (4.3.1) *has a nonnegative solution* $y \in AP(\mathbf{R})$ *with* $r \leq |y|_0 < R$.

4.4. Periodic solutions of nonlinear integral equations with applications.

In [45] the nonlinear integral equation

$$x(t) = \int_{t-\tau}^{t} f(s, x(s))\, ds \quad \text{for} \quad t \in \mathbf{R} \tag{4.4.1}$$

was interpreted as a model for the spread of certain infectious diseases with periodic contact rate that varies seasonally. In (4.4.1), $x(t)$ represents the proportion of infectives in the population at time t, $f(t, x(t))$ (which is periodic in t, that is, there exists $w > 0$ with $f(t + w, x) = f(t, x)$ for all $t \in \mathbf{R}$ and $x \geq 0$) is the proportion of new infectives per unit time and $\tau > 0$ is the length of time an individual remains infectious. Cooke and Kaplan in [45] were interested in obtaining a nontrivial periodic nonnegative solution to (4.4.1). They used Krasnoselskii's fixed point theorem in a cone to achieve this. Four years later Leggett and Williams [95] using their own fixed point theorem (which was a generalization of Krasnoselskii's) generalized the results in [45].

In this section we establish using a variety of fixed point theorems new existence results for the nonlinear integral equation

$$x(t) = \int_{t-\tau}^{t} k(t, s) f(s, x(s))\, ds, \text{ for } t \in \mathbf{R}. \tag{4.4.2}$$

We discuss the existence of one and multiple periodic solutions of (4.4.2). Moreover our results improve and extend those in [45,78,95].

We begin by using Krasnoselskii's fixed point theorem to establish a general existence result for (4.4.2).

Theorem 4.4.1. *Let* $1 \leq p \leq \infty$ *be a constant and* $q \geq 1$ *be such that* $\frac{1}{p} + \frac{1}{q} = 1$. *Suppose the following conditions are satisfied:*

$$k_t(s) = k(t, s) \in L^q[0, w] \textit{ for each } t \in [0, w], \tag{4.4.3}$$

$$\textit{the map } t \mapsto k_t \textit{ is continuous from } [0, w] \textit{ to } L^q[0, w], \tag{4.4.4}$$

$$\textit{for each } t \in [0, w],\ k(t, s) \geq 0 \textit{ for a.e. } s \in [0, w], \tag{4.4.5}$$

$$\textit{for every } t \in \mathbf{R},\ k(t, s + w) = k(t, s) \textit{ for a.e. } s \in \mathbf{R}, \tag{4.4.6}$$

$$\text{(4.4.7)} \qquad \textit{for a.e. } s \in \mathbf{R},\ k(t+w,s) = k(t,s) \textit{ for all } t \in \mathbf{R},$$

$$\text{(4.4.8)} \qquad f : [0,w] \times [0,\infty) \to [0,\infty) \textit{ is a Carathéodory function},$$

$$\text{(4.4.9)} \qquad f(s+w,u) = f(s,u) \textit{ for a.e. } s \in \mathbf{R} \textit{ and } u \geq 0,$$

$$\text{(4.4.10)} \qquad \begin{cases} \textit{for each } r > 0, \textit{ there exists } \mu_r \in L^p[0,w] \textit{ such that} \\ 0 \leq y \leq r \textit{ implies } |f(s,y)| \leq \mu_r(s) \textit{ for almost all } s \in [0,w], \end{cases}$$

$$\text{(4.4.11)} \qquad \begin{cases} \textit{there exists a function } \psi : [0,\infty) \to [0,\infty) \textit{ continuous} \\ \textit{and nondecreasing, a constant } A_0,\ 0 < A_0 \leq 1, \textit{ and} \\ q \in L^p[0,w],\ q(t+w) = q(t) \textit{ for a.e. } t \in \mathbf{R}, \textit{ with} \\ A_0 q(s)\psi(y) \leq f(s,y) \leq q(s)\psi(y) \\ \textit{for all } y \in [0,\infty), \textit{ and a.e. } s \in [0,w], \end{cases}$$

$$\text{(4.4.12)} \qquad \begin{cases} \textit{there exists a continuous function } \phi : (0,1) \to (0,\infty) \\ \textit{such that for any } 0 < m < 1 \textit{ and } u \geq 0, \\ \textit{we have } \psi(mu) \geq \phi(m)\psi(u), \end{cases}$$

$$\text{(4.4.13)} \qquad K_2 \equiv \inf_{t\in[0,w]} \int_{t-\tau}^{t} q(s)k(t,s)\,ds > 0,$$

$$\text{(4.4.14)} \qquad \begin{cases} \textit{there exists } 0 < M < 1 \textit{ with } \dfrac{M}{\phi(M)} \leq A_0 \dfrac{K_2}{K_1} \\ \textit{where } K_1 := \sup\limits_{t\in[0,w]} \displaystyle\int_{t-\tau}^{t} q(s)\,k(t,s)\,ds, \end{cases}$$

$$\text{(4.4.15)} \qquad \textit{there exists } \alpha > 0 \textit{ with } \alpha > K_1\psi(\alpha)$$

and

$$\text{(4.4.16)} \qquad \textit{there exists } \beta > 0,\ \beta \neq \alpha, \textit{ with } \beta < A_0 K_2 \psi(M\beta).$$

Then (4.4.2) *has at least one positive solution* $y \in A_\omega(\mathbf{R})$, *and either*

$$(A)\ 0 < \alpha < |y|_w < \beta \textit{ and } y(t) \geq M\alpha \textit{ for } t \in \mathbf{R} \textit{ if } \alpha < \beta$$

or

$$(B)\ 0 < \beta < |y|_w < \alpha \quad \textit{and } y(t) \geq M\beta \textit{ for } t \in \mathbf{R} \textit{ if } \beta < \alpha$$

holds.

PROOF: Let $E := (A_w, |.|_w)$ and

$$\begin{aligned} C &:= \{u \in C_w : u(t) \geq M|u|_w \text{ for } t \in [0, w]\} \\ &= \{u \in C_w : u(t) \geq M|u|_w \text{ for } t \in \mathbf{R}\} \end{aligned}$$

where M is as defined in (4.4.14). Let $K : C_w \to E$ be defined by

$$Ky(t) := \int_{t-\tau}^{t} k(t,s) f(s, y(s))\, ds, \text{ for } t \in \mathbf{R};$$

here $y \in A_w$. We <u>claim</u> that

$$K : C \to A_w \tag{4.4.17}$$

is true.

Let $y \in C$. Then for $t \in \mathbf{R}$, we have using (4.4.6), (4.4.7) and (4.4.9),

$$\begin{aligned} Ky(t+w) &= \int_{t+w-\tau}^{t+w} k(t+w, s) f(s, y(s))\, ds \\ &= \int_{t-\tau}^{t} k(t+w, x+w) f(x+w, y(x+w))\, dx \\ &= \int_{t-\tau}^{t} k(t+w, x+w) f(x+w, y(x))\, dx \\ &= \int_{t-\tau}^{t} k(t, x) f(x, y(x))\, dx = Ky(t). \end{aligned}$$

Also since $y \in C$, there exists $r > 0$ with $|y|_w \leq r$, and therefore (4.4.10) guarantees the existence of $\mu_r \in L^p[0, w]$, with $|f(s, y(s))| \leq \mu_r(s)$, for a.e. $s \in [0, w]$. This together with (4.4.9) implies that there exists $\eta_r \in L^p[-\tau, w]$ with $|f(s, y(s))| \leq \eta_r(s)$, for a.e. $s \in [-\tau, w]$ (alternatively we could assume without loss of generality that $\mu_r(t+w) = \mu_r(t)$ for a.e. $t \in \mathbf{R}$ and take $\eta_r = \mu_r$). Now for $t_1,\ t_2 \in [0, w]$, with $t_2 > t_1$ and $t_2 - \tau < t_1$, we have

$$\begin{aligned} Ky(t_1) \quad - \quad & Ky(t_2) = \int_{t_1-\tau}^{t_1} k(t_1, s) f(s, y(s)) ds - \int_{t_2-\tau}^{t_2} k(t_2, s) f(s, y(s)) ds \\ = \quad & \int_{t_1-\tau}^{t_2-\tau} k(t_1, s) f(s, y(s)) ds \\ + \quad & \int_{t_2-\tau}^{t_1} [k(t_1, s) - k(t_2, s)] f(s, y(s)) ds - \int_{t_1}^{t_2} k(t_2, s) f(s, y(s)) ds \end{aligned}$$

and therefore

$$\begin{aligned}
|Ky(t_1) &- Ky(t_2)| \le \left(\int_{t_1-\tau}^{t_2-\tau}[k_{t_1}(s)]^q\,ds\right)^{\frac{1}{q}}\left(\int_{t_1-\tau}^{t_2-\tau}[f(s,y(s))]^p\,ds\right)^{\frac{1}{p}}\\
&+ \left(\int_{t_2-\tau}^{t_1}[k_{t_1}(s)-k_{t_2}(s)]^q\,ds\right)^{\frac{1}{q}}\left(\int_{t_2-\tau}^{t_1}[f(s,y(s))]^p\,ds\right)^{\frac{1}{p}}\\
&+ \left(\int_{t_1}^{t_2}[k_{t_2}(s)]^q\,ds\right)^{\frac{1}{q}}\left(\int_{t_1}^{t_2}[f(s,y(s))]^p\,ds\right)^{\frac{1}{p}}.
\end{aligned}$$

Thus

$$\begin{aligned}
|Ky(t_1) &- Ky(t_2)|\\
&\le \sup_{t\in[0,w]}\left(\int_{-\tau}^{w-\tau}[k_t(s)]^q\,ds\right)^{\frac{1}{q}}\left(\int_{t_1-\tau}^{t_2-\tau}[\eta_r(s)]^p\,ds\right)^{\frac{1}{p}}\\
&+\left(\int_{-\tau}^{w}[k_{t_1}(s)-k_{t_2}(s)]^q\,ds\right)^{\frac{1}{q}}\left(\int_{-\tau}^{w}[\eta_r(s)]^p\,ds\right)^{\frac{1}{p}}\\
&+\sup_{t\in[0,w]}\left(\int_{0}^{w}[k_t(s)]^q\,ds\right)^{\frac{1}{q}}\left(\int_{t_1}^{t_2}[\mu_r(s)]^p\,ds\right)^{\frac{1}{p}}.
\end{aligned} \tag{4.4.18}$$

Remark 4.4.1. Note that (4.4.3) and (4.4.6) guarantee that $k_t \in L^q[-\tau,w]$ for $t\in[0,w]$.

Now (4.4.4) together with (4.4.18) guarantees that $Ky \in C[0,w]$, and hence (4.4.17) is true. Next we show that

$$K : C \to A_w \text{ is continuous.} \tag{4.4.19}$$

Let $y_n,\ y\in C$ $(n=1,2,...)$ with $y_n\to y$. Then there exists $r>0$ with $|y_n|_w \le r$ for $n=1,2,...$ and $|y|_w\le r$. Also (4.4.10) guarantees the existence of $\mu_r\in L^p[0,w]$, with $|f(s,y_n(s))|\le\mu_r(s)$ for a.e. $s\in[0,w]$ and $n=1,2,....$, and $|f(s,y(s))|\le\mu_r(s)$ for a.e. $s\in[0,w]$. This together with (4.4.9) implies that there exists $\eta_r\in L^p[-\tau,w]$ with $|f(s,y_n(s))|\le\eta_r(s)$ for a.e. $s\in[-\tau,w]$ and $n=1,2,....$, and $|f(s,y(s))|\le\eta_r(s)$ for a.e. $s\in[-\tau,w]$. Now (4.4.8) and (4.4.9) imply that

$$\rho_n(s)=|f(s,y_n(s))-f(s,y(s))|\to 0 \text{ pointwise for a.e. } s\in[-\tau,w]$$

and

$$\rho_n(s)\le 2\eta_r(s),\ \text{ for a.e. } s\in[-\tau,w].$$

Thus for $t \in [0, w]$ we have

$$
\begin{aligned}
|Ky_n(t) - Ky(t)| &\le \int_{t-\tau}^{t} k(t,s)\rho_n(s)\,ds \\
&\le \left(\int_{t-\tau}^{t} [k_t(s)]^q\,ds\right)^{\frac{1}{q}} \left(\int_{t-\tau}^{t} [\rho_n(s)]^p\,ds\right)^{\frac{1}{p}} \\
&\le \sup_{x\in[0,w]} \left(\int_{-\tau}^{w} [k_x(s)]^q\,ds\right)^{\frac{1}{q}} \left(\int_{-\tau}^{w} [\rho_n(s)]^p\,ds\right)^{\frac{1}{p}}
\end{aligned}
$$

and therefore

$$
\sup_{t\in[0,w]} |Ky_n(t) - Ky(t)| \to 0 \text{ as } n \to \infty.
$$

Consequently (4.4.19) is true.

Next we show

(4.4.20) $\qquad K : C \to A_w$ is completely continuous.

Let Ω be a bounded set in C, that is, there exists $r > 0$ with $|u|_w \le r$ for all $u \in \Omega$. We will show that $K\Omega$ is relatively compact in A_w. Let $(y_n)_1^\infty$ be a sequence in Ω, therefore $(Ky_n)_1^\infty$ is a sequence in $K\Omega$. Now (4.4.10) guarantees that there exists $\mu_r \in L^p[0,w]$ with $|f(s, y_n(s))| \le \mu_r(s)$ for a.e. $s \in [0,w]$ and $n = 1, 2, \ldots..$ This together with (4.4.9) implies that there exists $\eta_r \in L^p[-\tau, w]$ with $|f(s, y_n(s))| \le \eta_r(s)$ for a.e. $s \in [-\tau, w]$ and $n = 1, 2, \ldots..$ Now for $t \in [0, w]$ we have

$$
\begin{aligned}
|Ky(t)| &\le \int_{t-\tau}^{t} k(t,s) f(s, y_n(s))\,ds \\
&\le \left(\int_{t-\tau}^{t} [k_t(s)]^q\,ds\right)^{\frac{1}{q}} \left(\int_{t-\tau}^{t} [f(s, y_n(s))]^p\,ds\right)^{\frac{1}{p}} \\
&\le \sup_{x\in[0,w]} \left(\int_{-\tau}^{w} [k_x(s)]^q\,ds\right)^{\frac{1}{q}} \left(\int_{-\tau}^{w} [\eta_r(s)]^p\,ds\right)^{\frac{1}{p}},
\end{aligned}
$$

thus $(Ky_n)_1^\infty$ is a uniformly bounded sequence in $C[0,w]$. An argument similar to that in (4.4.18) guarantees that $(Ky_n)_1^\infty$ is equicontinuous on $[0,w]$. The Arzela–Ascoli theorem guarantees a $Ky \in C[0,w]$ and a subsequence $(Ky_{n_k})_{k=1}^\infty$ of $(Ky_n)_1^\infty$ which converges uniformly on $[0,w]$ to Ky. Let $Ky(t+w) = Ky(t)$, then since $Ky_{n_k}(t+w) = Ky_{n_k}(t)$, we have $Ky_{n_k} \to Ky$ in A_w. Thus (4.4.20) holds.

Let

$$\Omega_\alpha := \{y \in A_w : |y|_w < \alpha\}$$

and

$$\Omega_\beta := \{y \in A_w : |y|_w < \beta\}.$$

We wish to apply Krasnoselskii's fixed point theorem. We can do so if we show that

$$K : C \to C, \tag{4.4.21}$$

$$|Ky|_w < |y|_w \text{ for } y \in C \cap \partial\Omega_\alpha = S_\alpha \tag{4.4.22}$$

and

$$|Ky|_w > |y|_w \text{ for } y \in C \cap \partial\Omega_\beta = S_\beta \tag{4.4.23}$$

hold.

Let $y \in C$. Then (4.4.11) implies for $t \in [0, w]$ that

$$|Ky(t)| \le \psi(|y|_w) \int_{t-\tau}^{t} q(s)k(t,s)\,ds \le \psi(|y|_w) \sup_{t\in[0,w]} \int_{t-\tau}^{t} q(s)k(t,s)\,ds.$$

Remark 4.4.2. Fix $t \in [0, w]$. Note for $x \in [t - \tau, t]$, that there exists $s_x \in [0, w]$ and $n_x \in \mathbf{Z}$, with $x + n_x w = s_x$, and hence (4.4.9) and (4.4.11) imply for a.e. $x \in \mathbf{R}$ that

$$f(x, y(x)) = f(s_x, y(x)) \le q(s_x)\psi(y(x)) = q(x)\psi(y(x))$$

since $q(z + w) = q(z)$ for a.e. $z \in \mathbf{R}$.

Thus

$$|Ky|_w \le \psi(|y|_w) K_1. \tag{4.4.24}$$

On the other hand since $y \in C$, we have $y(t) \ge M|y|_w$ for $t \in [0, w]$, and therefore (4.4.11), (4.4.12), (4.4.24) and (4.4.14) give for $t \in [0, w]$,

$$\begin{aligned} Ky(t) &\ge A_0 \int_{t-\tau}^{t} q(s)k(t,s)\psi(y(s))\,ds \ge A_0\psi(M|y|_w) \int_{t-\tau}^{t} q(s)k(t,s)\,ds \\ &\ge A_0\phi(M)\psi(|y|_w) \int_{t-\tau}^{t} q(s)k(t,s)\,ds \ge K_2 A_0\phi(M)\psi(|y|_w) \\ &\ge \frac{K_2}{K_1} A_0\phi(M)|Ky|_w \ge M|Ky|_w. \end{aligned}$$

Therefore $Ky \in C$ and (4.4.21) holds.

To see (4.4.22), let $y \in C \cap \partial\Omega_\alpha$, therefore $|y|_w = \alpha$ and $y(t) \geq M\alpha$ for $t \in [0,w]$. Now for $t \in [0,w]$, we have

$$|Ky(t)| \leq \psi(|y|_w) \int_{t-\tau}^{t} q(s)k(t,s)\,ds \leq \psi(\alpha)K_1.$$

This together with (4.4.15) yields

$$|Ky|_w \leq \psi(\alpha)K_1 < \alpha = |y|_w, \tag{4.4.25}$$

thus (4.4.22) is satisfied.

Now let $y \in C \cap \partial\Omega_\beta$ therefore $|y|_w = \beta$ and $M\beta \leq y(t) \leq \beta$ for $t \in [0,w]$. Also for $t \in [0,w]$ we have

$$\begin{aligned} Ky(t) &\geq A_0 \int_{t-\tau}^{t} q(s)k(t,s)\psi(y(s))\,ds \\ &\geq A_0\psi(M\beta) \int_{t-\tau}^{t} q(s)k(t,s)\,ds \\ &\geq A_0K_2\psi(M\beta). \end{aligned}$$

This inequality together with (4.4.16) yields

$$Ky(t) \geq A_0K_2\psi(M\beta) > \beta = |y|_w \text{ for } t \in [0,w],$$

and thus $|Ky|_w > |y|_w$, that is, (4.4.23) holds.

Krasnoselskii's fixed point theorem guarantees that there is a solution $y \in A_w$ of (4.4.2) with $y \in C \cap \left(\overline{\Omega_\alpha} \backslash \Omega_\beta\right)$ if $\beta < \alpha$, whereas $y \in C \cap \left(\overline{\Omega_\beta} \backslash \Omega_\alpha\right)$ if $\alpha < \beta$. Finally note $|y|_w \neq \alpha$ and $|y|_w \neq \beta$. To see this suppose $|y|_w = \alpha$. Then since $y = Ky$ we have (follow the ideas used to prove (4.4.25)),

$$\alpha = |y|_w = |Ky|_w \leq \psi(\alpha)K_1 < \alpha = |y|_w;$$

– a contradiction. A similar argument shows that $|y|_w \neq \beta$. □

Example 4.4.1. Consider the nonlinear integral equation

$$y(t) = \int_{t-\tau}^{t} k(t,s)h(s)[y(s)]^\gamma\,ds \text{ for } t \in \mathbf{R} \tag{4.4.26}$$

with $0 < \gamma < 1$ and (4.4.3) – (4.4.7) holding. In addition assume

$$\begin{cases} \text{there exists } w > 0 \text{ with } h(t+w) = h(t) \\ \text{for a.e. } t \in \mathbf{R} \text{ and } h \in L^p[0,w] \end{cases} \tag{4.4.27}$$

and

$$\inf_{t\in[0,w]} \int_{t-\tau}^{t} k(t,s)h(s)\,ds > 0 \tag{4.4.28}$$

hold. Then (4.4.26) has at least one positive solution $y \in A_w$, with

$$\beta \equiv \frac{1}{2} M^{\frac{\gamma}{1-\gamma}} (K_2)^{\frac{1}{1-\gamma}} < |y|_w < 2\,(K_1)^{\frac{1}{1-\gamma}} \equiv \alpha \text{ and } y(t) \geq M\beta \text{ for } t \in \mathbf{R};$$

here

$$M := \frac{1}{2}\left(\frac{K_2}{K_1}\right)^{\frac{1}{1-\gamma}}$$

with

$$K_2 := \inf_{t\in[0,w]} \int_{t-\tau}^{t} h(s)k(t,s)\,ds \text{ and } K_1 := \sup_{t\in[0,w]} \int_{t-\tau}^{t} h(s)k(t,s)\,ds.$$

To see that the above is true we will apply Theorem 4.4.1 with

$$f(t,y) := q(t)y^{\gamma},\ \psi(y) := y^{\gamma},\ q := h,\ A_0 := 1 \text{ and } \phi(x) := x^{\gamma}.$$

Notice that (4.4.8) – (4.4.13) are clearly satisfied. To see (4.4.14), notice

$$\frac{M}{\phi(M)} = M^{1-\gamma} = \left(\frac{1}{2}\right)^{1-\gamma} \frac{K_2}{K_1} \leq \frac{K_2}{K_1} = A_0 \frac{K_2}{K_1}.$$

Also (4.4.15) holds since

$$\frac{\alpha}{\psi(\alpha)} = \alpha^{1-\gamma} = 2^{1-\gamma} K_1 > K_1.$$

Finally (4.4.16) is true since

$$\frac{\beta}{\psi(M\,\beta)} = \frac{1}{M^{\gamma}}\beta^{1-\gamma} = \frac{1}{M^{\gamma}}\left(\frac{1}{2}\right)^{1-\gamma} M^{\gamma} K_2 = \left(\frac{1}{2}\right)^{1-\gamma} K_2 < K_2 = A_0 K_2.$$

Now apply Theorem 4.4.1 (B).

It is easy to see that with additional conditions on the nonlinearity and kernel repeated application of Theorem 4.4.1 will yield additional positive periodic solutions of (4.4.2). For completeness we provide one *multiple* solution result.

Theorem 4.4.2. *Let $1 \le p \le \infty$ be a constant and $q \ge 1$ be such that $\frac{1}{p} + \frac{1}{q} = 1$. Suppose that (4.4.3)–(4.4.14) hold. In addition suppose there are constants $\gamma_0 < \gamma_1 < \gamma_2$ with the following satisfied:*

(4.4.29) $$\gamma_0 > 0 \text{ is such that } \gamma_0 < A_0 K_2 \psi(M\gamma_0),$$

(4.4.30) $$\gamma_1 > 0 \text{ is such that } \gamma_1 > K_1 \psi(\gamma_1)$$

and

(4.4.31) $$\gamma_2 > 0 \text{ is such that } \gamma_2 < A_0 K_2 \psi(M\gamma_2).$$

Then (4.4.2) has at least two positive solutions y_1, $y_2 \in A_w$ with $0 < \gamma_0 < |y_1|_w < \gamma_1 < |y_2|_w < \gamma_2$ and $y_1(t) \ge M\gamma_0$ for $t \in \mathbf{R}$, $y_2(t) \ge M\gamma_1$ for $t \in \mathbf{R}$.

PROOF: The existence of y_1 follows from Theorem 4.4.1 (B) with $\alpha = \gamma_1$ and $\beta = \gamma_0$, and the existence of y_2 follows from Theorem 4.4.1 (A) with $\alpha = \gamma_1$ and $\beta = \gamma_2$. □

It is possible to use the nonlinear alternative to obtain an existence result for (4.4.2).

Theorem 4.4.3. *Let $1 \le p \le \infty$ be a constant and $q \ge 1$ be such that $\frac{1}{p} + \frac{1}{q} = 1$. Suppose that (4.4.3) – (4.4.7) hold. In addition assume the following conditions are satisfied:*

(4.4.32) $$f : [0,w] \times \mathbf{R} \to [0,\infty) \text{ is a Carathéodory function,}$$

(4.4.33) $$f(s+w,u) = f(s,u), \text{ for a.e. } s \in \mathbf{R} \text{ and } u \in \mathbf{R},$$

(4.4.34) $$\begin{cases} \text{for each } r > 0 \text{ there exists } \mu_r \in L^p[0,w] \text{ such that } |y| \le r \\ \text{implies } |f(s,y)| \le \mu_r(s) \text{ for almost all } s \in [0,w], \end{cases}$$

(4.4.35) $$\begin{cases} \text{there exists a function } \psi : [0,\infty) \to [0,\infty) \text{ continuous} \\ \text{and nondecreasing, and } q \in L^p[0,w],\ q(t+w) = q(t) \\ \text{for a.e. } t \in \mathbf{R}, \text{ with } f(s,y) \le q(s)\psi(y) \\ \text{for all } y \in [0,\infty) \text{ and a.e. } s \in [0,w] \end{cases}$$

and

(4.4.36) $$\begin{cases} \text{there exists } \alpha > 0 \text{ with } \alpha > K_1 \psi(\alpha) \\ \text{where } K_1 := \sup_{t \in [0,w]} \int_{t-\tau}^{t} q(s)\, k(t,s)\, ds. \end{cases}$$

Then (4.4.2) *has a nonegative solution* $y \in A_w$ *with* $|y|_w < \alpha$.

Remark 4.4.3. In the above theorem it is possible for $|y|_w$ to be zero in some applications.

PROOF OF THEOREM 4.4.3: We will apply the nonlinear alternative. Let

$$E := (A_w, |.|_w) \text{ and } U := \{u \in C_w : |u|_w < \alpha\},$$

and $K : A_w \to A_w$ be defined by

$$Ky(t) := \int_{t-\tau}^{t} k(t,s) f(s, y(s))\, ds, \text{ for } t \in \mathbf{R};$$

here $y \in A_w$. Essentially the same reasoning as in Theorem 4.4.1 guarantees that

$$K : A_w \to A_w \text{ is continuous and completely continuous.}$$

Now let $y \in A_w$ be any solution of

$$y(t) = \lambda \left(\int_{t-\tau}^{t} k(t,s) f(s, y(s))\, ds \right), \quad t \in \mathbf{R},$$

for $0 < \lambda < 1$. Notice that (4.4.5) and (4.4.32) imply $y(t) \geq 0$ for $t \in \mathbf{R}$. Also for $t \in [0, w]$, we have

$$\begin{aligned} |y(t)| &\leq \int_{t-\tau}^{t} k(t,s) q(s) \psi(y(s))\, ds \\ &\leq \psi(|y|_w) \int_{t-\tau}^{t} k(t,s) q(s)\, ds \\ &\leq K_1 \psi(|y|_w) \end{aligned}$$

and therefore

$$|y|_w \leq K_1 \psi(|y|_w). \tag{4.4.37}$$

Now (4.4.36) and (4.4.37) imply $|y|_w \neq \alpha$. Apply the nonlinear alternative (notice (*ii*) cannot occur) to guarantee a nonnegative solution $y \in A_w$ with $|y|_w < \alpha$ (note $|y|_w \leq \alpha$ by the nonlinear alternative and $|y|_w \neq \alpha$ by an argument similar to the one above). □

Remark 4.4.4. It is also possible to combine Theorem 4.4.1 (A) and Theorem 4.4.3 to obtain a result which guarantees the existence of two nonnegative solutions of (4.4.2). We leave the details to the reader.

Next we use the Leggett Williams fixed point theorem to obtain two results for (4.4.2). In particular Theorem 4.4.4 improves results in [45] and Theorem 4.4.5 extends results in [95].

Theorem 4.4.4. *Let $1 \le p \le \infty$ be a constant and $q \ge 1$ be such that $\frac{1}{p} + \frac{1}{q} = 1$. Suppose that (4.4.3) – (4.4.10) hold. In addition assume the following conditions are satisfied:*

$$(4.4.38)\qquad \begin{cases} \text{there exists a nondecreasing continuous map} \\ \psi : [0,\infty) \to [0,\infty), \text{ a continuous map } \phi : [0,\infty) \to [0,\infty), \\ \text{and a } q \in L^p[0,w],\ q(t+w) = q(t) \text{ for a.e. } t \in \mathbf{R}, \\ \text{with } \phi(y)q(s) \le f(t,y) \le \psi(y)q(s) \\ \text{for all } y \in [0,\infty), \text{ and a.e. } s \in [0,w], \end{cases}$$

$$(4.4.39)\qquad K_2 := \inf_{t\in[0,w]} \int_{t-\tau}^{t} q(s)k(t,s)\,ds > 0,$$

$$(4.4.40)\qquad \text{there exists } r > 0 \text{ with } r < K_2\phi(r),$$

$$(4.4.41)\qquad \frac{\phi(x)}{x} \text{ is nonincreasing on } (0,r)$$

and

$$(4.4.42)\qquad \begin{cases} \text{there exists } R > r \text{ with } R > \psi(R)K_1, \\ \text{where } K_1 := \sup_{t\in[0,w]} \displaystyle\int_{t-\tau}^{t} q(s)k(t,s)\,ds. \end{cases}$$

Then (4.4.2) *has a nonnegative solution $y \in A_w$ with $r \le |y|_w < R$.*

Remark 4.4.5. If $R \ge \psi(R)K_1$ in (4.4.42) then the result in Theorem 4.4.4 is again true if $r \le |y|_w < R$ is replaced by $r \le |y|_w \le R$.

Remark 4.4.6. Suppose there exists $\epsilon > 0$ with $\phi(y) \ge y$ for $y \in [0,\epsilon]$. Then (4.4.40) is satisfied with $0 < r \le \epsilon$ fixed if $\inf_{t\in[0,w]} \int_{t-\tau}^{t} q(s)k(t,s)\,ds > 1$ (we need only note that $\frac{r}{\phi(r)} \le \frac{r}{r} = 1$).

PROOF OF THEOREM 4.4.4: Let $E := (a_w, |.|_w)$ and

$$\begin{aligned} C &:= \{u \in C_w :\ u(t) \ge 0 \ \text{ for } \ t \in [0,w]\} \\ &= \{u \in C_w :\ u(t) \ge 0 \ \text{ for } \ t \in \mathbf{R}\}. \end{aligned}$$

Also let $u_0 \equiv 1$ and note

$$C(u_0) = \{u \in C : \text{there exists } \lambda > 0 \text{ with } u(t) \geq \lambda \text{ for } t \in [0, w]\}.$$

Remark 4.4.7. Note that

$$C(u_0) = \{u \in C : u(t) > 0 \text{ for } t \in [0, w]\}$$

since if $u(t) > 0$ for $t \in [0, w]$ then $u \in C[0, w]$ guarantees that there exists $t_0 \in [0, w]$ with $\min_{t\in[0,w]} u(t) = u(t_0)$, and consequently $u(t) \geq u(t_0) > 0$ for $t \in [0, w]$.

Now let $K : C \to C$ be defined by

$$Ky(t) := \int_{t-T}^{t} k(t,s)f(s,y(s))\,ds, \text{ for } t \in \mathbf{R};$$

here $y \in C$. As in Theorem 4.4.1 we have

$$K : C \to C \text{ is continuous and completely continuous.}$$

We wish to apply Theorem 4.3.1, therefore we first show

$$|Ky|_w \leq |y|_w \text{ for } y \in S_R. \tag{4.4.43}$$

To see this notice if $y \in S_R$ then $|y|_w = R$ and thus for $t \in [0, w]$ we have

$$\begin{aligned} Ky(t) &\leq \int_{t-T}^{t} k(t,s)q(s)\psi(y(s))\,ds \\ &\leq \psi(|y|_w) \int_{t-T}^{t} k(t,s)q(s)\,ds \leq \psi(R)K_1. \end{aligned}$$

This together with (4.4.42) gives

$$|Ky|_w \leq \psi(R)K_1 < R = |y|_w, \tag{4.4.44}$$

and therefore (4.4.43) is true.

Next we show

$$Ky \not\leq y \text{ for } y \in S_r \cap C(u_0). \tag{4.4.45}$$

To see this let $y \in S_r \cap C(u_0)$, hence

$$|y|_w = r \quad \text{and} \quad r \geq y(t) > 0 \quad \text{for} \quad t \in [0, w].$$

Now for $t \in [0, w]$ we have

$$\begin{aligned} Ky(t) &\geq \int_{t-\tau}^{t} k(t,s)q(s)\phi(y(s))\,ds \\ &= \int_{t-\tau}^{t} k(t,s)q(s)\frac{\phi(y(s))}{y(s)}y(s)\,ds \\ &\geq \frac{\phi(r)}{r}\int_{t-\tau}^{t} k(t,s)q(s)\,y(s)\,ds. \end{aligned}$$

Let $t_0 \in [0, w]$ be such that $\min_{t\in[0,w]} y(t) = y(t_0)$ and this together with the previous inequality yields for $t \in [0, w]$,

$$Ky(t) \geq \frac{\phi(r)}{r}y(t_0)\int_{t-\tau}^{t} k(t,s)q(s)\,ds \geq \left(\frac{\phi(r)}{r}K_2\right)y(t_0).$$

Use (4.4.40) to obtain $Ky(t) > y(t_0)$ for $t \in [0, w]$. In particular $Ky(t_0) > y(t_0)$. Thus (4.4.45) is true.

Now apply Theorem 4.3.1 to deduce that (4.4.2) has a nonnegative solution $y \in A_w$ with $r \leq |y|_w \leq R$. Note that $|y|_w \neq R$ since if $|y|_w = R$ then from $y = Ky$ we have (follow the ideas from (4.4.43) to (4.4.44)),

$$R = |y|_w = |Ky|_w \leq \psi(R)K_1 < R = |y|_w,$$

– a contradiction. □

Example 4.4.2. Consider the nonlinear integral equation

$$y(t) = \int_{t-\tau}^{t} k(t,s)h(s)\left([y(s)]^\alpha + [y(s)]^\beta\right)ds, \text{ for } t \in \mathbf{R} \tag{4.4.46}$$

with $0 < \alpha < 1$, $\beta \geq 1$ and (4.4.3) – (4.4.7) satisfied. In addition assume

$$\left\{ \begin{array}{l} \text{there exists } w > 0 \text{ with } h(t+w) = h(t) \\ \text{for a.e. } t \in \mathbf{R} \text{ and } h \in L^p[0,w] \end{array} \right. \tag{4.4.47}$$

and

$$\inf_{t\in[0,w]}\int_{t-\tau}^{t} k(t,s)h(s)\,ds > 0 \text{ and } \sup_{t\in[0,w]}\int_{t-\tau}^{t} k(t,s)h(s)\,ds < \frac{1}{2} \tag{4.4.48}$$

hold. Then (4.4.46) has a nonegative solution $y \in A_w$ with

$$\left(\frac{1}{2}\inf_{t\in[0,w]}\int_{t-\tau}^{t} k(t,s)h(s)\,ds\right)^{\frac{1}{1-\alpha}} \leq |y|_w < 1.$$

Remark 4.4.8. Note $y \equiv 0$ is also a solution of (4.4.46).

To see this let

$f(t,y) := h(t)[y^\alpha + y^\beta]$, $\psi(y) := y^\alpha + y^\beta$, $\phi(y) := y^\alpha$ and $q(t) := h(t)$,

with

$$r := \left(\frac{1}{2} \inf_{t\in[0,w]} \int_{t-\tau}^{t} k(t,s)h(s)\,ds \right)^{\frac{1}{1-\alpha}} \quad \text{and } R := 1.$$

Clearly (4.4.8), (4.4.9), (4.4.10), (4.4.38) and (4.4.39) hold.

To see (4.4.40) notice

$$\frac{r}{\phi(r)} = r^{1-\alpha} = \frac{1}{2} \inf_{t\in[0,w]} \int_{t-\tau}^{t} k(t,s)h(s)\,ds < \inf_{t\in[0,w]} \int_{t-\tau}^{t} k(t,s)h(s)\,ds.$$

Also (4.4.41) is satisfied since $\frac{\phi(x)}{x} = \frac{1}{x^{1-\alpha}}$ and $0 < \alpha < 1$, and finally (4.4.42) holds with $R = 1$ (see (4.4.48)). Now apply Theorem 4.4.4.

Theorem 4.4.5. *Let $1 \le p \le \infty$ be a constant and $q \ge 1$ be such that $\frac{1}{p} + \frac{1}{q} = 1$. Suppose that* (4.4.3) – (4.4.10) *and* (4.4.38) *are satisfied. In addition assume the following conditions hold:*

$$\text{(4.4.49)} \qquad \begin{cases} \text{there exists } a \in L^q[0,w], \text{ with } a(t+w) = a(t) \\ \text{for a.e. } t \in \mathbf{R}, \text{ and with } k(t,s) \ge a(s) \\ \text{for a.e. } s \in [0,w], \text{ and all } t \in [0,w] \end{cases}$$

and

$$\text{(4.4.50)} \qquad \begin{cases} \text{there exists } r > 0 \text{ with } \left(\frac{\phi(r)}{r}\right)^N \prod_{j=1}^{N} \int_{I_j} a(s)q(s)\,ds > 1; \\ \text{here } N \text{ is the smallest positive integer such that} \\ \frac{w}{N} \le \frac{\tau}{2} \text{ and } I_j = \left[\frac{(j-1)w}{N}, \frac{j\,w}{N}\right],\ j = 0,1,....,N. \end{cases}$$

Finally suppose (4.4.41) *and* (4.4.42) *hold. Then* (4.4.2) *has a nonnegative solution $y \in A_w$ with $r \le |y|_w < R$.*

PROOF: Let E, C, u_0, $C(u_0)$ and K be as in Theorem 4.4.4. Essentially the same reasoning as in Theorem 4.4.4 establishes that

$$\text{(4.4.51)} \qquad K : C \to C \text{ is continuous and completely continuous.}$$

with

$$\text{(4.4.52)} \qquad |Ky|_w \le |y|_w \text{ for } y \in S_R.$$

Next we show

$$Ky \not\leq y \text{ for } y \in S_r \cap C(u_0). \tag{4.4.53}$$

To see this let $y \in S_r \cap C(u_0)$ therefore

$$|y|_w = r \text{ and } r \geq y(t) > 0 \text{ for } t \in [0, w].$$

Now for $t \in [0, w]$ we have

$$\begin{aligned} Ky(t) &\geq \int_{t-\tau}^{t} k(t,s)q(s)\phi(y(s))\,ds \\ &= \int_{t-\tau}^{t} k(t,s)q(s)\frac{\phi(y(s))}{y(s)}y(s)\,ds \\ &\geq \frac{\phi(r)}{r}\int_{t-\tau}^{t} k(t,s)q(s)y(s)\,ds \\ &\geq \frac{\phi(r)}{r}\int_{t-\tau}^{t} a(s)q(s)y(s)\,ds. \end{aligned}$$

We <u>claim</u> that there exists a $t_0 \in [0, w]$ with

$$\frac{\phi(r)}{r}\int_{t_0-\tau}^{t_0} a(s)q(s)y(s)\,ds > y(t_0). \tag{4.4.54}$$

If our claim is true then

$$Ky(t_0) \geq \frac{\phi(r)}{r}\int_{t_0-\tau}^{t_0} a(s)q(s)y(s)\,ds > y(t_0),$$

therefore (4.4.53) holds. Now apply Theorem 4.3.1 to deduce the result.

It remains to prove our claim. Suppose (4.4.54) is false. Then

$$y(t) \geq \frac{\phi(r)}{r}\int_{t-\tau}^{t} a(s)q(s)y(s)\,ds \text{ for all } t \in [0, w]. \tag{4.4.55}$$

Note if $t \in I_j$ for some $j \in \{1, 2, ..., N\}$ then clearly $I_j \subseteq [t-\tau, t]$ since $\frac{w}{N} \leq \frac{\tau}{2}$. This together with (4.4.55) gives for $j \in \{1, 2, ..., N\}$,

$$\begin{aligned} \int_{I_j} a(t)q(t)y(t)\,dt &\geq \frac{\phi(r)}{r}\int_{I_j} a(t)q(t)\int_{t-\tau}^{t} a(s)q(s)y(s)\,ds\,dt \\ &\geq \frac{\phi(r)}{r}\int_{I_j} a(t)q(t)\int_{I_{j-1}} a(s)q(s)y(s)\,ds\,dt \\ &= \frac{\phi(r)}{r}\left(\int_{I_j} a(t)q(t)\,dt\right)\left(\int_{I_{j-1}} a(s)q(s)y(s)\,ds\right). \end{aligned}$$

Thus we have

$$\int_{I_N} a(t)q(t)y(t)\,dt \geq \left(\frac{\phi(r)}{r}\right)^N \prod_{j=1}^{N} \left(\int_{I_j} a(t)\,q(t)\,dt\right) \left(\int_{I_0} a(s)q(s)y(s)\,ds\right)$$

and therefore

$$\begin{aligned} (4.4.56) \qquad & \int_{I_N} a(t)q(t)y(t)\,dt \\ & \geq \left(\frac{\phi(r)}{r}\right)^N \prod_{j=1}^{N} \left(\int_{I_j} a(t)q(t)\,dt\right) \left(\int_{I_N} a(s)q(s)y(s)\,ds\right). \end{aligned}$$

Note if $\int_{I_N} a(t)q(t)y(t)\,dt = 0$ then since $y(t) > 0$ for $t \in [0, w]$ we have $a(t)q(t) = 0$ for a.e. $t \in I_N$. This contradicts (4.4.50). We therefore see that $\int_{I_N} a(t)q(t)y(t)\,dt > 0$ and (4.4.56) yields

$$\left(\frac{\phi(r)}{r}\right)^N \prod_{j=1}^{N} \left(\int_{I_j} a(t)q(t)\,dt\right) \leq 1.$$

This contradicts (4.4.50), therefore (4.4.54) holds. □

CHAPTER 5

THREE SOLUTIONS TO INTEGRAL EQUATIONS VIA THE LEGGETT WILLIAMS FIXED POINT THEOREM

5.1. Introduction.

In this chapter we present results which guarantee the existence of three nonnegative solutions to integral equations. The strategy involves using the Leggett Williams fixed point theorem. For the remainder of the introduction we present some preliminaries which will be needed in Section 5.2. Let $E = (E, \|\,.\,\|)$ be a Banach space and $C \subset E$ be a cone. By a concave nonnegative continuous functional ψ on C we mean a continuous mapping $\psi : C \to [0, \infty)$ with

$$\psi(\,\lambda\, x + (1-\lambda)\, y\,) \geq \lambda\, \psi(x) + (1-\lambda)\, \psi(y)$$

for all $x,\, y \in C$ and $\lambda \in [0,1]$. Let $K,\, L,\, r > 0$ be constants with C and ψ as defined above. We let

$$C_K = \{y \in C :\; \|y\| < K\}$$

and

$$C(\psi, r, L) = \{y \in C :\; \psi(y) \geq r \;\text{ and }\; \|y\| \leq L\}.$$

We now state the Leggett Williams fixed point theorem [78, 96]. The proof of Theorem 5.1.1 can be found in Chapter 13.

Theorem 5.1.1. *Let $E = (E, \|\,.\,\|)$ be a Banach space, $C \subset E$ a cone of E and $R > 0$ a constant. Suppose there exists a concave nonnegative continuous functional ψ on C with $\psi(y) \leq \|y\|$ for $y \in \overline{C_R}$ and let $A : \overline{C_R} \to \overline{C_R}$ be a continuous, compact map. Assume there are numbers r, L and K with $0 < r < L < K \leq R$ such that*

(H1). $\{y \in C(\psi, L, K) :\; \psi(y) > L\} \neq \emptyset$ and $\psi(A\,y) > L$ for all $y \in C(\psi, L, K)$;

(H2). $\|A\,y\| < r$ for all $y \in \overline{C_r}$;

(H3). $\psi(A\,y) > L$ for all $y \in C(\psi, L, R)$ with $\|A\,y\| > K$.

Then A has at least three fixed points y_1, y_2 and y_3 in $\overline{C_R}$. Furthermore we have

$$y_1 \in C_r, \;\; y_2 \in \{y \in C(\psi, L, R) : \; \psi(y) > L\}$$

and

$$y_3 \in \overline{C_R} \setminus \left(C(\psi, L, R) \cup \overline{C_r}\right).$$

5.2. Three solutions to integral equations.

In this section we discuss the integral equation

$$y(t) = h(t) + \int_0^1 k(t,s)\, f(y(s))\, ds \quad \text{for} \quad t \in [0,1]. \tag{5.2.1}$$

The following conditions will be assumed:

$$f : [0,\infty) \to [0,\infty) \;\; \text{is continuous and nondecreasing} \tag{5.2.2}$$

$$\begin{cases} k_t(s) = k(t,s) \in L^1[0,1] \;\; \text{with} \;\; k_t \geq 0 \\ \text{a.e. on} \;\; [0,1], \;\; \text{for each} \;\; t \in [0,1] \end{cases} \tag{5.2.3}$$

$$\text{the map} \;\; t \mapsto k_t \;\; \text{is continuous from} \;\; [0,1] \;\; \text{to} \;\; L^1[0,1] \tag{5.2.4}$$

$$\begin{cases} \exists\, r > 0 \;\; \text{with} \;\; |h|_0 + f(r) \; \sup_{t \in [0,1]} \int_0^1 k(t,s)\, ds < r \\ (\text{here} \;\; |h|_0 = \sup_{t \in [0,1]} |h(t)|) \end{cases} \tag{5.2.5}$$

$$\begin{cases} \exists\, M, \; 0 < M < 1, \; \kappa \in L^1[0,1] \;\; \text{and an interval} \;\; [a,b] \subseteq [0,1], \\ a < b, \;\; \text{such that} \;\; k(t,s) \geq M\, \kappa(s) \geq 0 \;\; \text{for} \;\; t \in [a,b] \\ \text{and a.e.} \;\; s \in [0,1] \end{cases} \tag{5.2.6}$$

$$k(t,s) \leq \kappa(s), \;\; t \in [0,1], \;\; \text{a.e.} \;\; s \in [0,1] \tag{5.2.7}$$

$$\begin{cases} h \in C[0,1] \;\; \text{with} \;\; h(t) \geq 0 \;\; \text{for} \;\; t \in [0,1] \\ \text{and} \;\; \min_{t \in [a,b]} h(t) \geq M\, |h|_0 \end{cases} \tag{5.2.8}$$

$$\exists\, L > r \;\; \text{with} \;\; \min_{t \in [a,b]} \left[h(t) \, + \, f(L) \int_a^b k(t,s)\, ds \right] > L \tag{5.2.9}$$

and

$$(5.2.10) \qquad \exists\, R \geq L\, M^{-1} \quad \text{with} \quad |h|_0 + f(R) \sup_{t\in[0,1]} \int_0^1 k(t,s)\, ds \leq R.$$

Theorem 5.2.1. *Suppose* (5.2.2) – (5.2.10) *hold. Then* (5.2.1) *has three nonnegative solution* y_1, y_2 *and* y_3 *in* $C[0,1]$ *with*

$$|y_1|_0 < r, \quad y_2(t) > L \quad for \quad t \in [a,b]$$

and

$$|y_3|_0 > r \quad with \quad \min_{t\in[a,b]} y_3(t) < L.$$

PROOF: Let

$$E = (C[0,1], |\,.\,|_0) \quad \text{and} \quad C = \{u \in C[0,1]: \; u(t) \geq 0 \quad \text{for} \quad t \in [0,1]\}.$$

Now let $A : C \to C$ be defined by

$$(5.2.11) \qquad A\, y(t) = h(t) + \int_0^1 k(t,s)\, f(y(s))\, ds \quad \text{for} \quad t \in [0,1],$$

here $y \in C$. It is immediate (see (5.2.2), (5.2.3) and (5.2.4)) from the results in [137] that

$$A : C \to C \quad \text{is continuous and completely continuous.}$$

For $y \in C$ let

$$\psi(y) = min_{t\in[a,b]}\, y(t).$$

Next choose and <u>fix</u> K so that

$$(5.2.12) \qquad L\, M^{-1} \leq K \leq R;$$

this is possible since $R \geq L\, M^{-1}$. Let

$$C_r = \{y \in C: \; |y|_0 < r\},$$

$$C_R = \{y \in C: \; |y|_0 < R\},$$

$$C(\psi, L, K) = \{y \in C: \; \psi(y) \geq L \text{ and } |y|_0 \leq K\},$$

and

$$C(\psi, L, R) = \{y \in C: \; \psi(y) \geq L \text{ and } |y|_0 \leq R\}.$$

First notice condition $(H2)$ of Theorem 5.1.1 holds since for $y \in \overline{C_r}$ we have from (5.2.2), (5.2.5) and (5.2.11) that

$$|Ay|_0 \le |h|_0 + f(r) \sup_{t\in[0,1]} \int_0^1 k(t,s)\,ds < r.$$

Also $A : \overline{C_R} \to \overline{C_R}$ since if $y \in \overline{C_R}$,

$$|Ay|_0 \le |h|_0 + f(R) \sup_{t\in[0,1]} \int_0^1 k(t,s)\,ds \le R.$$

Next we show $(H1)$ of Theorem 5.1.1 holds. First notice if

$$u(t) = \frac{L+K}{2} \quad \text{for} \quad t \in [0,1]$$

then $u \in \{y \in C(\psi, L, K) : \psi(y) > L\}$. Also if $y \in C(\psi, L, K)$ then $\psi(y) = \min_{t\in[a,b]} y(t) \ge L$ and $|y|_0 \le K$, so

$$y(t) \in [L, K] \quad \text{for} \quad t \in [a,b].$$

This together with (5.2.9) yields

$$\begin{aligned} \psi(Ay) &= \min_{t\in[a,b]} \left(h(t) + \int_0^1 k(t,s)\, f(y(s))\,ds \right) \\ &\ge \min_{t\in[a,b]} \left(h(t) + \int_a^b k(t,s)\, f(y(s))\,ds \right) \\ &\ge \min_{t\in[a,b]} \left(h(t) \,+\, f(L) \int_a^b k(t,s)\,ds \right) > L, \end{aligned}$$

so condition $(H1)$ of Theorem 5.1.1 is satisfied. Finally to see that $(H3)$ of Theorem 5.1.1 holds let $y \in C(\psi, L, R)$ with $|Ay|_0 > K$. First notice (5.2.7) and (5.2.11) imply

$$|Ay|_0 \le |h|_0 + \int_0^1 \kappa(s)\, f(y(s))\,ds$$

and this together with (5.2.6), (5.2.8) and (5.2.12) yields

$$\begin{aligned} \psi(Ay) &= \min_{t\in[a,b]} \left(h(t) + \int_0^1 k(t,s)\, f(y(s))\,ds \right) \\ &\ge M\,|h|_0 + M \int_0^1 \kappa(s)\, f(y(s))\,ds \\ &\ge M\,|Ay|_0 > M\,K \ge L. \end{aligned}$$

Thus condition $(H3)$ of Theorem 5.1.1 holds. Now apply Theorem 5.1.1 to deduce the result. □

Remark. Notice (5.2.3) and (5.2.4) can be replaced by any conditions which guarantee that the map $A : C \to C$ is continuous and completely continuous.

To illustrate how Theorem 5.2.1 can be applied to n^{th} $(n \geq 2)$ order boundary value problems we consider the Lidstone boundary value problem

$$(5.2.13) \qquad \begin{cases} (-1)^n y^{(2n)} = \phi(t)\, f(y), & t \in [0,1] \\ y^{(2i)}(0) = 0, \ \ y^{(2i)}(1) = 0, & 0 \leq i \leq n-1. \end{cases}$$

The Green's function $g_n(t,s)$ for the boundary value problem

$$(5.2.14) \qquad \begin{cases} y^{(2n)} = 0 \quad \text{on} \ [0,1] \\ y^{(2i)}(0) = 0, \ \ y^{(2i)}(1) = 0, \ \ 0 \leq i \leq n-1 \end{cases}$$

satisfied (see [8, 164]),

$$(5.2.15) \qquad (-1)^n g_n(t,s) \leq \frac{1}{6^{n-1}}\, s\,(1-s)$$

for $(t,s) \in [0,1] \times [0,1]$, and

$$(5.2.16) \qquad (-1)^n g_n(t,s) \geq \frac{1}{4^n} \left(\frac{3}{32}\right)^{n-1} s\,(1-s)$$

for $(t,s) \in \left[\frac{1}{4}, \frac{3}{4}\right] \times [0,1]$.

Theorem 5.2.2. *Assume the following conditions hold:*

$$(5.2.17) \qquad f : [0,\infty) \to [0,\infty) \ \textit{is continuous and nondecreasing}$$

$$(5.2.18) \qquad \begin{cases} \phi \in C(0,1) \ \textit{with} \ \phi > 0 \ \textit{on} \ (0,1) \\ \textit{and} \ \int_0^1 t(1-t)\,\phi(t)\,dt < \infty \end{cases}$$

$$(5.2.19) \qquad \exists\, r > 0 \ \textit{with} \ f(r) \sup_{t\in[0,1]} \int_0^1 (-1)^n g_n(t,s)\,\phi(s)\,ds < r$$

$$(5.2.20) \qquad \exists\, L > r \ \textit{with} \ f(L) \min_{t\in\left[\frac{1}{4},\frac{3}{4}\right]} \int_{\frac{1}{4}}^{\frac{3}{4}} (-1)^n g_n(t,s)\,\phi(s)\,ds > L$$

and

$$\text{(5.2.21)} \qquad \begin{cases} \exists\, R \ge L \left[4^n \left(\frac{32}{3}\right)^{n-1} \frac{1}{6^{n-1}}\right] \quad \textit{with} \\ f(R)\ \sup_{t\in[0,1]} \int_0^1 (-1)^n\, g_n(t,s)\, \phi(s)\, ds \ \le\ R. \end{cases}$$

Then (5.2.13) *has three nonnegative solution* y_1, y_2 *and* y_3 *in* $C[0,1]$ *with*

$$|y_1|_0 < r, \quad y_2(t) > L \quad \textit{for} \quad t \in \left[\frac{1}{4}, \frac{3}{4}\right]$$

and

$$|y_3|_0 > r \quad \textit{with} \quad \min_{t\in[\frac{1}{4},\frac{3}{4}]} y_3(t) < L.$$

PROOF: Let $A : C \to C$ (here C is as defined in Theorem 2.1) be defined by

$$A\, y(t) = \int_0^1 (-1)^n\, g_n(t,s)\, \phi(s)\, f(y(s))\, ds \quad \text{for} \quad y \in C.$$

The results in [8] guarantee that

$$A : C \to C \quad \text{is continuous and completely continuous.}$$

Now apply Theorem 5.2.1 (with Remark) with

$$k(t,s) = (-1)^n\, g_n(t,s)\, \phi(s), \ \ h \equiv 0, \ \ a = \frac{1}{4}, \ b = \frac{3}{4}$$

and

$$\kappa(s) = \frac{1}{6^{n-1}}\ s\,(1-s), \quad M = \frac{1}{4^n}\left(\frac{3}{32}\right)^{n-1} 6^{n-1}.$$

Notice (5.2.15) and (5.2.16) guarantee that (5.2.6) and (5.2.7) hold. □

CHAPTER 6

FIXED POINT SETS FOR ABSTRACT VOLTERRA OPERATORS

6.1. Introduction.

In this chapter we study the topological structure of the solution set for the equation

$$y = F(y). \tag{6.1.1}$$

Here $F : E \to E$ is a nonlinear abstract Volterra operator and $E = E(I, \mathbf{R}^M)$ is a topological space of functions defined on the subinterval I of $[0, \infty)$, with values in $\mathbf{R}^M$ (M is a positive integer), to be specified later.

Recall that an operator $F : E \to E$, where $E = E(I, \mathbf{R}^M)$ is a function space and $I = [0, T]$, $0 \leq T < \infty$ or $I = [0, T)$, $0 \leq T \leq \infty$, is an *abstract Volterra operator* if for an arbitrary $T^\star \in I$, F satisfies the following condition:

$$x(t) = y(t), \;\; t \in [0, T^\star] \text{ implies that } Fx(t) = Fy(t), \;\; t \in [0, T^\star].$$

(When considering spaces of measurable functions, the relevant equalities are replaced by equalities almost everywhere.)

The solution set of the given equation coincides with the set of all fixed points of F, i.e.

$$Fix(F) = \{x \in E : x = F(x)\}.$$

We also recall that a set S in E is called an R_δ set if S is homeomorphic with a decreasing sequence of compact absolute retracts [25, 35]. If S is an R_δ set, then S is also a continuum (i.e. a nonempty, compact and connected set).

The case where E is a Banach space is discussed in Section 6.2. In particular we show that if $E = C([0, a], \mathbf{R}^M)$ (the Banach space of continuous functions on $[0, a]$, $a > 0$ given) or if $E = L^p([0, a], \mathbf{R}^M)$, $1 < p < \infty$ (the Banach space of p-integrable functions on $[0, a]$), and F is an abstract Volterra operator that satisfies appropriate conditions, then $Fix(F)$ is an R_δ set.

In Section 6.3 we first consider the functional equation

$$y(t) = Vy(t), \quad t \in [0,\infty), \tag{6.1.2}$$

where it is assumed that $V : C[0,\infty) \to C[0,\infty)$ is an abstract Volterra operator. In Section 6.3 we present a set of conditions that $V : C[0,\infty) \to C[0,\infty)$ must satisfy in order that the solution set of (6.1.2) is a continuum, that is, nonempty, compact and connected, in the Fréchet space $C[0,\infty)$.

Notation. $C[0,\infty)$ denotes the Fréchet space of functions that are continuous on each compact subinterval of $[0,\infty)$, while analogously, $L^p_{loc}[0,\infty)$ denotes the Fréchet space of functions that are L^p-integrable on each compact subinterval of $[0,\infty)$. Two other function spaces that we will encounter in this chapter are $W^{1,p}[0,n]$ and $W^{1,p}_{loc}[0,\infty)$. We denote by $W^{1,p}[0,n]$, the Sobolev space of all absolutely continuous functions $u : [0,n] \to \mathbf{R}$ with $u' \in L^p[0,n]$. $W^{1,p}[0,n]$ is a Banach space with norm $||u|| = ||u||_{L^p[0,n]} + ||u'||_{L^p[0,n]}$. Finally, $W^{1,p}_{loc}[0,\infty)$ is the Fréchet space of all functions u that are absolutely continuous and whose derivatives are L^p-integrable on each compact subinterval of $[0,\infty)$.

Since Aronszajn's discussion in 1942 [25] regarding the R_δ nature of solution sets of Volterra equations of the form

$$y(t) = Vy(t), \quad t \in [0,T], \tag{6.1.3}$$

much is known about the existence, compactedness and connectedness of solution sets of such equations in the Banach space $C[0,T]$. In particular, the results in Section 6.2 details the conditions that one requires on the operator $V : C[0,T] \to C[0,T]$ in order that the set of solutions of (6.1.3) is a continuum in $C[0,T]$. In our first main result of Section 6.3, we show that if for each $n \in \mathbf{N}$, the solution set of

$$y(t) = Vy(t), \quad t \in [0,n], \tag*{$(6.1.4)_n$}$$

is a continuum in $C[0,n]$, then (6.1.2) has a continuum of solutions in the Fréchet space $C[0,\infty)$. With the results of Section 6.2 we can then readily determine what conditions to impose on $V : C[0,\infty) \to C[0,\infty)$ in order that the continuum condition on the solution set of $(6.1.4)_n$ be satisfied.

In the second part of Section 6.3, we discuss the functional-differential equation

$$\begin{cases} y'(t) = \tilde{V}y(t), \ \text{a.e. } t \in [0,\infty) \\ y(0) = y_0. \end{cases} \tag{6.1.5}$$

Here it is assumed that the abstract Volterra operator $\tilde{V}$ is such that $\tilde{V}: C[0,\infty) \to L^p_{loc}[0,\infty)$, $1 \le p \le \infty$ is satisfied, and a continuum of solutions of (6.1.5) is sought in the Fréchet space $W^{1,p}_{loc}[0,\infty)$. By examining (6.1.5)'s counterpart on the compact interval $[0,n]$, namely

$$(6.1.6)_n \qquad \begin{cases} y'(t) = \tilde{V}y(t), & \text{a.e. } t \in [0,n] \\ y(0) = y_0, \end{cases}$$

and requesting that for each $n \in \mathbf{N}$, the solution set of $(6.1.6)_n$ be a continuum in $W^{1,p}[0,n]$, an analogous argument to the one used for (6.1.2) ensures that the solution set of (6.1.5) is nonempty, compact and connected in $W^{1,p}_{loc}[0,\infty)$. We also show that if a continuity condition is imposed on $\tilde{V}: C[0,\infty) \to L^p_{loc}[0,\infty)$, the fact that for each $n \in \mathbf{N}$, the solution set of $(6.1.6)_n$ is a continuum in $C[0,n]$, will suffice to give the desired result for (6.1.5).

Therefore we must further address the question of what conditions

$$\tilde{V}: C[0,n] \to L^p[0,n]$$

must satisfy in order that the set of solutions of $(6.1.6)_n$ be a continuum in $C[0,n]$ and indeed $W^{1,p}[0,n]$. Of course by considering the representation

$$(6.1.7)_n \qquad y(t) = Vy(t) := y_0 + \int_0^t \tilde{V}y(s)\,ds, \quad t \in [0,n]$$

of $(6.1.6)_n$, one can apply the results of Section 6.2 to ensure that the solution set of $(6.1.7)_n$, and hence $(6.1.6)_n$, is nonempty, compact and connected in $C[0,n]$. Further elaboration on this result, will give us a continuum of solutions of $(6.1.6)_n$ in $W^{1,p}[0,n]$.

We remark that if one is content with seeking a continuum of solutions of (6.1.5) in the Fréchet space $C[0,\infty)$, the result follows easily from the results obtained for (6.1.2) if we use $(6.1.7)_n$ and

$$(6.1.8) \qquad y(t) = Vy(t) := y_0 + \int_0^t \tilde{V}y(s)\,ds, \quad t \in [0,\infty),$$

to represent $(6.1.6)_n$ and (6.1.5) respectively.

Applications of the results obtained in Section 6.3 are discussed in Section 6.4. In particular, results obtained for (6.1.2) can be applied to the Volterra integral equation

$$(6.1.9) \qquad y(t) = Vy(t) = h(t) + \int_0^t k(t,s)f(s,y(s))\,ds, \quad t \in [0,\infty),$$

whereas information on the solution sets of the Cauchy problem

$$(6.1.10)\qquad \begin{cases} y'(t) = \tilde{V}y(t) = f(t, y(t)), \text{ a.e. } t \in [0, \infty) \\ y(0) = y_0, \end{cases}$$

and the Volterra integrodifferential equation

$$(6.1.11)\qquad \begin{cases} y'(t) = \tilde{V}y(t) \\ \qquad = h(t) + \int_0^t k(t,s)g(s,y(s))\, ds, \text{ a.e. } t \in [0, \infty), \\ y(0) = y_0, \end{cases}$$

in $W^{1,p}_{loc}[0, \infty)$ can be extracted from the results established for (6.1.5).

In Section 6.5 and 6.6 we prove a stronger result that in Section 6.3. The methods presented in these sections are different than those seen in Section 6.3. In particular in Section 6.5 (respectively Section 6.6) we give appropriate conditions, if $E = C([0, \infty), \mathbf{R}^M)$ (respectively $E = L^p_{loc}([0, \infty), \mathbf{R}^M)$, $1 < p < \infty$), when $Fix(F)$ is an R_δ set.

6.2. Fixed points for operators on Banach spaces.

Let $E = C([0, a], \mathbf{R}^M)$ or $E = L^p([0, a], \mathbf{R}^M)$, $1 < p < \infty$, where a is given. Consider an operator $F : E \to E$. Let X be a closed subset of E such that int (X) is nonempty. Assume that $F : X \to E$ is continuous and compact.

Define the sequence of operators $\{F_n\}_n$, $F_n : E \to E$, as follows:

$$(6.2.1)\qquad F_n(x)(t) = F(x)(r_n(t)), \text{ for } x \in E \text{ and } t \in [0, a],$$

where

$$(6.2.2)\qquad r_n(t) = \begin{cases} 0, \text{ if } t \in [0, a/n]; \\ t - \frac{a}{n}, \text{ if } t \in (a/n, a]. \end{cases}$$

Consider the following conditions:

(i) $\exists u_0 \in \mathbf{R}^M$ such that $F(x)(0) = u_0$, all $x \in X$;

(ii) $\forall \epsilon \in (0, a)$, $\forall x, y \in X$, if $x(t) = y(t)$, $\forall t \in [0, \epsilon]$, then $F(x)(t) = F(y)(t)$, $\forall t \in [0, \epsilon]$ (i.e. F is an abstract Volterra operator);

(iii) if $y \in E$ satisfies $y = F(y)$, then $y \in X$;

(iv) $\forall n$, if $y \in E$ satisfies $y = F_n(y)$, then $y \in$ int (X).

Let $|.|$ be the usual norm in $\mathbf{R}^M$. Let $C([0, a], \mathbf{R}^M)$ denote the Banach space of all continuous functions on $[0, a]$, with the norm $\|x\| =$

$\sup\{|x(t)|, t \in [0,a]\}$. Let $L^p([0,a], \mathbf{R}^M)$ be the Banach space of all strongly measurable functions on $[0,a]$, such that $(\int_0^a |x(t)|^p dt)^{1/p} < \infty$, with the norm $\|x\|_p = (\int_0^a |x(t)|^p dt)^{1/p}$; assume that $1 < p < \infty$.

Theorem 6.2.1. *Assume that* $E = L^p([0,a], \mathbf{R}^M)$ *and that (i), (ii), (iii) and (iv) hold. Then* $Fix(F)$ *is an* R_δ *set.*

Theorem 6.2.2. *Assume that* $E = C([0,a], \mathbf{R}^M)$ *and that (i), (ii), (iii) and (iv) hold. Then* $Fix(F)$ *is an* R_δ *set.*

To prove these two theorems, we shall use the following result [35]:

Lemma 6.2.3. *Let* X *be a metric space,* E *a Banach space, and let* $\{f_n\}_n$ *be a sequence of continuous maps,* $f_n : X \to E$*, which converge uniformly on* X *to a map* $f : X \to E$*. Assume* f_n *and* f *are proper, and that for all* y *in a neighborhood of* 0 *in* E *there exists exactly one solution of the equation* $f_n(x) = y$*. Then* $f^{-1}(0)$ *is an* R_δ *set.*

PROOF OF THEOREM 6.2.1: Let f and f_n be defined by: $f = I - F$, $f_n = I - F_n$, where I is the identity operator. Note from (iii) that $f^{-1}(0) = Fix(F)$. We now show that the following hold:

(6.2.3) $$\lim_{n\to\infty} f_n(x) = f(x), \text{ uniformly for } x \in X;$$

(6.2.4) $$f \text{ and } f_n \text{ are proper for all } n;$$

and

(6.2.5) $$f_n \text{ is a homeomorphism from } X \text{ to } f_n(X), \text{ for all } n.$$

Notice first that each F_n is continuous and compact. To see that F_n is continuous, let $\epsilon > 0$; since F is continuous, $\exists \delta > 0$ such that $\|x - y\|_p < \delta$ implies $\|F(x) - F(y)\|_p < \epsilon$. Then for x, y in X with $\|x - y\|_p < \delta$ we also have

$$\begin{aligned} \|F_n(x) - F_n(y)\|_p^p &= \int_{a/n}^a \|F(x)(t - \frac{a}{n}) - F(y)(t - \frac{a}{n})\|^p dt \\ &\le \int_0^a \|F(x)(s) - F(y)(s)\|^p ds < \epsilon^p. \end{aligned}$$

To see that F_n is compact, let B be a bounded subset of X. Since F is compact, $F(B)$ is relatively compact, therefore there exists $M > 0$ with

(6.2.6) $$\|F(x)(t)\|_p \le M, \text{ all } x \in B,$$

and

(6.2.7) $\int_0^a \|F(x)(t+\tau) - F(x)(t)\|^p dt \to 0$ as $\tau \to 0$, uniformly for $x \in B$.

Now if $y_n \in F_n(B)$, then $y_n = F_n(x)$, for some $x \in B$, and

$$\|y_n\|_p^p = \int_0^{a/n} \|u_0\|^p dt + \int_0^{a-a/n} \|F(x)(t)\|^p dt$$

so that $F_n(B)$ is bounded by $(a\|u_0\|^p + M^p)^{1/p}$. Also, for $\tau \to 0$ we may assume $\tau < \frac{a}{n}$, and then, if we let G be defined by

$$G(t) = F_n(x)(t+\tau) - F_n(x)(t),$$

we have

$$\int_0^a \|G(t)\|^p dt = \int_0^{a/n-\tau} \|G(t)\|^p dt + \int_{a/n-\tau}^{a/n} \|G(t)\|^p dt + \int_{a/n}^a \|G(t)\|^p dt.$$

The first integral is zero; the third integral has the limit 0 as $\tau \to 0$, uniformly for $x \in X$ (since (6.2.7) holds); the second integral is bounded above by $\tau(\|u_0\|^p + M^p)$, which has the limit 0 as $\tau \to 0$, uniformly for $x \in X$. Thus $F_n(B)$ is relatively compact.

We now show that $\lim_{n\to\infty} F_n(x) = F(x)$, uniformly for $x \in X$. Since F is compact, (6.2.6) and (6.2.7) hold for $B = X$. Then

$$\begin{aligned}\int_0^a \|F_n(x)(t) - F(x)(t)\|^p dt &= \int_0^{a/n} \|F(x)(0) - F(x)(t)\|^p dt \\ &+ \int_{a/n}^a \|F(x)(t - \frac{a}{n}) - F(x)(t)\|^p dt\end{aligned}$$

The first term is bounded above by $\frac{a}{n}(\|u_0\|^p + M^p)$, therefore its limit is 0 as $n \to \infty$, uniformly for $x \in X$; the second term has the limit 0 as $n \to \infty$, uniformly for $x \in X$, since (6.2.7) holds.

To prove that $I - F$ is proper, let Y be a compact set in E. If $\{x_j\}_j$ is a sequence in $(I-F)^{-1}(Y)$, then $x_j - F(x_j) \in Y$ for all j. Since Y is compact, there exists a subsequence of $\{x_j\}_j$, again denoted $\{x_j\}_j$, and there exists a $y \in Y$, such that

(6.2.8) $$x_j - F(x_j) \to y \text{ in } E \text{ as } j \to \infty.$$

Since F is compact, $F(X)$ is relatively compact, so there exists a subsequence of $\{x_j\}_j$, again denoted $\{x_j\}_j$, and there exists a $z \in E$, such that

(6.2.9) $$F(x_j) \to z \text{ in } E \text{ as } j \to \infty.$$

Now (6.2.8) and (6.2.9), imply that $x_j \to y+z$ in E as $j \to \infty$. In addition to this, F is continuous and Y is compact, so $y+z \in (I-F)^{-1}(Y)$. Thus $I-F$ is proper. Using the same arguments, one concludes that $I-F_n$ is proper for all n.

Next we show that each $I-F_n$ is a homeomorphism, since F_n is continuous, $I-F_n$ is one-to-one, and $(I-F_n)^{-1}$ is continuous. We provide a few details. Let $x, y \in X$ be such that $(I-F_n)(x) = (I-F_n)(y)$. Then for $t \in [0, \frac{a}{n}]$,

$$x(t) - y(t) = F(x)(r_n(t)) - F(y)(r_n(t)) = u_0 - u_0 = 0;$$

for $t \in [\frac{a}{n}, \frac{2a}{n}]$,

$$x(t) - y(t) = F(x)(t - \frac{a}{n}) - F(y)(t - \frac{a}{n})$$

with $t - \frac{a}{n} \in [0, \frac{a}{n}]$, so $x(t) - y(t) = 0$, since (ii) holds; repeat the procedure for each $[\frac{ia}{n}, \frac{(i+1)a}{n}]$, $i = 1, \ldots n-1$, and conclude that $I - F_n$ is one-to-one (see [6, pp. 161] also). For the continuity of $(I-F_n)^{-1}$, let $\{x_j\}_j$ be a sequence in X, and let x be in X, such that

$$\lim_{j\to\infty} \|(x_j - F_n(x_j)) - (x - F_n(x))\|_p = 0. \tag{6.2.10}$$

For $t \in [0, \frac{a}{n}]$, we have $F_n(x_j)(t) = F_n(x)(t) = u_0$, thus (6.2.10) gives

$$\lim_{j\to\infty} \|(x_j - x)\chi_{[0,\frac{a}{n}]}\|_p = 0,$$

where $\chi_{[0,\frac{a}{n}]}$ is the characteristic function of the interval $[0, \frac{a}{n}]$. Also, since F is continuous, it follows that

$$\lim_{j\to\infty} \|(F(x_j) - F(x))\chi_{[0,\frac{a}{n}]}\|_p = 0. \tag{6.2.11}$$

For $t \in [\frac{a}{n}, \frac{2a}{n}]$, we have $t - \frac{a}{n} \in [0, \frac{a}{n}]$, and $F_n(x_j)(t - \frac{a}{n}) = u_0$,

$$F_n(x_j)(t) = F(x_j)(t - \frac{a}{n}) \quad \text{and} \quad F_n(x)(t) = F(x)(t - \frac{a}{n}),$$

thus (6.2.10) and (6.2.11) give

$$\lim_{j\to\infty} \|(F(x_j) - F(x))\chi_{[0,\frac{2a}{n}]}\|_p = 0.$$

Repeat the procedure on each interval of length $\frac{a}{n}$, and obtain

$$\lim_{j\to\infty} \|F(x_j) - F(x)\|_p = 0,$$

thus $\lim_{j\to\infty} \|x_j - x\|_p = 0$, so $(I - F_n)^{-1}$ is continuous (see [13, p.169] also).

Now we claim that $\forall n, \exists x \in E$ with $x - F_n(x) = 0$. To see this, define x as follows:

$$x(t) = u_0 \quad \text{for} \quad t \in [0, a/n];$$

$$x(t) = F_n(x)(t) \quad \text{for} \quad t \in [(i-1)a/n, ia/n],\ i = 1, ..., n.$$

Then, by (iv), this x is in $\text{int}\,(X)$, and $0 \in f_n(\text{int}\,(X))$. Since f_n^{-1} is continuous, there exists a neighbourhood U of 0 in E such that $U \subseteq f_n(\text{int}\,(X)) \subseteq f_n(X)$. Now apply Lemma 6.2.3 with $f : X \to E$, $f_n : X \to E$, $f = I - F$, $f_n = I - F_n$, to conclude that $Fix(F)$ is an R_δ set. $\square$

PROOF OF THEOREM 6.2.2: Let f and f_n be defined as above. For $E = C([0, a], \mathbf{R}^M)$, it is easy to check that (6.2.3), (6.2.4) and (6.2.5) hold. As in the previous proof, one can show that $f_n(X)$ contains a neighbourhood of the origin. Therefore, according to Lemma 6.2.3, we have that $Fix(F)$ is an R_δ set. $\square$

To show how to apply Theorem 6.2.1 in practice for $E = L^p([0, a], \mathbf{R})$ $(1 < p < \infty)$, we consider the integral equation

$$y(t) = \int_0^t k(t, s) g(s, y(s)) ds \tag{6.2.12}$$

for $t \in [0, a]$; here $0 < a < \infty$ is fixed.

Theorem 6.2.4. *Assume $p, q > 1$ with $\frac{1}{p} + \frac{1}{q} = 1$. Suppose the following conditions hold:*

$$\textit{the map } t \mapsto g(t, y) \textit{ is measurable for all } y \in \mathbf{R}; \tag{6.2.13}$$

$$\textit{the map } y \mapsto g(t, y) \textit{ is continuous for almost all } t \in [0, a]; \tag{6.2.14}$$

$$\begin{cases} \textit{there exists a function } b : [0, a] \to [0, \infty],\ b \in L^q([0, a], \mathbf{R}), \\ \textit{and a } c \geq 0 \textit{ with } |g(t, y)| \leq b(t) + c|y|^{p/q} \end{cases} \tag{6.2.15}$$

$$\begin{cases} k : [0, a] \times [0, t] \to \mathbf{R} \textit{ is such that } (t, s) \mapsto k(t, s) \\ \textit{is measurable, } \int_0^a \int_s^a |k(t, s)|^p dt ds < \infty, \end{cases} \tag{6.2.16}$$

$$\begin{cases} \textit{there exists a function } \psi : [0, a] \to [0, \infty] \textit{ with} \\ \psi \in L^p([0, a], \mathbf{R}) \textit{ and } |k(t, s)| \leq \psi(s) \textit{ for } 0 \leq s \leq t. \end{cases} \tag{6.2.17}$$

Let $I : [0, \infty] \to [0, \infty]$ *be an increasing map given by*

$$I(z) = \int_0^z \frac{dx}{1 + x^{p/q}}$$

and let

$$e(t) = 2^{p-1} max\Big\{ \int_0^t \psi(s)b(s)ds, c^p \int_0^t [\psi(s)]^p ds \Big\}.$$

Also assume

(6.2.18) $$\int_0^a e(t)dt \in dom(I^{-1})$$

holds.

Then the solution set of (6.2.12) *is a* R_δ *set.*

PROOF: Let

(6.2.19) $$r = \Big[I^{-1}\Big(\int_0^a e(t)dt \Big) \Big]^{1/p}, \; X = \{ x \in E : \|x\|_p \le r + 1 \}$$

and $E = L^p([0, a], \mathbf{R})$. Also let $F : X \to E$ be given by

(6.2.20) $$Fy(t) = \int_0^t k(t, s) g(s, y(s)) ds \text{ for } y \in X \text{ and } t \in [0, a].$$

Now [137, Chapter 4] implies $F : X \to E$ is a continuous, compact map. Notice (i) and (ii) in Theorem 6.2.1 hold. Next we show (iii) in Theorem 6.2.1. To show (iii), let $y \in E$ satisfy $y = F(y)$. We must show $\|y\|_p| \le r+1$. First notice for $t \in [0, a]$ that

$$\begin{aligned} |y(t)| &\le \int_0^t \psi(s)b(s)ds + c \int_0^t \psi(s)|y(s)|^{p/q} ds \\ &\le \int_0^t \psi(s)b(s)ds + c\Big(\int_0^t [\psi(s)]^p ds \Big)^{1/p} \Big(\int_0^t |y(s)|^p ds \Big)^{1/q}, \end{aligned}$$

and so

$$|y(t)|^p \le 2^{p-1} \Big\{ \Big(\int_0^t \psi(s)b(s)ds \Big)^p + c^p \Big(\int_0^t [\psi(s)]^p ds \Big) \Big(\int_0^t |y(s)|^p ds \Big)^{p/q} \Big\}.$$

Let

$$u(t) = \int_0^t |y(s)|^p ds$$

and so

$$u'(t) \le 2^{p-1} \Big\{ \Big(\int_0^t \psi(s)b(s)ds \Big)^p + c^p \Big(\int_0^t [\psi(s)]^p ds \Big) \Big(u(t) \Big)^{p/q} \Big\} \text{ for } t \in [0, a].$$

As a result,

$$u'(t) \le e(t)\Big(1 + \big(u(t)\big)^{p/q}\Big)$$

and integration from 0 to a yields

$$I(u(a)) = \int_0^{u(a)} \frac{dx}{1 + x^{p/q}} \le \int_0^a e(t)dt.$$

Consequently,

$$\int_0^a |y(s)|^p ds \le I^{-1}\Big(\int_0^a e(t)dt\Big)$$

and as a result

$$\|y\|_p \le r < r+1$$

(where r is as defined in (6.2.19)). A similar argument shows that if $n \in \{1, 2, ...\}$ and $y \in E$ satisfies

$$y(t) = \int_0^{r_n(t)} k(r_n(t), s)\, g(s, y(s))\, ds$$

then $\|y\|_p < r + 1$. Thus the conditions of Theorem 6.2.1 are satisfied, and the result follows. □

6.3. Continuums of solutions in Fréchet spaces.

The following technical result [148] which will be needed in this section.

Theorem 6.3.1. *Let (X, d) be a complete metric space and $\{S_m : m = 1, 2, \ldots\}$ a sequence of nonempty, compact and connected sets such that $S_{m+1} \subseteq S_m$ for $m = 1, 2, \ldots$. Then $\bigcap_{m=1}^{\infty} S_m$ is a nonempty, compact and connected set.*

The first result in this section examines the relationship between the solution set of

$$(6.3.1) \qquad y(t) = Vy(t), \;\; t \in [0, \infty)$$

in the Fréchet space $C[0, \infty)$, and for each $n \in \mathbf{N}$, the solution set of

$$(6.3.2)_n \qquad y(t) = Vy(t), \;\; t \in [0, n]$$

in the Banach space $C[0, n]$.

Theorem 6.3.2. *Suppose that*

(6.3.3) $\quad V : C[0, \infty) \to C[0, \infty)$ *is an abstract Volterra operator*

and

$$(6.3.4) \qquad \begin{cases} \textit{for each } n \in \mathbf{N}, \textit{ the solution set of } (2.2)_n \textit{ is} \\ \textit{nonempty, compact and connected in } C[0,n] \end{cases}$$

hold. Then the solution set of (6.3.1) *is nonempty, compact and connected in* $C[0,\infty)$.

Notation. Let S denote the solution set of (6.3.1) and S_n the solution set of $(6.3.2)_n$ for each $n \in \mathbf{N}$, that is,

$$S := \{y \in C[0,\infty) : y \text{ solves } (6.3.1)\}$$

and

$$S_n := \{y \in C[0,n] : y \text{ solves } (6.3.2)_n\}, \quad n \in \mathbf{N}.$$

Let $y|_{[a,b]}$ denote the restriction of a function y to the interval $[a,b]$, and for any $m \in \mathbf{N}$ and $n \in \{m, m+1, \ldots\}$, we use $S_n|_{[0,m]}$ to denote the restriction of S_n to the interval $[0,m]$, that is,

$$S_n|_{[0,m]} := \{y|_{[0,m]} : y \in S_n\}.$$

PROOF OF THEOREM 6.3.2: From (6.3.4) and the fact that the mapping from $C[0,n]$ to $C[0,1]$ which assigns to each function $y \in C[0,n]$ the restriction $y|_{[0,1]}$ is continuous, it is clear that

$$(6.3.5) \qquad \begin{cases} S_n|_{[0,1]} \text{ is nonempty, compact and connected} \\ \text{in } C[0,1] \text{ for each } n \in \mathbf{N} \end{cases}$$

holds. Also for any $n \in \mathbf{N}$, if $y \in S_{n+1}$ (that is, $y \in C[0,n+1]$ solves $(6.3.2)_{n+1}$) then as a result of the Volterra nature of the operator $V : C[0,\infty) \to C[0,\infty)$, we have that $y|_{[0,n]}$ is a solution of $(6.3.2)_n$, that is, $y|_{[0,n]} \in S_n$. Consequently for $n \in \mathbf{N}$,

$$S_{n+1}|_{[0,n]} \subseteq S_n.$$

Indeed it follows trivially that for $n \in \mathbf{N}$,

$$(6.3.6) \qquad S_{n+1}|_{[0,1]} \subseteq S_n|_{[0,1]}.$$

Using (6.3.5) and (6.3.6), Theorem 6.3.1 implies that

$$\bigcap_{n=1}^{\infty} (S_n|_{[0,1]})$$

is a continuum in $C[0,1]$. For convenience define

$$I_1 := \bigcap_{n=1}^{\infty} (S_n|_{[0,1]})$$

In general, for any $k \in \mathbf{N}$ (fixed), using an argument similar to the one just used above, we see that for each $n \in \{k, k+1, \ldots\}$,

$$S_n|_{[0,k]} \text{ is a continuum in } C[0,k]$$

and

$$S_{n+1}|_{[0,k]} \subseteq S_n|_{[0,k]}.$$

Consequently Theorem 6.3.1 implies that

$$I_k := \bigcap_{n=k}^{\infty} (S_n|_{[0,k]})$$

is a continuum in $C[0,k]$. We now make the following claim:

Claim. For each $k \in \{1, 2, \ldots\}$,

$$I_{k+1}|_{[0,k]} = I_k. \tag{6.3.7}$$

PROOF OF CLAIM: Fix $k \in \{1, 2, \ldots\}$. Let $\tilde{z}_{k+1} \in I_{k+1}|_{[0,k]}$. Then there exists

$$z_{k+1} \in I_{k+1} \text{ such that } z_{k+1}(t) = \tilde{z}_{k+1}(t), \;\; t \in [0,k].$$

Since $z_{k+1} \in I_{k+1} \subseteq S_{k+1}$ and $V : C[0,\infty) \to C[0,\infty)$ is a Volterra operator, we note that

$$z_{k+1}|_{[0,k]} = \tilde{z}_{k+1} \in S_k.$$

In addition, since $z_{k+1} \in I_{k+1}$, for each $n \in \{k+1, \ldots\}$, there exists $y_n \in S_n$ such that $y_n(t) = z_{k+1}(t)$, $t \in [0, k+1]$. Hence $\tilde{z}_{k+1} \in S_k$ and is the restriction of $y_n \in S_n$ to $[0,k]$, for each $n \in \{k+1, k+2, \ldots\}$. Consequently $\tilde{z}_{k+1} \in I_k$ and we have that

$$I_{k+1}|_{[0,k]} \subseteq I_k.$$

Conversely we want to show that

$$I_k \subseteq I_{k+1}|_{[0,k]}.$$

Let $z_k \in I_k$. Then we must find

$$z_{k+1} \in I_{k+1} \text{ such that } z_{k+1}(t) = z_k(t), \quad t \in [0, k].$$

Since $z_k \in I_k$, for each $n \in \{k, k+1, \ldots\}$, there exists

$$y_n \in S_n \text{ such that } y_n(t) = z_k(t), \quad t \in [0, k].$$

Consider the sequence $\{y_n\}_{n=k+1}^{\infty}$. Note that due to the Volterra nature of the operator $V : C[0, \infty) \to C[0, \infty)$, the restriction of this sequence to the interval $[0, k+1]$ is contained in S_{k+1}, that is,

$$\{y_n|_{[0,k+1]}\}_{n=k+1}^{\infty} \subseteq S_{k+1}.$$

Now S_{k+1} is compact, therefore there exists a subsequence of integers, say $N_{k+1} \subseteq \{k+1, \ldots\}$, and a function $z_{k+1} \in S_{k+1}$ such that

$$y_n|_{[0,k+1]} \to z_{k+1} \text{ uniformly on } [0, k+1] \text{ as } n \to \infty \text{ in } N_{k+1}.$$

Note that $z_{k+1}(t) = z_k(t)$, $t \in [0, k]$ and therefore if we show that $z_{k+1} \in I_{k+1}$, we have proved the claim. In other words we want to show that for each $m \in \{k+1, k+2, \ldots\}$, there exists

$$z_m \in S_m \text{ such that } z_m(t) = z_{k+1}(t), \quad t \in [0, k+1].$$

Let $N_{k+2}^{\star} = N_{k+1} \backslash \{k+1\}$ and consider the sequence $\{y_n\}_{n \in N_{k+2}^{\star}}$. The restriction of this sequence to the interval $[0, k+2]$ is contained in S_{k+2}, that is,

$$\{y_n|_{[0,k+2]}\}_{n \in N_{k+2}^{\star}} \subseteq S_{k+2},$$

and since S_{k+2} is compact there exists a subsequence N_{k+2} of $N_{k+2}^{\star}$, and a function $z_{k+2} \in S_{k+2}$ such that

$$y_n|_{[0,k+2]} \to z_{k+2} \text{ uniformly on } [0, k+2] \text{ as } n \to \infty \text{ in } N_{k+2}$$

and

$$z_{k+2}(t) = z_{k+1}(t), \quad t \in [0, k+1].$$

Proceeding inductively we can show that for any $m \in \{k+1, k+2, \ldots\}$, there exists a subsequence of integers

$$N_m \subseteq \{m, m+1, \ldots\} \text{ with } N_{k+1} \supseteq N_{k+2} \supseteq \ldots \supseteq N_m,$$

and a function $z_m \in S_m$ such that

$$y_n|_{[0,m]} \to z_m \text{ uniformly on } [0, m] \text{ as } n \to \infty \text{ in } N_m$$

and

$$z_m(t) = z_{k+1}(t), \quad t \in [0, k+1].$$

Thus we have shown that for each $m \in \{k+1, \ldots\}$, z_{k+1} is the restriction of a function $z_m \in S_m$ to the interval $[0, k+1]$. Hence $z_k \in I_{k+1}|_{[0,k]}$ and the claim is proved.

Now define

$$S^\star := \{y \in C[0,\infty) : y|_{[0,k]} \in I_k \text{ for each } k \in \mathbf{N}\}.$$

We assert that $S = S^\star$. Certainly if $y \in S \subseteq C[0,\infty)$, then since $V : C[0,\infty) \to C[0,\infty)$ is a Volterra operator, for any fixed $k \in \mathbf{N}$,

$$y|_{[0,n]} \in S_n \text{ for } n \in \{k, k+1, \ldots\}.$$

Consequently $y \in C[0,\infty)$ and $y|_{[0,k]} \in I_k$ for each $k \in \mathbf{N}$. Hence $S \subseteq S^\star$.

Conversely suppose that $y \in S^\star$. Then

$$y|_{[0,k]} \in I_k \subseteq S_k \text{ for each } k \in \mathbf{N},$$

therefore once again using the fact that $V : C[0,\infty) \to C[0,\infty)$ is Volterra, y must solve (6.3.1), that is, $y \in S$. Hence $S^\star \subseteq S$ and the assertion is true.

With this alternate description of S we are now in a position to show that S is nonempty, compact and connected in $C[0,\infty)$. Since $I_k \neq \emptyset$ for each $k \in \mathbf{N}$, there exists $z_1 \in I_1$. Now by (6.3.7),

$$I_1 = I_2|_{[0,1]},$$

therefore there exists $z_2 \in I_2$ such that $z_2(t) = z_1(t)$, $t \in [0,1]$. However

$$I_2 = I_3|_{[0,2]},$$

therefore there exists $z_3 \in I_3$ such that $z_3(t) = z_2(t)$, $t \in [0,2]$. Continuing inductively, for any $k \in \mathbf{N}$, we can find $z_{k+1} \in I_{k+1}$ such that $z_{k+1}|_{[0,k]} = z_k$. Define a function $y \in C[0,\infty)$ by

$$y(t) = z_k(t), \quad t \in [0,k] \text{ for each } k \in \mathbf{N}.$$

Note that $y|_{[0,k]} \in I_k$ for each $k \in \mathbf{N}$ and hence $y \in S$. Therefore $S \neq \emptyset$.

To show that S is compact, let $\{y_n\}_{n=1}^\infty$ be a sequence in S. Now

$$\{y_n|_{[0,1]}\}_{n=1}^\infty \subseteq I_1,$$

and since I_1 is compact, there exists a subsequence N_1 of $\mathbf{N}$ and a function $z_1 \in I_1$ such that

$$y_n|_{[0,1]} \to z_1 \text{ uniformly on } [0,1] \text{ as } n \to \infty \text{ in } N_1.$$

As in a previous argument, one can proceed inductively so that for any $k \in \mathbf{N}$, there exists a subsequence

$$N_{k+1} \subseteq \{k+1, \ldots\} \text{ with } N_1 \supseteq N_2 \supseteq \ldots \supseteq N_{k+1},$$

and a function $z_{k+1} \in I_{k+1}$ such that

$$y_n|_{[0,k+1]} \to z_{k+1} \text{ uniformly on } [0,k+1] \text{ as } n \to \infty \text{ in } N_{k+1},$$

and

$$z_{k+1}(t) = z_k(t), \quad t \in [0,k].$$

Defining $y : [0,\infty) \to \mathbf{R}$ by

$$y(t) = z_k(t), \quad t \in [0,k] \text{ for each } k \in \mathbf{N}$$

we see that $y \in C[0,\infty)$ and $y|_{[0,k]} \in I_k$ for each $k \in \mathbf{N}$. In conclusion we have shown that $\{y_n\}_{n=1}^{\infty}$ contains a subsequence which converges in $C[0,\infty)$ to $y \in S$. Hence S is compact.

Finally we show that S is connected. Suppose not. Then we can decompose S into the union of two nonempty, disjoint, compact sets, S_A and S_B, that is,

$$S = S_A \cup S_B.$$

Fix $k \in \mathbf{N}$ and consider the restriction of S_A and S_B to $[0,k]$. Then

$$S_A|_{[0,k]} \cup S_B|_{[0,k]} = S|_{[0,k]},$$

but since $S|_{[0,k]} = I_k$ which is connected, we see that $S_A|_{[0,k]}$ and $S_B|_{[0,k]}$ (which are nonempty and compact) cannot be disjoint. Let $y_k \in S_A|_{[0,k]} \cap S_B|_{[0,k]}$. Consequently there exists $y_{k,A} \in S_A$ and $y_{k,B} \in S_B$ such that

$$y_{k,A}(t) = y_{k,B}(t) = y_k(t), \quad t \in [0,k]$$

and

$$y_{k,A} \neq y_{k,B}$$

since S_A and S_B are assumed to be disjoint. Since this argument holds for each $k \in \mathbf{N}$, we obtain sequences

$$\{y_{n,A}\}_{n=1}^{\infty} \subseteq S_A \text{ and } \{y_{n,B}\}_{n=1}^{\infty} \subseteq S_B$$

where for each $n \in \mathbf{N}$,

$$y_{n,A}(t) = y_{n,B}(t), \quad t \in [0, n].$$

The compactedness of S_A and S_B means that there exists subsequences $\mathbf{N}_A$, $\mathbf{N}_B \subseteq \mathbf{N}$ and functions $y_A \in S_A$, $y_B \in S_B$, $y_A \neq y_B$ such that

$$y_{n,A} \to y_A \text{ and } y_{n,B} \to y_B$$

uniformly on each compact subinterval of $[0, \infty)$ as $n \to \infty$ in $\mathbf{N}_A$ and $\mathbf{N}_B$ respectively. Observe however that for any fixed $k \in \mathbf{N}$,

$$y_{n,A}(t) = y_{n,B}(t), \quad t \in [0, k], \text{ for all } n \in \{k, k+1, \ldots\}.$$

Consequently

$$y_A(t) = y_B(t), \quad t \in [0, k]$$

and since this argument holds for all $k \in \mathbf{N}$ we have that $y_A = y_B$ – a contradiction. Therefore S must be connected. □

Looking at the hypotheses of Theorem 6.3.2, it is clear that condition (6.3.4) requires further discussion. However we have discussed this in Section 6.2, and we restate the result here for the convenience of the reader.

Theorem 6.3.3. *Let $n, M_n > 0$ and $X_n = \{x \in C[0,n] : |x|_{0,n} = \sup_{t \in [0,n]} |x(t)| \leq M_n\}$. Suppose that*

(6.3.8) $\quad V : C[0, n] \to C[0, n]$ *is an abstract Volterra operator,*

(6.3.9) $\quad V : X_n \to C[0, n]$ *is a continuous, compact operator,*

(6.3.10) *there exists $y_0 \in \mathbf{R}$ such that $Vy(0) = y_0$ for all $y \in C[0, n]$*

and

$$(6.3.11) \qquad \begin{cases} \textit{if } y \in C[0,n] \textit{ satisfies } y(t) = Vy(t), \ t \in [0,n], \\ \textit{then } y \in X_n; \ \textit{ for } m \in \{1,2,...\} \ \textit{ if } y \in C[0,n] \\ \textit{satisfies } y(t) = V_m y(t), \ t \in [0,n], \\ \textit{then } y \in \textit{int}\,(X_n) \end{cases}$$

hold; here $V_m y(t) = V y(r_m(t))$ where

$$r_m(t) = \begin{cases} 0, & \textit{if } t \in [0, \frac{n}{m}] \\ t - \frac{n}{m}, & \textit{if } t \in (\frac{n}{m}, n] \,. \end{cases}$$

Then Fix(V) is an R_δ set, therefore in particular, is nonempty, compact and connected in $C[0, n]$.

Combining Theorem 6.3.2 and Theorem 6.3.3, we now can state the desired result for (6.3.1).

Theorem 6.3.4. *Suppose that* (6.3.3),

(6.3.12) *there exists* $y_0 \in \mathbf{R}$ *such that* $Vy(0) = y_0$ *for all* $y \in C[0, \infty)$,

and for each $n \in \mathbf{N}$, (6.3.8), (6.3.9) *and* (6.3.11) *hold. Then the solution set of* (6.3.1) *is nonempty, compact and connected in* $C[0, \infty)$.

PROOF: By Theorem 6.3.3, (6.3.8), (6.3.9), (6.3.11) and (6.3.12) imply that condition (6.3.4) is satisfied. Since we have assumed (6.3.3) is true, the result follows directly from Theorem 6.3.2. □

We now consider the abstract Volterra functional-differential equation

(6.3.13) $$\begin{cases} y'(t) = \tilde{V}y(t), & \text{a.e. } t \in [0, \infty) \\ y(0) = y_0, \end{cases}$$

where

(6.3.14) $$\begin{cases} \tilde{V} : C[0, \infty) \to L^p_{loc}[0, \infty) \ (1 \le p \le \infty) \\ \text{is an abstract Volterra operator} \end{cases}$$

holds. By examining (6.3.13)'s counterpart on the compact interval, namely

$(6.3.15)_n$ $$\begin{cases} y'(t) = \tilde{V}y(t), & \text{a.e. } t \in [0, n] \\ y(0) = y_0, \end{cases}$$

and imposing certain conditions on the solution set of $(6.3.15)_n$, we have the following result concerning the solution set of (6.3.13).

Theorem 6.3.5. *Suppose that* (6.3.14) *and*

(6.3.16) $$\begin{cases} \textit{for each } n \in \mathbf{N}, \textit{ the solution set of } (6.3.15)_n \textit{ is} \\ \textit{nonempty, compact and connected in } W^{1,p}[0, n] \end{cases}$$

hold. Then the solution set of (6.3.13) *is nonempty, compact and connected in* $W^{1,p}_{loc}[0, \infty)$.

Notation. We denote the solution sets of (6.3.13) and $(6.3.15)_n$ by

$$S(\tilde{V}) := \{y \in W^{1,p}_{loc}[0, \infty) : y \text{ solves } (6.3.13)\}$$

and

$$S_n(\tilde{V}) := \{y \in W^{1,p}[0,n] : y \text{ solves } (6.3.15)_n\}$$

respectively.

PROOF OF THEOREM 6.3.5: The proof follows using an argument that is almost identical to the one used in the proof of Theorem 6.3.2. In fact, in the proof of Theorem 6.3.2, replace (6.3.4) with (6.3.16), S and S_l with $S(\tilde{V})$ and $S_l(\tilde{V})$ respectively, and $V : C[0,\infty) \to C[0,\infty)$ with $\tilde{V} : C[0,\infty) \to L^p_{loc}[0,\infty)$. Elsewhere replace $C[0,l]$ and $C[0,\infty)$ with $W^{1,p}[0,l]$ and $W^{1,p}_{loc}[0,\infty)$ respectively, and where we have uniform convergence, consider it now to be convergence in $W^{1,p}[0,l]$ or $W^{1,p}_{loc}[0,\infty)$, as appropriate. (Note above that $l \in \{1, k, k+1, k+2, m, n, n+1\}$.) Then we have the proof of Theorem 6.3.5. □

In fact, imposing a continuity condition on $\tilde{V} : C[0,\infty) \to L^p_{loc}[0,\infty)$, we can restate Theorem 6.3.5 with a slightly weaker requirement on the solution set of $(6.3.15)_n$.

Theorem 6.3.6. *Suppose that*

$$\left\{ \begin{array}{l} \tilde{V} : C[0,\infty) \to L^p_{loc}[0,\infty) \ (1 \le p \le \infty) \textit{ is a} \\ \textit{continuous, abstract Volterra operator} \end{array} \right. \tag{6.3.17}$$

and

$$\left\{ \begin{array}{l} \textit{for each } n \in \mathbf{N}, \textit{ the solution set of } (6.3.15)_n \textit{ is} \\ \textit{nonempty, compact and connected in } C[0,n] \end{array} \right. \tag{6.3.18}$$

hold. Then the solution set of (6.3.13) *is nonempty, compact and connected in* $W^{1,p}_{loc}[0,\infty)$.

PROOF: From (6.3.18) we know that $S_n(\tilde{V})$ is a continuum in $C[0,n]$ for each $n \in \mathbf{N}$. If we show that for each $n \in \mathbf{N}$, $S_n(\tilde{V})$ is nonempty, compact and connected in $W^{1,p}[0,n]$, then the result will follow from Theorem 6.3.5.

Fix $n \in \mathbf{N}$. By (6.3.17), $S_n(\tilde{V})$ is trivially nonempty in $W^{1,p}[0,n]$. To show that $S_n(\tilde{V})$ is compact in $W^{1,p}[0,n]$, let $\{y_m\}_{m=1}^{\infty} \subseteq S_n(\tilde{V})$. The compactness of $S_n(\tilde{V})$ in $C[0,n]$ guarantees the existence of a subsequence $\tilde{N}$ of $\mathbf{N}$, and a function $y \in S_n(\tilde{V})$, such that

$$y_m \to y \text{ in } C[0,n], \text{ as } m \to \infty \text{ in } \tilde{N}.$$

However the continuity of $\tilde{V} : C[0,n] \to L^p[0,n]$ (a consequence of (6.3.17)) further implies that

$$y_m' = \tilde{V} y_m \to \tilde{V} y \text{ in } L^p[0,n], \text{ as } m \to \infty \text{ in } \tilde{N}.$$

In summary, we see that

$$y_m \to y \text{ in } W^{1,p}[0,n], \text{ as } m \to \infty \text{ in } \tilde{N},$$

and as a result, $S_n(\tilde{V})$ is compact in $W^{1,p}[0,n]$.

We argue by contradiction to illustrate that $S_n(\tilde{V})$ is a connected subset of $W^{1,p}[0,n]$. Suppose

$$S_n(\tilde{V}) = A \cup B,$$

where A and B are nonempty, closed, disjoint subsets of $W^{1,p}[0,n]$. A and B are clearly nonempty, disjoint subsets of $C[0,n]$, therefore if we show that A and B are also closed subsets of $C[0,n]$, then we have a contradiction since $S_n(\tilde{V})$ is connected in $C[0,n]$. Let

$$\{y_m\}_{m=1}^{\infty} \subseteq A \text{ with } y_m \to y \text{ in } C[0,n].$$

We will show that $y \in A$. By (6.3.17) we see that $y_m \in W^{1,p}[0,n]$, and as above the continuity of $\tilde{V} : C[0,n] \to L^p[0,n]$ yields

$$y_m' = \tilde{V} y_m \to \tilde{V} y \text{ in } L^p[0,n] \text{ as } n \to \infty.$$

Consequently, $y_m \to y$ in $W^{1,p}[0,n]$, and since A is a closed subset of $W^{1,p}[0,n]$, we have that $y \in A$. Thus A, and similarly B, are closed in $C[0,n]$. We have therefore shown that $S_n(\tilde{V})$ is a connected subset of $W^{1,p}[0,n]$.

Since the above arguments are true for any $n \in \mathbf{N}$, the result follows from Theorem 6.3.5. □

Observe that (6.3.14) (and (6.3.17)) enables us to integrate (6.3.13) and $(6.3.15)_n$ from 0 to t to obtain

$$\text{(6.3.19)} \qquad y(t) = Vy(t) := y_0 + \int_0^t \tilde{V} y(s)\, ds, \quad t \in [0,\infty)$$

and

$$(6.3.20)_n \qquad y(t) = Vy(t) := y_0 + \int_0^t \tilde{V} y(s)\, ds, \quad t \in [0,n].$$

respectively. If one was content with having the solution set of (6.3.13) be a continuum in $C[0,\infty)$, then using the above representations for (6.3.13) and $(6.3.15)_n$, the following theorem is an easy consequence of Theorem 6.3.2.

Theorem 6.3.7. *Suppose that* (6.3.14) *and* (6.3.18) *hold. Then the solution set of* (6.3.13) *is nonempty, compact and connected in* $C[0,\infty)$.

Notation. Denote the solution sets of (6.3.19) and $(6.3.20)_n$ by

$$S(V) := \{y \in C[0,\infty) : y \text{ solves } (6.3.19)\}$$

and

$$S_n(V) := \{y \in C[0,n] : y \text{ solves } (6.3.20)_n\}$$

respectively.

PROOF OF THEOREM 6.3.7: Since $S_n(\tilde{V}) = S_n(V)$, we have by (6.3.18), that $S_n(V)$ is a continuum in $C[0,n]$ for each $n \in \mathbf{N}$. In addition, (6.3.14) implies that $V : C[0,\infty) \to C[0,\infty)$ as defined in (9.3.19) is an abstract Volterra operator. Therefore Theorem 6.3.2 guarantees that $S(V)$ is a continuum in $C[0,\infty)$. Since $S(V) = S(\tilde{V})$, we have the desired result for (6.3.13). □

It is necessary now to examine what conditions one requires on $\tilde{V} : C[0,n] \to L^p[0,n]$ to ensure that (6.3.16) and/or (6.3.18) hold(s). We first concentrate on the slightly weaker condition (6.3.18). If we hope to make use of Theorem 6.3.3, then we must examine the representation of $(6.3.15)_n$ given by $(6.3.20)_n$. Showing that $S_n(V)$ is a continuum in $C[0,n]$ is equivalent to showing that $S_n(\tilde{V})$ is a continuum in $C[0,n]$. Hence we are interested in the conditions that $\tilde{V} : C[0,n] \to L^p[0,n]$ must satisfy in order that $V : C[0,n] \to C[0,n]$ satisfies the hypotheses of Theorem 6.3.3. These are detailed in the following result.

Theorem 6.3.8. *Let* $n, M_n > 0$, $1 \le p \le \infty$, *and* $X_n = \{x \in C[0,n] : |x|_{0,n} \le M_n\}$. *Suppose that*

(6.3.21) $\quad \tilde{V} : C[0,n] \to L^p[0,n]$ *is an abstract Volterra operator,*

(6.3.22) $\quad \tilde{V} : X_n \to L^p[0,n]$ *is a continuous operator,*

(6.3.23) $\quad \left\{ \begin{array}{l} \textit{there exists } \mu_{M_n} \in L^p[0,n] \textit{ such that for any} \\ y \in X_n \textit{ we have } |\tilde{V}y(t)| \le \mu_{M_n}(t), \ \textit{a.e. } t \in [0,n] \end{array} \right.$

and

(6.3.24) $\quad \left\{ \begin{array}{l} \textit{if } y \in C[0,n] \textit{ satisfies } (6.3.20)_n, \textit{ then} \\ y \in X_n; \ \textit{for } m \in \{1,2,...\} \textit{ if } y \in C[0,n] \\ \textit{satisfies } y(t) = V_m y(t), \ t \in [0,n], \\ \textit{then } y \in \textit{int}\,(X_n) \end{array} \right.$

hold; here $V_m\, y(t) = V\, y(r_m(t))$ *where*

$$r_m(t) = \begin{cases} 0, & \text{if } t \in [0, \frac{n}{m}] \\ t - \frac{n}{m}, & \text{if } t \in (\frac{n}{m}, n] \,. \end{cases}$$

Then the solution set of $(6.3.20)_n$ *is nonempty, compact and connected in* $C[0,n]$.

PROOF: One can show [137,Chapter 2] that $V : X_n \to C[0,n]$ as defined in $(6.3.20)_n$ is a continuous, compact, abstract Volterra operator, and consequently V satisfies conditions (6.3.8) and (6.3.9) of Theorem 6.3.3. By definition of $V : C[0,n] \to C[0,n]$, (6.3.10) is true and (6.3.24) above ensures that (6.3.11) holds. Therefore from Theorem 6.3.3 it follows that $S_n(V)$ is a continuum in $C[0,n]$. □

Remark. In [137, Chapter 2] it is assumed that $\tilde{V} : C[0,n] \to L^p[0,n]$ is a continuous operator that satisfies

$$(6.3.25) \qquad \begin{cases} \text{for any constant } r > 0, \text{ there exists } \mu_r \in L^p[0,n] \\ \text{such that for any } \ y \in C[0,n] \text{ with } |y|_{0,n} \leq r, \\ \text{we have } \ |\tilde{V}y(t)| \leq \mu_r(t), \text{ a.e. } t \in [0,n], \end{cases}$$

and it is shown that this implies that $V : C[0,n] \to C[0,n]$ as defined in $(6.3.20)_n$ is a continuous and *completely continuous* operator. Recall that an operator $T : X \to Y$, where X and Y are normed spaces, is *compact* if $T(X)$ is relatively compact in Y, whereas $T : X \to Y$ is *completely continuous* if for every bounded set $M \subseteq X$, $T(M)$ is relatively compact in Y. Therefore if $V : C[0,n] \to C[0,n]$ is a completely continuous operator, then $V : X_n \to C[0,n]$ is certainly compact, as was required in the above proof.

Remark. Condition (6.3.22) in Theorem 6.3.8 can be replaced by

$$(6.3.26) \qquad \begin{cases} \text{for any sequence } \{y_k\}_{k=1}^{\infty} \subseteq X_n \text{ and } y \in X_n \\ \text{such that } \ y_k \to y \ \text{ uniformly on } [0,n], \text{ we} \\ \text{have that } \ \tilde{V}y_k(t) \to \tilde{V}y(t), \text{ a.e. on } [0,n]. \end{cases}$$

In fact (6.3.23) and (6.3.26) along with the Lebesgue Dominated Convergence Theorem, imply that (6.3.22) is true.

Combining Theorem 6.3.6 and Theorem 6.3.8, we have the following existence principle for (6.3.13).

Theorem 6.3.9. *Suppose that* (6.3.17) *is true and assume that for all* $n \in \mathbf{N}$, (6.3.23) *and* (6.3.24) *hold. Then the solution set of* (6.3.13) *is nonempty, compact and connected in* $W^{1,p}_{loc}[0,\infty)$.

PROOF: Since $\tilde{V} : C[0,\infty) \to L^p_{loc}[0,\infty)$ is a continuous, Volterra operator we have that for any $n \in \mathbf{N}$, $\tilde{V} : C[0,n] \to L^p[0,n]$ is also a continuous, Volterra operator. Hence for any $n \in \mathbf{N}$, the hypotheses of Theorem 6.3.8 are satisfied, thus implying that (6.3.18) in Theorem 6.3.6 is true. Since we have assumed that (6.3.17) holds, the result follows immediately from Theorem 6.3.6. □

Finally we remark that in the proof of Theorem 6.3.6, the continuity of the operator $\tilde{V} : C[0,\infty) \to L^p_{loc}[0,\infty)$ ensured that uniform convergence in $S_n(\tilde{V})$, implied convergence with respect to the $W^{1,p}[0,n]$ norm in $S_n(\tilde{V})$. Exploiting this fact we see that in fact the hypotheses of Theorem 6.3.8 ensure that (6.3.16) holds.

Theorem 6.3.10. *Suppose that the hypotheses of Theorem 6.3.8 hold. Then the solution set of* $(6.3.15)_n$ *is nonempty, compact and connected in* $W^{1,p}[0,n]$.

PROOF: By Theorem 6.3.8 we have that $S_n(V) = S_n(\tilde{V})$ is a continuum in $C[0,n]$. The exact argument used in Theorem 6.3.6 now gives us the desired result for $(6.3.15)_n$. □

We can now restate the existence principle for (6.3.13) as follows.

Theorem 6.3.11. *Suppose that* (6.3.14) *and the hypotheses of Theorem 6.3.8 hold. Then the solution set of* (6.3.13) *is nonempty, compact and connected in* $W^{1,p}_{loc}[0,\infty)$.

PROOF: The proof follows directly from Theorem 6.3.10 and Theorem 6.3.5. □

6.4. Applications.

Consider the nonlinear Volterra integral equation

$$(6.4.1) \qquad y(t) = h(t) + \int_0^t k(t,s)f(s,y(s))\,ds, \quad t \in [0,\infty)$$

defined on the half-line $[0,\infty)$. First we recall the definition of a locally L^p-Carathéodory function:

Definition 6.4.1. Let I be an interval in $\mathbf{R}$. A function $g : I \times \mathbf{R} \to \mathbf{R}$ is an L^p-*Carathéodory function* if the following conditions hold:

(i) the map $t \mapsto g(t,y)$ is measurable for all $y \in \mathbf{R}$,
(ii) the map $y \mapsto g(t,y)$ is continuous for almost all $t \in I$,
(iii) for any $r > 0$ there exists $\mu_r \in L^p(I)$ such that $|y| \leq r$ implies that $|g(t,y)| \leq \mu_r(t)$ for almost all $t \in I$.
A function $g : I \times \mathbf{R} \to \mathbf{R}$ is a *locally L^p-Carathéodory function,* if the above conditions hold when g is restricted to $J \times \mathbf{R}$, with J any compact subinterval of I.

Let $1 \leq p_1 \leq \infty$ and p_2 be such that $1/p_1 + 1/p_2 = 1$. Suppose that

$$h \in C[0,\infty), \tag{6.4.2}$$

$$f : [0,\infty) \times \mathbf{R} \to \mathbf{R} \text{ is a locally } L^{p_2}\text{-Carathéodory function,} \tag{6.4.3}$$

$$k_t(s) = k(t,s) \in L^{p_1}[0,t], \text{ for each } t \in [0,n], \text{ and each } n \in \mathbf{N} \tag{6.4.4}$$

and

$$\begin{cases} \text{for any } t, t' \in [0,\infty), \\ \int_0^{t^\star} |k_t(s) - k_{t'}(s)|^{p_1}\, ds \to 0 \text{ as } t \to t', \\ \text{where } t^\star = \min\{t,t'\} \end{cases} \tag{6.4.5}$$

hold, and define

$$Vy(t) := h(t) + \int_0^t k(t,s) f(s,y(s))\, ds, \quad t \in [0,\infty). \tag{6.4.6}$$

To see that $V : C[0,\infty) \to C[0,\infty)$, choose t_1, $t_2 \in [0,\infty)$ with $t_1 < t_2$, and let $y \in C[0,\infty)$. Then

$$\begin{aligned} |Vy(t_2) - Vy(t_1)| &\leq |h(t_2) - h(t_1)| + \int_0^{t_1} |k_{t_2}(s) - k_{t_1}(s)||f(s,y(s))|\, ds \\ &+ \int_{t_1}^{t_2} |k_{t_2}(s)||f(s,y(s))|\, ds. \end{aligned}$$

Since $f : [0,\infty) \times \mathbf{R} \to \mathbf{R}$ satisfies (6.4.3), there exists $\mu_r \in L^{p_2}[0,t_2]$ such that $|f(t,y(t))| \leq \mu_r(t)$, a.e. $t \in [0,t_2]$ where $r > 0$ is such that $|y|_{0,t_2} \leq r$. Hence

$$\begin{aligned} |Vy(t_2) - Vy(t_1)| &\leq |h(t_2) - h(t_1)| \\ &+ \left\{\int_0^{t_1} |k_{t_2}(s) - k_{t_1}(s)|^{p_1}\, ds\right\}^{\frac{1}{p_1}} ||\mu_r||_{p_2} \\ &+ \left\{\int_{t_1}^{t_2} |k_{t_2}(s)|^{p_1}\, ds\right\}^{\frac{1}{p_1}} ||\mu_r||_{p_2} \\ &\to 0 \quad \text{as} \quad t_1 \to t_2, \end{aligned}$$

illustrating that $V : C[0,\infty) \to C[0,\infty)$ is well-defined. Therefore $V : C[0,\infty) \to C[0,\infty)$ is a Volterra operator for which

$$Vy(0) = h(0) \text{ for all } y \in C[0,\infty),$$

and thus satisfies (6.3.3) and (6.3.12) of Theorem 6.3.4. Also, on restricting V to operate on $C[0,n]$, it can be shown [137,Chapter 4] that

$$\text{(6.4.7)} \qquad \left\{ \begin{array}{l} V : C[0,n] \to C[0,n] \text{ is a continuous and} \\ \text{completely continuous Volterra operator} \end{array} \right.$$

is true.

In addition to the hypotheses listed above, suppose we insist that

$$\text{(6.4.8)} \qquad \left\{ \begin{array}{l} \text{there exists } \alpha \in L^1_{loc}[0,\infty) \text{ and } \psi : [0,\infty) \to (0,\infty), \\ \text{a nondecreasing, continuous function such that} \\ |k(t,s)f(s,y(s))| \le \alpha(s)\psi(|y|), \text{ for a.e. } t \in [0,\infty), \\ \text{a.e. } s \in [0,t] \text{ and for all } y \in C[0,\infty) \end{array} \right.$$

and

$$\text{(6.4.9)} \qquad \int_0^n \alpha(s)\,ds < \int_{|h|_{0,n}}^{\infty} \frac{ds}{\psi(s)}, \text{ for each } n \in \mathbf{N}$$

hold. Let

$$\tilde{M}_n = I_n^{-1}\left(\int_0^n \alpha(s)\,ds\right) \text{ where } I_n(z) = \int_{|h|_{0,n}}^{z} \frac{ds}{\psi(s)}$$

and

$$M_n = \tilde{M}_n + 1.$$

Then if $y \in C[0,n]$ is a solution of

$$y(t) = h(t) + \int_0^t k(t,s)f(s,y(s))\,ds, \quad t \in [0,n],$$

we have that

$$|y(t)| \le |h|_{0,n} + \int_0^t \alpha(s)\psi(|y(s)|)\,ds \equiv w(t), \quad t \in [0,n].$$

Now

$$w'(t) = \alpha(t)\psi(|y(t)|) \le \alpha(t)\psi(w(t)), \text{ a.e. } t \in [0,n],$$

and therefore

$$\int_{|h|_{0,n}}^{w(t)} \frac{ds}{\psi(s)} = \int_0^t \frac{w'(s)}{\psi(w(s))}\,ds \le \int_0^t \alpha(s)\,ds \le \int_0^n \alpha(s)\,ds, \quad t \in [0,n].$$

Thus $w(t) \le \tilde{M}_n$, for any $t \in [0,n]$ and therefore

$$|y(t)| \le \tilde{M}_n < M_n, \quad \text{for all } t \in [0,n].$$

A similar argument shows if $m \in \{1,2,...\}$ is fixed and $y \in C[0,n]$ satisfies

$$y(t) = h(r_n(t)) + \int_0^{r_n(t)} k(r_n(t),s) f(s,y(s))\,ds,$$

then $|y(t)| < M_n$ for $t \in [0,n]$. Letting $X_n = \{y \in C[0,n] : |y|_{0,n} \le M_n\}$, we have just shown above that (6.3.11) of Theorem 6.3.4 is satisfied. Finally from (6.4.7), it is clear that $V : X_n \to C[0,n]$ fulfills conditions (6.3.8) and (6.3.9). In summary we have the following application of Theorem 6.3.4:

Theorem 6.4.1. *Let* $1 \le p_1 \le \infty$ *and suppose* p_2 *is such that* $1/p_1 + 1/p_2 = 1$. *Assume that* (6.4.2) – (6.4.5), (6.4.8) *and* (6.4.9) *hold. Then the integral equation* (6.4.1) *has a nonempty, compact and connected solution set in* $C[0,\infty)$.

To complete the section we consider two applications of Theorem 6.3.11. Two special cases of (6.3.13) occur when the abstract Volterra operator $\tilde{V}$ is given by

$$\tilde{V}y(t) = Fy(t) := f(t,y(t)), \ \text{a.e. } t \in [0,\infty) \tag{6.4.10}$$

– a Niemytzki operator, and

$$\tilde{V}y(t) = Ky(t) := h(t) + \int_0^t k(t,s) g(s,y(s))\,ds, \ \text{a.e. } t \in [0,\infty) \tag{6.4.11}$$

– a Volterra integral operator. These reduce (6.3.13) to the Cauchy problem on the half-line,

$$\begin{cases} y'(t) = Fy(t) = f(t,y(t)), \ \text{a.e. } t \in [0,\infty) \\ y(0) = y_0 \end{cases} \tag{6.4.12}$$

and the integrodifferential equation on the half-line,

$$\begin{cases} y'(t) = Ky(t) \\ \qquad = h(t) + \int_0^t k(t,s) g(s,y(s))\,ds, \ \text{a.e. } t \in [0,\infty) \\ y(0) = y_0, \end{cases} \tag{6.4.13}$$

respectively.

It can be shown that if

$$(6.4.14)\qquad \begin{cases} f : [0,\infty)\times\mathbf{R}\to\mathbf{R} \text{ is a locally} \\ L^p\text{-Carathéodory function } (1\le p\le\infty), \end{cases}$$

holds, then $F : C[0,\infty) \to L^p_{loc}[0,\infty)$ is a Volterra operator and satisfies (6.3.21) and (6.3.22) for each $n \in \mathbf{N}$. (Here the operator F is as defined in (6.4.10)).

For some $1 \le p \le \infty$ and $1/p + 1/q = 1$, suppose that

$$(6.4.15)\qquad h \in L^1_{loc}[0,\infty),$$

$$(6.4.16)\qquad \begin{cases} \text{the map } (t,s)\mapsto k(t,s) \text{ is measurable, and } k(t,s) \text{ is finite} \\ \text{for almost all } (t,s)\in[0,n]\times[0,t], \text{ and each } n\in\mathbf{N}, \end{cases}$$

$$(6.4.17)\qquad k_t(s) = k(t,s) \in L^p[0,n] \text{ for a.e. } t\in[0,n], \text{ for each } n\in\mathbf{N},$$

$$(6.4.18)\qquad \sup_{t\in[0,n]} \left\{ \int_0^t |k_t(s)|^p\, ds \right\}^{\frac{1}{p}} < \infty, \text{ for each } n\in\mathbf{N}$$

and

$$(6.4.19)\qquad g : [0,\infty)\times\mathbf{R}\to\mathbf{R} \text{ is a locally } L^q\text{-Carathéodory function}$$

are true, then one can check (see [137, Chapter 2]) that $K : C[0,\infty) \to L^1_{loc}[0,\infty)$ is a Volterra operator and satisfies (6.3.21) and (6.3.22) for each $n \in \mathbf{N}$. (Here the operator K is as defined in (6.4.11)).

With appropriate growth conditions on f and g, one can show using Theorem 6.3.11 and the ideas outlined above for (6.4.1) that the solution sets of (6.4.12) and (6.4.13) respectively are nonempty, compact and connected in $W^{1,p}_{loc}[0,\infty)$. We state the theorems but omit the proofs as the details are similar to those described above and also to those found in [137, Chapter 3].

Theorem 6.4.2. *Let $1 \le p \le \infty$ and suppose that* (6.4.14),

$$(6.4.20)\qquad \begin{cases} \textit{there exists } \alpha\in L^1_{loc}[0,\infty) \textit{ and } \psi:[0,\infty)\to[0,\infty) \\ \textit{such that } \ \psi:(0,\infty)\to(0,\infty), \textit{ is a Borel} \\ \textit{measurable function such that for a.e. } t\in[0,\infty) \\ \textit{and all } y\in\mathbf{R}, \textit{ we have } |f(t,y)|\le\alpha(t)\psi(|y|) \end{cases}$$

and

$$\int_0^n \alpha(s)\, ds < \int_{|y_0|}^{\infty} \frac{ds}{\psi(s)}\, ds \text{ for each } n \in \mathbf{N} \tag{6.4.21}$$

hold. Then the solution set of (6.4.12) *is nonempty, compact and connected in* $W^{1,p}_{loc}[0,\infty)$.

Theorem 6.4.3. *Let* $1 \le p \le \infty$ *and* q *satisfy* $1/p + 1/q = 1$. *Suppose that* (6.4.15) – (6.4.19), (6.4.21) *and*

$$\begin{cases} \text{there exists } \alpha \in L^1_{loc}[0,\infty) \text{ and } \psi : [0,\infty) \to [0,\infty) \\ \text{such that } \psi : (0,\infty) \to (0,\infty), \text{ is a Borel} \\ \text{measurable function such that for a.e. } t \in [0,\infty), \\ \text{a.e } s \in [0,t], \text{ and all } y \in \mathbf{R},\ |k(t,s)f(t,y)| \le \alpha(t)\psi(|y|) \end{cases} \tag{6.4.22}$$

hold. Then the solution set of (6.4.13) *is nonempty, compact and connected in* $W^{1,p}_{loc}[0,\infty)$.

We finish with a simple example.

Example. Consider the initial-value problem

$$\begin{cases} y'(t) = \tilde{V}y(t) = y^{\frac{1}{n}}(t), \text{ a.e. } t \in [0,\infty), \\ y(0) = 0, \end{cases} \tag{6.4.23}$$

where $1 < n \in \mathbf{N}$ and n is odd.

Clearly,

$$f(t,y) = y^{\frac{1}{n}}$$

satisfies the conditions of Theorem 6.4.2, and in particular, (6.4.20) and (6.4.21) are true with $\alpha(t) = 1$ and $\psi(y) = y^{\frac{1}{n}}$, since

$$n = \int_0^n 1\, ds < \int_0^\infty \frac{ds}{\psi(s)} = \int_0^\infty s^{-\frac{1}{n}}\, ds = \infty, \text{ for each } n \in \mathbf{N}.$$

Therefore we have that the solution set of (6.4.23) is a continuum in $W^{1,p}_{loc}[0,\infty)$.

Note that $y = 0$ is a solution of the initial-value problem, and for any $\tilde{t} \in [0,\infty)$, it is easy to check that the following are also two solutions of (6.4.23) :

$$y_{\tilde{t}}(t) = \begin{cases} 0 & ,0 \le t \le \tilde{t} \\ \pm\left[\left(\dfrac{n-1}{n}\right)(t-\tilde{t})\right]^{\frac{n}{n-1}} & ,\tilde{t} \le t < \infty. \end{cases}$$

6.5. Fixed points for operators on $C([0,\infty),\mathbf{R}^M)$.

In Sections 6.5 and 6.6, $E = E([0,\infty),\mathbf{R}^M)$ will be a space of functions defined on the interval $[0,\infty)$ in $\mathbf{R}$, with values in $\mathbf{R}^M$. Assume that E is a Fréchet space with the topology given by the complete family of seminorms $\{p_m\}_{m\geq 1}$, or, equivalently, by the distance d defined by

$$d(x,y) = \sum_{m=1}^{\infty} \frac{1}{2^m} \frac{p_m(x-y)}{1+p_m(x-y)},$$

for $x, y \in E$.

Let X be a nonempty closed subset of E. We consider an operator $F: E \to E$.

Define the sequence of operators $\{F_n\}_n$, $F_n : E \to E$, as follows:

$$F_n(x)(t) = F(x)(r_n(t)), \text{ for } x \in E \text{ and } t \geq 0, \tag{6.5.1}$$

where

$$r_n(t) = \begin{cases} 0, \text{ if } t \in [0,1/n]; \\ t - \frac{1}{n}, \text{ if } t > 1/n. \end{cases} \tag{6.5.2}$$

The main theorems of Sections 6.5 and 6.6 will be proved using the following result [91].

Lemma 6.5.1. *Let X be a closed set in a Fréchet space (E,d), and $F: X \to E$ a continuous compact operator. Assume there exists a sequence $\{U_n\}_n$ of closed convex sets in E such that*

$$\forall n \in \mathbf{N},\ 0 \in U_n; \tag{6.5.3}$$

$$\lim_{n\to\infty} diam(U_n) = 0, \tag{6.5.4}$$

and there exists a sequence $\{F_n\}_n$ of operators $F_n : X \to E$, such that

$$\forall n \in \mathbf{N},\ \forall x \in X,\ F(x) - F_n(x) \in U_n; \tag{6.5.5}$$

(6.5.6) *$I - F_n$ is a homeomorphism of the set $(I-F_n)^{-1}(U_n)$ onto U_n.*

Then $\{x \in X : x = Fx\}$ is an R_δ set.

The rest of Section 6.5 considers the case when E is the space of all continuous functions on $[0,\infty)$, with the topology given by uniform convergence on compacts in $[0,\infty)$.

Theorem 6.5.2. *Assume that* $E = C([0,\infty), \mathbf{R}^M)$, *with*

$$p_m(x) = \sup\{|x(t)| : t \in [0,m]\},$$

and that $F : X \to E$ *is a continuous and compact operator. Also assume that the following conditions hold:*

(i) $\exists u_0 \in \mathbf{R}^M$ *such that* $F(x)(0) = u_0$, *for all* $x \in X$;

(ii) $\forall \epsilon > 0$, $\forall x, y \in X$, *if* $x(t) = y(t)$, $\forall t \in [0,\epsilon]$, *then* $F(x)(t) = F(y)(t)$, $\forall t \in [0,\epsilon]$ *(i.e. F is an abstract Volterra operator);*

(iii) if $y \in E$ *satisfies* $y = F(y)$, *then* $y \in X$;

(iv) $\exists \eta > 0$ *such that* $\forall n$, *if* $y \in E$ *satisfies* $y = F_n(y) + z$, *where* $p_m(z) \leq \eta, \forall m$, *then* $y \in X$.

Then $Fix(F)$ *is an* R_δ *set.*

PROOF: Consider the sequence $\{F_n\}_n$ defined by (6.5.1)-(6.5.2). We show that there exists a sequence $\{U_n\}_n$ of closed convex sets in E and there exists a subsequence $\{G_n\}_n$ of $\{F_n\}_n$ such that $\{U_n\}_n$ and $\{G_n\}_n$ satisfy conditions (6.5.3) – (6.5.6) in Lemma 6.5.1.

First, we show that $\forall n$, $I - F_n$ is a homeomorphism from X onto $(I - F_n)(X)$.

The continuity of $I - F_n$ follows immediately from the continuity of F and the fact that

$$\begin{aligned}
p_m(F_n(x) - F_n(y)) &= \sup\{|F_n(x)(t) - F_n(y)(t)|, t \in [0,m]\} \\
&= \sup\{|F(x)(t) - F(y)(t)|, t \in [0, m - 1/n]\} \\
&\leq \sup\{|F(x)(t) - F(y)(t)|, t \in [0,m]\} \\
&= p_m(F(x) - F(y)).
\end{aligned}$$

To see that $I - F_n$ is one-to-one, let $x, y \in X$ be such that $(I - F_n)(x) = (I - F_n)(y)$. Then, for $t \in [0, 1/n]$,

$$x(t) - y(t) = F(x)(r_n(t)) - F(y)(r_n(t)) = F(x)(0) - F(y)(0) = u_0 - u_0 = 0.$$

If $t \in [1/n, 2/n]$, then $t - \frac{1}{n} \in [0, 1/n]$ and

$$x(t) - y(t) = F(x)\left(t - \frac{1}{n}\right) - F(y)\left(t - \frac{1}{n}\right) = F(x)(s) - F(y)(s),$$

with $s \in [0, 1/n]$, therefore the difference is 0, because (ii) holds. Proceed by induction. Assume we know that $x(t) - y(t) = 0$ for $t \in [0, (k-1)/n]$, for some positive integer k. Since (ii) holds, this means that $F(x)(t)-F(y)(t) = 0$ for $t \in [0, (k-1)/n]$. Then, for $t \in [(k-1)/n, k/n]$,

$$x(t) - y(t) = F(x)\left(t - \frac{1}{n}\right) - F(y)\left(t - \frac{1}{n}\right) = F(x)(s) - F(y)(s),$$

with $s \in [0, (k-1)/n]$, so the difference is 0, and $I - F_n$ is one-to-one.

To see that $(I - F_n)^{-1}$ is continuous, let $\{x_j\}_j$ be a sequence in X. Let $x \in X$ be such that

$$\lim_{j\to\infty} ((x_j - F_n(x_j)) - (x - F_n(x))) = 0 \text{ in } E, \tag{6.5.7}$$

which implies that

$$\lim_{j\to\infty} ((x_j(t) - F_n(x_j)(t)) - (x(t) - F_n(x)(t))) = 0, \tag{6.5.8}$$

uniformly for t in every compact in $[0, \infty)$. We show that

$$\lim_{j\to\infty} (x_j - x) = 0 \text{ in } E.$$

For $t \in [0, 1/n]$, $F_n(x_j)(t) = F(x)(t) = u_0$, so

$$(x_j(t) - F_n(x_j)(t)) - (x(t) - F_n(x)(t)) = x_j(t) - x(t),$$

and since (6.5.8) holds uniformly for $t \in [0, 1/n]$, we have

$$\lim_{j\to\infty} \sup\{|x_j(t) - x(t)| : t \in [0, 1/n]\} = 0. \tag{6.5.9}$$

Now if $\overline{x}$ is the extension of x defined by

$$\overline{x} = \begin{cases} x(t) \text{ for } t \in [0, 1/n] \\ x\left(\frac{1}{n}\right) \text{ for } t > \frac{1}{n} \end{cases}$$

and $\overline{x}_j$ is the extension of x_j defined similarly, then, by (6.5.9) and the continuity of F, we have

$$\lim_{j\to\infty} (F(\overline{x}_j) - F(\overline{x})) = 0 \text{ in } E. \tag{6.5.10}$$

In particular, for $t \in [1/n, 2/n]$, (6.5.10) and the definition of F_n give

$$\begin{aligned} &\limsup_{j\to\infty}\{|F_n(x_j)(t) - F_n(x)(t)| : t \in [1/n, 2/n]\} \\ &\quad = \limsup_{j\to\infty}\{|F(\overline{x}_j)(t) - F(\overline{x})(t)| : t \in [0, 1/n]\} = 0. \end{aligned} \tag{6.5.11}$$

Now (6.5.8) and (6.5.11) imply

$$\lim_{j\to\infty}\sup\{|x_j(t)-x(t)| : t\in[1/n,2/n]\}=0. \tag{6.5.12}$$

By induction, assume that for some k,

$$\lim_{j\to\infty}\sup\{|x_j(t)-x(t)| : t\in[(k-1)/n,k/n]\}=0. \tag{6.5.13}$$

Repeating the extension argument above, this also implies

$$\lim_{j\to\infty}\sup\{|F_n(x_j)(t)-F_n(x)(t)| : t\in[k/n,(k+1)/n]\}=0. \tag{6.5.14}$$

Now (6.5.8) and (6.5.14) imply

$$\lim_{j\to\infty}\sup\{|x_j(t)-x(t)| : t\in[k/n,(k+1)/n]\}=0.$$

Thus, we showed by induction that

$$\lim_{j\to\infty}\sup\{|x_j(t)-x(t)| : t\in[(k-1)/n,k/n]\}=0, \forall k,$$

which implies that $\lim_{j\to\infty}p_m(x_j-x)=0, \forall m$, or $\lim_{j\to\infty}(x_j-x)=0$ in E.

To find the sequence $\{U_n\}_n$, for a given n, consider the equation

$$x=F_n(x)+z, \tag{6.5.15}$$

where $z\in E$ is such that $p_m(z)\le\eta, \forall m$. Let $x\in E$ be defined by induction as follows:

$$x(t)=u_0+z(t), \text{ for } t\in[0,1/n]. \tag{6.5.16}$$

This x is continuous on $[0,1/n]$ because z is continuous on $[0,\infty)$ (by "x is continuous on $[0,1/n]$" we mean that x is continuous on $(0,1/n)$, it is continuous at 0 from the right, and it is continuous at $1/n$ from the left). Now let x_1 be a continuous extension of x from $[0,1/n]$ to $[0,\infty)$, given by Tietze's theorem. Let

$$x(t)=F(x)\left(t-\frac{1}{n}\right)+z(t), \text{ for } t\in(1/n,2/n]. \tag{6.5.17}$$

Since $t-\frac{1}{n} \in (0, 1/n]$, we have $F(x)\left(t-\frac{1}{n}\right) = F(x_1)\left(t-\frac{1}{n}\right)$, and therefore x is continuous on $(1/n, 2/n]$, because $F(x_1)$ and z are continuous on $[0, \infty)$. To see that x is also continuous at $1/n$, we have

$$\begin{aligned}
\lim_{t\to 1/n, t>1/n} x(t) &= \lim_{t\to 1/n, t>1/n} \left(F(x_1)\left(t-\frac{1}{n}\right) + z(t)\right) \\
&= F(x_1)(0) + z\left(\frac{1}{n}\right) \\
&= F(x)(0) + z\left(\frac{1}{n}\right) \\
&= u_0 + z\left(\frac{1}{n}\right)
\end{aligned}$$

from the continuity of $F(x_1)$ and z, and

$$\lim_{t\to 1/n, t<1/n} x(t) = u_0 + z\left(\frac{1}{n}\right)$$

from (6.5.16). Now x is constructed on $[0, 2/n]$ and is continuous on $[0, 2/n]$. By induction, assume x is defined and continuous on $[0, k/n]$. Let x_k be a continuous extension of x from $[0, k/n]$ to $[0, \infty)$, given by Tietze's theorem. Define

$$x(t) = F(x)\left(t-\frac{1}{n}\right) + z(t), \text{ for } t \in (k/n, (k+1)/n], \tag{6.5.18}$$

Since $t-\frac{1}{n} \in (0, k/n]$, we have $F(x)\left(t-\frac{1}{n}\right) = F(x_k)\left(t-\frac{1}{n}\right)$, and therefore x is continuous on $(k/n, (k+1)/n]$, because $F(x_k)$ and z are continuous on $[0, \infty)$. To see that x is also continuous at k/n, we have

$$\begin{aligned}
\lim_{t\to k/n, t>k/n} x(t) &= \lim_{t\to k/n, t>k/n} \left(F(x_k)\left(t-\frac{1}{n}\right) + z(t)\right) \\
&= F(x_k)\left(\frac{k-1}{n}\right) + z\left(\frac{k}{n}\right)
\end{aligned}$$

from the continuity of $F(x_k)$ and z, and

$$\lim_{t\to k/n, t<k/n} x(t) = F(x_k)\left(\frac{k-1}{n}\right) + z\left(\frac{k}{n}\right)$$

from (6.5.18). Now this x, constructed inductively, satisfies (6.5.18) and is continuous on $[0, \infty)$. Let V be defined by:

$$V = \{y \in E : p_m(y) \leq \eta, \forall m\}.$$

This x is in E and satisfies the equation $x = F_n(x) + z$. According to (iv), it follows that $x \in X$. This implies that $V \subseteq (I - F_n)(X)$. Notice $0 \in V$, V is closed and convex. Now for each n, let

$$U_n = \{y \in E : p_m(y) \leq \eta/n, \forall m\}.$$

Then $0 \in U_n$, U_n is closed and convex. In addition,

$$U_n \subseteq V \subseteq (I - F_n)(X),$$

and so $I - F_n$ is a homeomorphism from $(I - F_n)^{-1}(U_n)$ onto U_n. Also, $\operatorname{diam}(U_n) \to 0$ as $n \to \infty$, because for $x, y \in U_n$,

$$d(x,y) = \sum_{m=1}^{\infty} \frac{1}{2^m} \frac{p_m(x-y)}{1 + p_m(x-y)} \leq \sum_{m=1}^{\infty} \frac{1}{2^m} p_m(x-y) \leq \frac{2\eta}{n} \to 0$$

as $n \to \infty$.

We now show that F_n converges to F in E, uniformly for $x \in X$. Since F is compact, $F(X)$ is relatively compact in $C([0,\infty), \mathbf{R}^M)$. Then $F(X)$ is relatively compact in $C([0,m], \mathbf{R}^M)$, $\forall m$. By the Arzelà-Ascoli theorem, $F(X)$ is bounded and equicontinuous in $C([0,m], \mathbf{R}^M)$, $\forall m$, i.e. $\forall m, \exists M_m > 0$ such that $p_m(F(x)) \leq M_m, \forall x \in X$, and $\forall \epsilon > 0, \exists \delta_m > 0$ such that if $t, s \in [0,m]$ with $|t - s| < \delta_m$, then

$$|F(x)(t) - F(x)(s)| < \epsilon, \forall x \in X. \tag{6.5.19}$$

Let $\epsilon > 0$ and let n be sufficiently large, such that $\frac{1}{n} < \delta_m$. We have

$$F_n(x)(t) - F(x)(t) = \begin{cases} F(x)(0) - F(x)(t), \text{ if } t \in [0, 1/n] \\ F(x)\left(t - \frac{1}{n}\right) - F(x)(t), \text{ if } t \geq \frac{1}{n}, \end{cases} \tag{6.5.20}$$

therefore $|F_n(x)(t) - F(x)(t)| = |F(x)(t) - F(x)(s)|$, where

$$s = \begin{cases} 0, \text{ if } t \in [0, 1/n) \\ t - \frac{1}{n}, \text{ if } t \in [1/n, \infty), \end{cases}$$

so $|t - s| < \frac{1}{n} < \delta_m$. Then (6.5.19) and (6.5.20) show that for n sufficiently large, we have

$$|F_n(x)(t) - F(x)(t)| < \epsilon, \forall t \in [0,m], \forall x \in X, \forall m,$$

or

$$p_m(F_n(x) - F(x)) < \epsilon, \forall x \in X, \forall m. \tag{6.5.21}$$

Thus F_n converges to F in $C([0,\infty),\mathbf{R}^M)$, uniformly for $x \in X$.

To finish the proof, we show that there exists a subsequence $\{G_n\}_n$ of $\{F_n\}_n$, such that for all n

$$G_n(x) - F(x) \in U_n, \forall x \in X, \tag{6.5.22}$$

since in that case $\{U_n\}_n$ and $\{G_n\}_n$ satisfy the hypotheses in Lemma 6.5.1, and so $Fix(F)$ is an R_δ set. For this, apply (6.5.21) for $\epsilon = \eta/n$, to construct $\{G_n\}_n$ inductively. For $\epsilon = \eta/1$, $\exists n_1$ such that for $n \geq n_1$

$$p_m(F_n(x) - F(x)) < \eta/1, \forall x \in X, \forall m.$$

For $\epsilon = \eta/2$, $\exists n_2 > n_1$ such that for $n \geq n_2$

$$p_m(F_n(x) - F(x)) < \eta/2, \forall x \in X, \forall m.$$

By induction, for $\epsilon = \eta/k$, $\exists n_k > n_{k-1}$ such that for $n \geq n_k$

$$p_m(F_n(x) - F(x)) < \eta/k, \forall x \in X, \forall m.$$

Now define $G_k := F_{n_k}, \forall k$, and $\{G_k\}_k$ is such that for all k

$$p_m(G_k(x) - F(x)) < \eta/k, \forall x \in X, \forall m,$$

i.e. (6.5.22) holds true, and the proof is complete. □

Corollary 6.5.3. *Under the same hypotheses as Theorem 6.5.2, $Fix(F)$ is nonempty, compact and connected in $C([0,\infty),\mathbf{R}^M)$.*

This is an immediate, since any R_δ set is nonempty, compact and connected.

Notice that a result similar to that of Corollary 6.5.3 has also been obtained in Section 6.3, by studying the connection between the solution set of equation

$$y(t) = F(y)(t),\ t \in [0,\infty) \tag{6.5.23}$$

in $C([0,\infty),\mathbf{R}^M)$, and the solution set of equations

$$y(t) = F(y)(t),\ t \in [0,m] \tag*{(6.5.24)$_m$}$$

in $C([0,m],\mathbf{R}^M)$, for all positive integers m. Notice in Section 6.3 we assumed $M = 1$, however the extension to arbitrary M is immediate. More precisely, the following results were derived in Section 6.3.

Theorem 6.5.4. *Suppose that the following two conditions hold:*

$$\left\{\begin{array}{l} F : C([0,\infty),\mathbf{R}^M) \to C([0,\infty),\mathbf{R}^M) \\ \textit{is an abstract Volterra operator} \end{array}\right. \tag{6.5.25}$$

$$\left\{\begin{array}{l} \forall m, \textit{ the solution set of } (6.5.24)_m \textit{ is nonempty,} \\ \textit{compact and connected in } C([0,m],\mathbf{R}^M). \end{array}\right. \tag{6.5.26}$$

Then the solution set of (6.5.23) *is nonempty, compact and connected in* $C([0,\infty),\mathbf{R}^M)$.

Theorem 6.5.5 *Suppose that* $F : C([0,\infty),\mathbf{R}^M) \to C([0,\infty),\mathbf{R}^M)$ *is an abstract Volterra operator satisfying:*

$$\exists u_0 \in \mathbf{R}^M \textit{ such that } F(x)(0) = u_0, \forall x \in C([0,\infty),\mathbf{R}^M). \tag{6.5.27}$$

For each positive integer m, *let* $M_m > 0$, *and let*

$$X_m = \{x \in C([0,m],\mathbf{R}^M) : \|x\|_m = \sup\{|x(t)| : t \in [0,m]\} \leq M_m\}.$$

Suppose that for each m

$$F : X_m \to C([0,\infty),\mathbf{R}^M) \textit{ is an abstract Volterra operator;} \tag{6.5.28}$$

$$F : X_m \to C([0,\infty),\mathbf{R}^M) \textit{ is a continuous, compact operator;} \tag{6.5.29}$$

$$\left\{\begin{array}{l} \textit{if } y \in C([0,m],\mathbf{R}^M) \textit{ satisfies } y(t) = F(y)(t), \\ t \in [0,m], \textit{ then } y \in X_m; \end{array}\right. \tag{6.5.30}$$

$$\left\{\begin{array}{l} \forall n, \textit{ if } y \in C([0,m],\mathbf{R}^M) \textit{ satisfies } y(t) = F_{n,m}(y)(t), \\ t \in [0,m], \textit{ then } y \in \textit{int}\,(X_m), \end{array}\right. \tag{6.5.31}$$

where $\forall n, \forall m$, $F_{n,m}$ *is defined by*

$$F_{n,m}(y)(t) = \left\{\begin{array}{l} u_0, \textit{ if } t \in [0,m/n]; \\ F(y)\left(t - \frac{m}{n}\right), \textit{ if } t \in [m/n,m]. \end{array}\right. \tag{6.5.32}$$

Then $Fix(F)$ *is nonempty, compact and connected in* $C([0,\infty),\mathbf{R}^M)$.

Consider the integral equation

$$y(t) = h(t) + \int_0^t k(t,s)g(s,y(s))ds \quad \text{for} \quad t \in [0,\infty). \tag{6.5.33}$$

Assume $p, q > 1$ with $\frac{1}{p} + \frac{1}{q} = 1$.

Theorem 6.5.6. *Assume that $g : [0,\infty) \times \mathbf{R} \to \mathbf{R}$ is a locally L^q-Carathéodory function, i.e. for each compact subinterval I of $[0,\infty)$, the following three conditions hold:*

$$\text{the map } t \mapsto g(t,y) \text{ is measurable for all } y \in \mathbf{R}; \tag{6.5.34}$$

$$\text{the map } y \mapsto g(t,y) \text{ is continuous for almost all } t \in I; \tag{6.5.35}$$

$$\begin{cases} \text{for all } r > 0 \text{ there exists } \mu_r \in L^q(I,\mathbf{R}) \text{ such that} \\ |y| < r \text{ implies that } |g(t,y)| \le \mu_r(t), \text{ for a.a. } t \in I. \end{cases} \tag{6.5.36}$$

Assume that

$$h \in C([0,\infty),\mathbf{R}); \tag{6.5.37}$$

$$k_t(s) = k(t,s) \in L^p([0,t],\mathbf{R}), \forall t \in [0,m], \forall m \in \mathbf{N}; \tag{6.5.38}$$

$$\begin{cases} \forall t, t' \in [0,\infty), \\ \int_0^{t^*} |k_t(s) - k_{t'}(s)|^p ds \to 0 \text{ as } t \to t', \\ \text{where } t^* = \min\{t,t'\}. \end{cases} \tag{6.5.39}$$

Also assume that

$$\begin{cases} \text{there exists an } \alpha \in L^1_{loc}([0,\infty),\mathbf{R}), \text{ and there exists} \\ \text{a nondecreasing, continuous function } \psi : [0,\infty) \to [0,\infty), \\ \text{such that } |k(t,s)g(s,y(s))| \le \alpha(s)\psi(|y(s)|), \\ \text{a.e. } t \in [0,\infty), \text{ a.e. } s \in [0,t], \ \forall y \in C([0,\infty),\mathbf{R}); \end{cases} \tag{6.5.40}$$

$$\int_0^m \alpha(s)ds < \int_{p_m(h)}^\infty \frac{ds}{\psi(s)}, \ \forall m \in \mathbf{N}. \tag{6.5.41}$$

Let

$$I_m(z) = \int_{p_m(h)}^z \frac{ds}{\psi(s)}, \tag{6.5.42}$$

(6.5.43) $$\overline{M}_m = I_m^{-1}\left(\int_0^m \alpha(s)ds\right),$$

(6.5.44) $$M_m = \overline{M}_m + 1,$$

(6.5.45) $$X_m = \{y \in C([0,m],\mathbf{R}) : p_m(y) \leq M_m\},$$

and

(6.5.46) $$X = \{y \in C([0,\infty) : p_m(y) \leq M_m, \forall m \in \mathbf{N}\}.$$

Finally assume

(6.5.47) $$\sup_{m\in\mathbf{N}} \overline{M}_m \equiv R < \infty.$$

Then the solution set of equation (6.5.33) *is an* R_δ *set.*

PROOF: Let $E = C([0,\infty),\mathbf{R})$, and let $F : E \to E$ be defined by

$$F(y)(t) = h(t) + \int_0^t k(t,s)g(s,y(s))ds \text{ for } t \in [0,\infty).$$

Conditions (6.5.34) – (6.5.39) ensure that F is well defined, F is a Volterra operator such that $F(y)(0) = h(0)$, $\forall y \in C([0,\infty),\mathbf{R})$, and the restriction $F : C([0,m],\mathbf{R}) \to C([0,m],\mathbf{R})$ is continuous. In fact, $F : X \to C([0,\infty),\mathbf{R})$ is continuous, because if $\{y_j\}_{j\in\mathbf{N}}$ is a sequence in X and $y_0 \in C([0,\infty),\mathbf{R})$ is such that $y_j \to y_0$ in $C([0,\infty),\mathbf{R})$ as $j \to \infty$, then $y_j \to y_0$ in $C([0,m],\mathbf{R})$ as $j \to \infty$, for all m. Since $F : C([0,m],\mathbf{R}) \to C([0,m],\mathbf{R})$ is continuous, we then have that $F(y_j) \to F(y_0)$ in $C([0,m],\mathbf{R})$ as $j \to \infty$, for all m. This implies that $F(y_j) \to F(y_0)$ in $C([0,\infty),\mathbf{R})$ as $j \to \infty$.

We show now that if all hypotheses are satisfied, then $F : X \to C([0,\infty),\mathbf{R})$ is compact, and that (iii) and (iv) of Theorem 6.5.2 hold true.

To see that

$$F : X \to C([0,\infty),\mathbf{R}) \text{ is compact,}$$

let B be a bounded set in X. Let $\{y_j\}_{j\in\mathbf{N}}$ be a sequence in B and consider the sequence $\{F(y_j)\}_{j\in\mathbf{N}}$ in $F(B)$. Since B is bounded in $C([0,\infty),\mathbf{R})$, $B|_{[0,m]}$ is bounded in $C([0,m],\mathbf{R})$, for all m, where

$$B|_{[0,m]} = \{y|_{[0,m]} : y \in B\}.$$

The restriction $F : B|_{[0,m]} \to C([0,m], \mathbf{R})$, is compact, therefore $F(B|_{[0,m]})$ is relatively compact in $C([0,m], \mathbf{R})$. For $m = 1$, there exists a subsequence N_1 of $\mathbf{N}$, and there exists a $z_1 \in C([0,1], \mathbf{R})$, such that

$$F(y_j)|_{[0,1]} \to z_1 \text{ in } C([0,1], \mathbf{R}) \text{ as } j \to \infty \text{ in } N_1.$$

Now consider the sequence $\{F(y_j)\}_{j \in N_1}$, restricted to $[0, 2]$. Since $F(B|_{[0,2]})$ is relatively compact in $C([0,2], \mathbf{R})$, there exists a subsequence N_1 of N_2, and there exists a $z_2 \in C([0,2], \mathbf{R})$, such that

$$F(y_j)|_{[0,2]} \to z_2 \text{ in } C([0,2], \mathbf{R}) \text{ as } j \to \infty \text{ in } N_2.$$

In addition,

$$z_2|_{[0,1]} = z_1 \text{ on } [0,1].$$

By induction, assume the sequence $\{F(y_j)\}_{j \in N_k}$ and $z_k \in C([0,k], \mathbf{R})$ are found such that $N_k \subseteq N_{k-1} \subseteq ... \subseteq N_1 \subseteq \mathbf{N}$,

$$F(y_j)|_{[0,k]} \to z_k \text{ in } C([0,k], \mathbf{R}) \text{ as } j \to \infty \text{ in } N_k,$$

and

$$z_k|_{[0,1]} = z_{k-1} \text{ on } [0, k-1].$$

Since $F(B|_{[0,k+1]})$ is relatively compact in $C([0, k+1], \mathbf{R})$, there exists a subsequence N_{k+1} of N_k, and there exists a $z_{k+1} \in C([0, k+1], \mathbf{R})$, such that

$$F(y_j)|_{[0,k+1]} \to z_{k+1} \text{ in } C([0, k+1], \mathbf{R}) \text{ as } j \to \infty \text{ in } N_{k+1}.$$

In addition,

$$z_{k+1}|_{[0,k]} = z_k \text{ on } [0,k].$$

Now define $z \in C[0, \infty)$ by

$$z(t) = z_k(t),\ t \in [k-1, k),\ k = 1, 2, \ldots.$$

The induction above shows that the sequence $\{F(y_j)\}_{j \in \mathbf{N}}$ contains a subsequence which converges in $C([0,\infty), \mathbf{R})$ to $z \in C[0,\infty)$. Therefore $F(B)$ is relatively compact in $C([0,\infty), \mathbf{R})$, and the operator $F : X \to C([0,\infty), \mathbf{R})$ is compact.

Condition (iii) of Theorem 6.5.2 is satisfied, because if $y \in C([0,\infty), \mathbf{R})$ satisfies equation $y = F(y)$ on $[0, \infty)$, then $y|_{[0,m]}$ satisfies this equation on $[0, m]$, for all m, so (like in Section 6.4), $y \in X_m$, for all m, i.e. $y \in X$.

To see that (iv) of Theorem 6.5.2 is satisfied, let $n \in \mathbf{N}$, let $\eta > 0$, and let $y \in C([0,\infty),\mathbf{R})$ be such that $y(t) = F_n(y)(t) + z(t)$, $t \in [0,\infty)$, where z is such that $p_m(z) \le \eta$, $\forall m$, and

$$F_n(y)(t) = \begin{cases} h(0), \text{ if } t \in [0,1/n) \\ h\left(t-\frac{1}{n}\right) + \int_0^{t-1/n} k\left(t-\frac{1}{n},s\right) g(s,y(s))ds, \text{ if } t \in [1/n,\infty). \end{cases}$$

Let $m \in \mathbf{N}$ be arbitrary. Then, we have

$$\begin{aligned} |y(t)| &\le |F_n(y)(t)| + |z(t)| \\ &\le p_m(h) + \int_0^t \alpha(s)\psi(|y(s)|)ds + \eta, \ t \in [0,m]. \end{aligned}$$

Let

$$w(t) = p_m(h) + \int_0^t \alpha(s)\psi(|y(s)|)ds + \eta, \ t \in [0,m]. \tag{6.5.48}$$

Then, like in Section 6.4, we have

$$w'(t) \le \alpha(t)\psi(w(t)), \ \ t \in [0,m],$$

which implies that (see Section 6.4)

$$\int_{p_m(h)+\eta}^{w(t)} \frac{ds}{\psi(s)} \le \int_0^m \alpha(s)ds, \ \ t \in [0,m],$$

so that

$$\int_{p_m(h)}^{w(t)} \frac{ds}{\psi(s)} \le \int_0^m \alpha(s)ds + \int_{p_m(h)}^{p_m(h)+\eta} \frac{ds}{\psi(s)}, \ \ t \in [0,m].$$

Then for $t \in [0,m]$,

$$\begin{aligned} w(t) &\le I_m^{-1}\left(\int_0^m \alpha(s)ds + \int_{p_m(h)}^{p_m(h)+\eta} \frac{ds}{\psi(s)}\right) \\ &= I_m^{-1}\left(\int_0^m \alpha(s)ds + \int_0^\eta \frac{du}{\psi(u+p_m(h))}\right) \\ &\le I_m^{-1}\left(\int_0^m \alpha(s)ds + \int_0^\eta \frac{du}{\psi(u)}\right) \\ &= I_m^{-1}\left(I_m(M_m) + \int_0^\eta \frac{du}{\psi(u)} - \int_{\overline{M}_m}^{\overline{M}_m+1} \frac{du}{\psi(u)}\right) \\ &\le I_m^{-1}\left(I_m(M_m) + \int_0^\eta \frac{du}{\psi(u)} - \int_R^{R+1} \frac{du}{\psi(u)}\right) \\ &\le M_m \end{aligned}$$

if η is small enough (recall ψ is nondecreasing). Thus for all m we have $p_m(y) \leq M_m$, i.e. $y \in X$.

Now, according to Theorem 6.5.2, the solution set of equation (6.5.33) is an R_δ set. □

Remark. It is possible, using more subtle arguments (we need to obtain a new theorem of Theorem 6.5.2 type), to remove assumption (6.5.47) in the statement of Theorem 6.5.6.

6.6. Fixed points for operators on $L^p_{loc}([0,\infty), \mathbf{R}^M)$.

Let $E = E([0,\infty), \mathbf{R}^M)$, $\{p_m\}_{m\geq 1}$, d, X and F be as in the beginning of Section 6.5, and let $\{F_n\}_n$ be the sequence defined by (6.5.1) – (6.5.2).

This section will discuss the case when $E = L^p_{loc}([0,\infty), \mathbf{R}^M)$, the space of locally p-integrable functions on $[0,\infty)$, $1 < p < \infty$.

Theorem 6.6.1. *Assume that* $E = L^p_{loc}([0,\infty), \mathbf{R}^M)$, *with*

$$p_m(x) = \left(\int_0^m |x(t)|^p dt \right)^{1/p},$$

$1 < p < \infty$, *and that* $F : X \to E$ *is a continuous and compact operator. Also assume that the following conditions hold:*

(i) $\exists u_0 \in \mathbf{R}^M$ *such that* $F(x)(0) = u_0$, *for all* $x \in X$;

(ii) $\forall \epsilon > 0$, $\forall x, y \in X$, *if* $x(t) = y(t)$, $\forall t \in [0,\epsilon]$, *then* $F(x)(t) = F(y)(t)$, $\forall t \in [0,\epsilon]$ *(i.e. F is an abstract Volterra operator);*

(iii) if $y \in E$ *satisfies* $y = F(y)$, *then* $y \in X$;

(iv) $\exists \eta > 0$ *such that* $\forall n$, *if* $y \in E$ *satisfies* $y = F_n(y) + z$, *where* $p_m(z) \leq \eta, \forall m$, *then* $y \in X$.

Then $Fix(F)$ *is an* R_δ *set.*

PROOF: Consider the sequence $\{F_n\}_n$ defined by (6.5.1) – (6.5.2). We show that there exists a sequence $\{U_n\}_n$ of closed convex sets in E and there exists a subsequence $\{G_n\}_n$ of $\{F_n\}_n$ such that $\{U_n\}_n$ and $\{G_n\}_n$ satisfy conditions (6.5.3) – (6.5.6) in Lemma 6.5.1.

First, we show that $\forall n$, $I - F_n$ is a homeomorphism from X onto $(I - F_n)(X)$. To see that $I - F_n$ is continuous, let $\epsilon > 0$. Since F is continuous, $\exists \delta > 0$ such that if $x, y \in X$ with $p_m(x-y) < \delta$, then $p_m(F(x) - F(y)) < \epsilon$, $\forall m$. Then, if $x, y \in X$ are such that $p_m(x-y) < \delta$, $\forall m$, we also have

$$p^p_m(F_n(x) - F_n(y)) \quad = \quad \int_0^m |F_n(x)(t) - F_n(y)(t)|^p dt$$

$$
\begin{aligned}
&= \int_{1/n}^{m} |F_n(x)(t) - F_n(y)(t)|^p dt \\
&= \int_{1/n}^{m} \left| F(x)\left(t - \frac{1}{n}\right) - F(y)\left(t - \frac{1}{n}\right) \right|^p dt \\
&= \int_{0}^{m-1/n} |F(x)(t) - F(y)(t)|^p dt \\
&\le \int_{0}^{m} |F(x)(t) - F(y)(t)|^p dt \\
&= p_m^p(F(x) - F(y)) < \epsilon^p, \ \forall m
\end{aligned}
$$

Therefore, F_n is continuous, and thus $I - F_n$ is continuous.

The fact that $I - F_n$ is one-to-one follows exactly like in the proof of Theorem 6.5.2.

To see that $(I - F_n)^{-1}$ is continuous, let $\{x_j\}_j$ be a sequence in X. Let $x \in X$ be such that

$$
\lim_{j\to\infty} p_m((x_j - F_n(x_j)) - (x - F_n(x))) = 0, \forall m. \tag{6.6.1}
$$

For $t \in [0, 1/n]$ we have $F_n(x_j)(t) = F(x_j)(0) = u_0 = F_n(x)(t)$, so (6.6.1) implies that

$$
\lim_{j\to\infty} p_m((x_j - x)\chi_{[0,1/n]}) = 0, \forall m,
$$

where $\chi_{[0,1/n]}$ is the characteristic function of the interval $[0, 1/n]$. Since F is continuous, it also follows that

$$
\lim_{j\to\infty} p_m((F(x_j) - F(x))\chi_{[0,1/n]}) = 0, \forall m. \tag{6.6.2}
$$

For $t \in [1/n, 2/n]$, we have $t - \frac{1}{n} \in [0, 1/n]$ and $F_n(x_j)\,(t) = F(x_j)\left(t - \frac{1}{n}\right)$, $F_n(x)\,(t) = F(x)\left(t - \frac{1}{n}\right)$, therefore (6.6.1) and (6.6.2) give

$$
\lim_{j\to\infty} p_m((F(x_j) - F(x))\chi_{[0,2/n]}) = 0, \forall m.
$$

and therefore

$$
\lim_{j\to\infty} p_m((x_j - x)\chi_{[0,2/n]}) = 0, \forall m.
$$

Assume by induction that for some positive integer k we have

$$
\lim_{j\to\infty} p_m((x_j - x)\chi_{[0,(k-1)/n]}) = 0, \forall m. \tag{6.6.3}
$$

By the continuity of F, we also have

$$\lim_{j\to\infty} p_m((F(x_j) - F(x))\chi_{[0,(k-1)/n]}) = 0, \forall m. \tag{6.6.4}$$

Then, as above, (6.6.3) and (6.6.4) imply that

$$\lim_{j\to\infty} p_m((x_j - x)\chi_{[0,k/n]}) = 0, \forall m. \tag{6.6.5}$$

Therefore, (6.6.5) is true for all k, which shows that

$$\lim_{j\to\infty} p_m(x_j - x) = 0, \forall m.$$

and therefore $(I - F_n)^{-1}$ is continuous.

Now to find the sequence $\{U_n\}_n$, for a given n, consider the equation

$$x = F_n(x) + z, \tag{6.6.6}$$

where $z \in E$ is such that $p_m(z) \leq \eta, \forall m$. Let $x \in E$ be defined by induction as follows:

$$x(t) = u_0 + z(t), \text{ for } t \in [0, 1/n);$$

$$x(t) = F(x)\left(t - \frac{1}{n}\right) + z(t), \text{ for } t \in [1/n, 2/n).$$

Given $x(t)$ for $t \in [(k-1)/n, k/n)$, define

$$x(t) = F(x)\left(t - \frac{1}{n}\right) + z(t), \text{ for } t \in [k/n, (k+1)/n),$$

and let V be defined by:

$$V = \{y \in E : p_m(y) \leq \eta, \forall m\}.$$

Then this x is in E and it satisfies the equation $x = F_n(x) + z$. According to (iv), it follows that $x \in X$. This implies that $V \subseteq (I - F_n)(X)$. Notice $0 \in V$, V is closed and convex. Now for each n, let

$$U_n = \{y \in E : p_m(y) \leq \eta/n, \forall m\}.$$

Then $0 \in U_n$, U_n is closed and convex. In addition,

$$U_n \subseteq V \subseteq (I - F_n)(X),$$

and so $I - F_n$ is a homeomorphism from $(I - F_n)^{-1}(U_n)$ onto U_n. Also, $\operatorname{diam}(U_n) \to 0$ as $n \to \infty$ (see the proof of Theorem 6.5.2).

Like in the proof of Theorem 6.5.2, the proof is complete if we show that F_n converges to F uniformly for $x \in X$. Since F is compact, $F(X)$ is relatively compact in $L^p_{loc}([0,\infty), \mathbf{R}^M)$, which implies that $F(X)$ is also relatively compact in $L^p([0,m], \mathbf{R}^M)$, $\forall m$. Therefore, $\forall m$, $\exists M_m > 0$ such that

$$p_m(F(x)) \le M_m, \forall x \in X \tag{6.6.7}$$

and

$$\int_0^m |F(x)(t+\tau) - F(x)(t)|^p dt \to 0 \text{ as } \tau \to 0, \text{ uniformly for } x \in X. \tag{6.6.8}$$

Thus we have

$$\begin{aligned} p_m^p(F_n(x) - F(x)) &= \int_0^m |F_n(x)(t) - F(x)(t)|^p dt \\ &= \int_0^{1/n} |F(x)(0) - F(x)(t)|^p dt \\ &+ \int_{1/n}^m \left| F(x)\left(t - \frac{1}{n}\right) - F(x)(t) \right|^p dt. \end{aligned}$$

The first term is bounded above by $\frac{1}{n}(|u_0|^p + M_m^p)$, therefore its limit is 0 as $n \to \infty$, uniformly for $x \in X$. The second term has the limit 0 as $n \to \infty$, uniformly for $x \in X$, since (6.6.8) holds. Therefore,

$$\lim_{n\to\infty} p_m(F_n(x) - F(x)) = 0, \text{ uniformly for } x \in X, \forall m,$$

and the proof is complete. □

Like in Section 6.5, we have the following immediate consequence:

Corollary 6.6.2. *Under the same hypotheses as Theorem 6.6.1, $Fix(F)$ is nonempty, compact and connected in $L^p_{loc}([0,\infty), \mathbf{R}^M)$.*

Some of the results in Section 6.3 can be extended for the space of locally p-integrable functions, by studying the connection between the solution set of (6.5.23) in $L^p_{loc}([0,\infty), \mathbf{R}^M)$, and the solution sets of $(6.5.24)_m$ in $L^p([0,m], \mathbf{R}^M)$, for all m. The following holds true.

Theorem 6.6.3. *Suppose that the following two conditions hold:*

$$\left\{ \begin{array}{l} F : L^p_{loc}([0,\infty), \mathbf{R}^M) \to L^p_{loc}([0,\infty), \mathbf{R}^M) \\ \textit{is an abstract Volterra operator} \end{array} \right. \tag{6.6.9}$$

(6.6.10) $$\begin{cases} \forall m, \text{ the solution set of } (6.5.24)_m \text{ is nonempty,} \\ \text{compact and connected in } L^p([0,m],\mathbf{R}^M). \end{cases}$$

Then the solution set of (6.5.23) is nonempty, compact and connected in $L^p_{loc}([0,\infty),\mathbf{R}^M)$.

PROOF: To prove this result, follow the proof of Theorem 6.3.2, and replace the uniform convergence in $C([0,m],\mathbf{R}^M)$ by the convergence in $L^p([0,m],\mathbf{R}^M)$. □

Theorem 6.6.4. *Suppose that*

$$F : L^p_{loc}([0,\infty),\mathbf{R}^M) \to L^p_{loc}([0,\infty),\mathbf{R}^M)$$

is an abstract Volterra operator satisfying:

(6.6.11) $\exists u_0 \in \mathbf{R}^M$ *such that* $F(x)(0) = u_0, \forall x \in L^p_{loc}([0,\infty),\mathbf{R}^M)$.

For each positive integer m, let $M_m > 0$, and let

$$X_m = \{x \in L^p([0,m],\mathbf{R}^M) : \|x\|_{m,p} = \int_0^m (|x(t)|^p dt)^{1/p} \le M_m\}.$$

Suppose that for each m

(6.6.12) $F : X_m \to L^p([0,\infty),\mathbf{R}^M)$ *is an abstract Volterra operator;*

(6.6.13) $F : X_m \to L^p([0,\infty),\mathbf{R}^M)$ *is a continuous, compact operator;*

(6.6.14) $$\begin{cases} \text{if } y \in L^p([0,m],\mathbf{R}^M) \text{ satisfies } y(t) = F(y)(t), \\ t \in [0,m], \text{ then } y \in X_m; \end{cases}$$

(6.6.15) $$\begin{cases} \forall n, \text{ if } y \in L^p([0,m],\mathbf{R}^M) \text{ satisfies } y(t) = F_{n,m}(y)(t), \\ t \in [0,m], \text{ then } y \in int(X_m), \end{cases}$$

where $\forall n, \forall m$, $F_{n,m}$ is defined by

$$F_{n,m}(y)(t) = \begin{cases} u_0, \text{ if } t \in [0,m/n]; \\ F(y)\left(t - \frac{m}{n}\right), \text{ if } t \in [m/n,m]. \end{cases}$$

Then $Fix(F)$ is nonempty, compact and connected in $L^p_{loc}([0,\infty),\mathbf{R}^M)$.

CHAPTER 7

WEAKLY CONDENSING OPERATORS AND OPERATOR EQUATIONS

7.1. Introduction.

This chapter presents a general continuation theory for weakly condensing operators between Banach spaces. Our theory is based on ideas in O'Regan [130, 131], which in turn were motivated by the papers of Banas and Rivero [31], De Blasi [53] and Emmanuele [61]. The continuation theorem established in Section 7.2 is used in Section 7.3 to establish very general existence principles for nonlinear operator equations. In addition we show, in the case of second order boundary value problems, that the theory developed here for weakly continuous operators can lead to the same conclusions as the theory of compact (strong) operators.

For the remainder of this section we gather together some preliminaries that will be needed in Sections 7.2 and 7.3. Let Ω_E be the bounded subsets of a Banach space E and let K^w be the family of all weakly compact subsets of E. Also let B be the unit ball of E. The DeBlasi [31, 53, 61] measure of weak noncompactness is the map $w : \Omega_E \to [0, \infty)$ defined by (here $X \in \Omega_E$),

$$w(X) = \inf\{t > 0 : \text{ there exists } Y \in K^w \text{ with } X \subseteq Y + tB\}.$$

For convenience we recall some properties of w:
Let $S, T \in \Omega_E$. Then
(i). $S \subseteq T$ implies $w(S) \leq w(T)$.
(ii). $w(S) = 0$ iff $\overline{S^w} \in K^w$; here $\overline{S^w}$ is the weak closure of S in E.
(iii). $w(\overline{S^w}) = w(S)$.
(iv). $w(S \cup T) = \max\{w(S), w(T)\}$.
(v). $w(rS) = rw(S)$ for all $r > 0$.
(vi). $w(co(S)) = w(S)$.
(vii). $w(S + T) \leq w(S) + w(T)$.

Suppose $F : Y \subseteq E \to E$ maps bounded sets into bounded sets and is weakly continuous. We call F a w–Lipschitzian map if there exists a constant $k \geq 0$ with $w(F(X)) \leq kw(X)$ for all bounded sets $X \subseteq Y$. We call F a w-condensing map if F is w-Lipschitzian with $k = 1$ and $w(F(X)) < w(X)$ for all bounded sets X with $w(X) \neq 0$.

Theorem 7.1.1. *(Emmanuele) [61].*
Let E be a Banach space, X a nonempty, bounded, closed, convex subset of E and $F : X \to X$ a w-condensing map. Then F has a fixed point.

Also the following results will be used in Section's 7.2 and 7.3.

Theorem 7.1.2. *[88]. Every topological Hausdorff linear space is Tychonoff ($T_{3\frac{1}{2}}$).*

Theorem 7.1.3. *[62]. If A is a compact subset of a Tychonoff space X then for every closed set B disjoint from A there exists a continuous function $\mu : X \to [0,1]$ such that $\mu(x) = 1$ for $x \in A$ and $\mu(x) = 0$ for $x \in B$.*

Theorem 7.1.4. *[82, pg. 368]. Let E be a Banach space whose dual space $E^\star$ is separable. Let $A \subseteq E$ be weakly compact. Then there exists a weakly continuous retraction onto A.*

Theorem 7.1.5. *[44, 57]. A convex subset of a normed space is closed iff it is weakly closed.*

Theorem 7.1.6. *[44, 57]. A subset of a reflexive Banach space is weakly compact iff it is closed in the weak topology and bounded in the norm topology.*

Theorem 7.1.7. *[66, pg. 65]. In a Banach space, a bounded weakly closed set A is weakly compact iff for every weakly closed B which is disjoint from A,*

$$d(A, B) = \inf\{\|a - b\| : a \in A,\, b \in B\} > 0.$$

Theorem 7.1.8. *(Eberlein Šmulian) [44, 57].*
Suppose K is weakly closed in a Banach space E. Then the following are equivalent:

(i). K is weakly compact

(ii). K is weakly sequentially compact i.e. any sequence in K has a subsequence which converges weakly.

Theorem 7.1.9. *(Krein Šmulian) [44, 57].*
If E is a Banach space and Q is a weakly compact subset of E then $\overline{co}(Q)$ is weakly compact.

7.2. Fixed point theory.

In this section we present a continuation theory for w–condensing operators. Let E be a Banach space and let Q and C be closed, bounded, convex subsets of E with $Q \subseteq C$. Now let $X \subseteq Q$ and $A \subseteq X$ with A weakly closed in X and X weakly closed in Q.

Definition 7.2.1. We let $P_A(X, C)$ denote the set of all w–condensing mappings $F : X \to C$ such that F is fixed point free on A.

We call $N : X \times [0, 1] \to C$ a w–condensing mapping if N is weakly continuous, $w(N(Y)) \leq w(\pi Y)$ for all sets $Y \subseteq X \times [0, 1]$ and $w(N(Z)) < w(\pi Z)$ for all sets $Z \subseteq X \times [0, 1]$ with $w(\pi Z) \neq 0$. Here $\pi : X \times [0, 1] \to X$ is the natural projection.

Definition 7.2.2. A map $F \in P_A(X, C)$ is **essential** if every map in $P_A(X, C)$ which agrees with F on A has a fixed point. Otherwise F is **inessential** i.e. there exists a fixed point free map $G \in P_A(X, C)$ with $G = F$ on A.

Definition 7.2.3. We say two maps $F, G \in P_A(X, C)$ are homotopic in $P_A(X, C)$ and we write $F \cong G$ in $P_A(X, C)$ if there is a w–condensing mapping $N : X \times [0, 1] \to C$ with $N_t(u) = N(u, t) : X \to C$ belonging to $P_A(X, C)$ for each $t \in [0, 1]$ and $N_0 = F$, $N_1 = G$.

Remark. Notice $\cong$ is an equivalence relation in $P_A(X, C)$.

Theorem 7.2.1. *Let C, X, A, Q, E be as above with $F \in P_A(X, C)$. Then the following are equivalent:*
(i). F is inessential
(ii). there is a fixed point free $G \in P_A(X, C)$ such that $F \cong G$ in $P_A(X, C)$.

PROOF: We first show that (i) implies (ii). Let $G \in P_A(X, C)$ be a fixed point free map with $G = F$ on A. Define $N : X \times [0, 1] \to C$ by

$$N(x, t) = tG(x) + (1 - t)F(x).$$

Remark. (E, w), the space E endowed with the weak topology, is a locally convex linear topological space (in particular Hausdorff).

Now $N : X \times [0, 1] \to C$ is weakly continuous. To see this let $(x_\alpha, \lambda_\alpha)$ be a net in $X \times [0, 1]$ with $x_\alpha \rightharpoonup x$ and $\lambda_\alpha \to \lambda$; here $\rightharpoonup$ denotes weak convergence. Then

$$N(x_\alpha, \lambda_\alpha) \quad = \quad \lambda_\alpha G(x_\alpha) + (1 - \lambda_\alpha)F(x_\alpha)$$

$$\rightharpoonup \lambda G(x) + (1-\lambda)F(x) = N(x,\lambda),$$

so N is M–S (Moore–Smith) sequentially weakly continuous and consequently $N : X \times [0,1] \to C$ is weakly continuous. Also N is a w–condensing map. To see this let Y be a subset of $X \times [0,1]$ with $w(\pi Y) \neq 0$. Then

$$N(Y) \subseteq co\ (G(\pi Y) \cup F(\pi Y))$$

since if $(x,t) \in Y$ then

$$N(x,t) = tG(x) + (1-t)F(x) \subseteq co\ (G(\pi Y) \cup F(\pi Y)).$$

This together with the properties of w implies

$$\begin{aligned} w(N(Y)) &\leq w\,(co\ (G(\pi Y) \cup F(\pi Y))) = w\,(G(\pi Y) \cup F(\pi Y)) \\ &= \max\{w(G(\pi Y)),\, w(F(\pi Y))\} \\ &< \max\{w(\pi Y), w(\pi Y)\} = w(\pi Y). \end{aligned}$$

Thus $N : X \times [0,1] \to C$ is a w–condensing map. Also since $F = G$ on A and G is fixed point free on A we have for $x \in A$ that

$$N_t(x) = tG(x) + (1-t)F(x) = G(x) \neq x,$$

so N_t is fixed point free on A for each $t \in [0,1]$. It remains to show $N_t \in P_A(X,C)$ for each $t \in [0,1]$. Fix $t \in [0,1]$ and let Z be a bounded subset of X with $w(Z) \neq 0$. Then

$$w(N_t(Z)) = w(N(Z \times \{t\})) < w(\pi(Z \times \{t\})) = w(Z)$$

since $\pi(Z \times \{t\}) = Z$.

Thus for each $t \in [0,1]$ we have $N_t \in P_A(X,C)$. Finally $N_0 = F$ and $N_1 = G$ so $F \cong G$ in $P_A(X,C)$.

We next show that (ii) implies (i). Let $N : X \times [0,1] \to C$ be the w–condensing mapping from $G \in P_A(X,C)$ to F with $N_0 = G$ and $N_1 = F$. In particular N_t is fixed point free on A for each $t \in [0,1]$. Let

$$B = \{x \in X : x = N(x,t) \text{ for some } t \in [0,1]\}.$$

If B is empty then in particular $F = N_1$ has no fixed points and so F is inessential. So it remains to consider the case when B is nonempty. First note $B \cap A = \emptyset$. Also B is weakly closed. To see this let $x \in \overline{B^w}$. Then there is a M–S sequence (x_α) (i.e. a net) in B (i.e. $x_\alpha = N(x_\alpha, t_\alpha)$)

which converges to x (i.e. $x_\alpha \rightharpoonup x$). Without loss of generality assume t_α converges to $t \in [0,1]$. Then since $N : X \times [0,1] \to C$ is weakly continuous it follows that $x = N(x,t)$ and thus B is weakly closed. Next we claim that B is weakly compact. If $w(B) \neq 0$ then since $B \subseteq N(B \times [0,1])$ we have

$$w(B) \leq w(N(B \times [0,1])) < w(\pi(B \times [0,1])) = w(B)$$

since $\pi(B \times [0,1]) = B$. This contradiction implies B is weakly compact.

Now $E = (E, w)$, the space E endowed with the weak topology, is a locally convex Hausdorff topological space. Thus E (and hence X, with subspace topology) is Tychonoff by Theorem 7.1.2. This together with Theorem 7.1.3 implies that there is a continuous (weakly) function $\mu : X \to [0,1]$ with $\mu(A) = 1$ and $\mu(B) = 0$. Define $J : X \to C$ by $J(x) = N(x, \mu(x))$. Now J is weakly continuous. To see this let (x_α) be a net in X with $x_\alpha \rightharpoonup x$. Then $\mu(x_\alpha) \rightharpoonup \mu(x)$ and so $\mu(x_\alpha) \to \mu(x)$ since strong and weak convergence are the same in finite dimensional spaces. Consequently since $N : X \times [0,1] \to C$ is weakly continuous we have that $J(x_\alpha) = N(x_\alpha, \mu(x_\alpha)) \rightharpoonup N(x, \mu(x)) = J(x)$ so J is weakly continuous. We claim that $J : X \to C$ is a w–condensing fixed point free map with $J = F$ on A. If this is true then $J \in P_A(X,C)$ and J is a fixed point free map which agrees with F on A. Consequently F is inessential and we are finished.

It remains to prove the claim. J is fixed point free since $J(x) = x$ means $N(x, \mu(x)) = x$ which implies $x \in B$ and so $\mu(x) = 0$ (i.e. $N(x,0) = x$), a contradiction since $N(x,0) = G(x)$ is fixed point free. To see that $J = F$ on A notice if $x \in A$ then $\mu(x) = 1$ and so $J(x) = N(x, \mu(x)) = N(x,1) = F(x)$. It remains to show J is a w–condensing map. Let Ω be a bounded subset of X with $w(\Omega) \neq 0$. Now $\Omega^\star = \{\, (x, \mu(x)) : x \in \Omega \,\} \subseteq X \times [0,1]$. Then since $J(\Omega) = N(\Omega^\star)$ and $\pi(\Omega^\star) = \Omega$ we have

$$w(J(\Omega)) = w(N(\Omega^\star)) < w(\pi(\Omega^\star)) = w(\Omega).$$

Thus J is a w–condensing map. □

Theorem 7.2.2. *Let C, X, A, Q, E be as above. Suppose F and G are two maps in $P_A(X,C)$ such that $F \cong G$ in $P_A(X,C)$. Then F is essential iff G is essential.*

PROOF: If F is inessential then Theorem 7.2.1 guarantees a fixed point free map $T \in P_A(X,C)$ with $F \cong T$ in $P_A(X,C)$. Thus $G \cong T$ in

$P_A(X, C)$ and so G is inessential by Theorem 7.2.1. Symmetry will now imply that F is inessential iff G is inessential. □

Theorem 7.2.3. *Let Q and C be closed, bounded, convex subsets of a Banach space E with $Q \subseteq C$. In addition let U be a weakly open subset of Q with $u_0 \in U$. The constant map $F(\overline{U^w}) = u_0$ is essential in $P_{\partial_C U}(\overline{U^w}, C)$; here $\overline{U^w}$ denotes the weak closure of U in Q (which equals the weak closure of U in C) and $\partial_C U$ denotes the weak boundary of U in C.*

PROOF: Let $G : \overline{U^w} \to C$ be any w–condensing map with $G|_{\partial_C U} = F|_{\partial_C U} = u_0$. Define

$$I(x) = \begin{cases} G(x), & x \in \overline{U^w} \\ u_0, & x \in C \setminus \overline{U^w}. \end{cases}$$

It is easy to see that $I : C \to C$ is weakly continuous. In fact $I : C \to C$ is a w–condensing map. To see this let Ω be a bounded subset of C with $w(\Omega) \neq 0$. Then since

$$I(\Omega) \subseteq G(\Omega \cap \overline{U^w}) \cup \{0\}$$

we have

$$w(I(\Omega)) \leq w(G(\Omega \cap \overline{U^w}) \cup \{0\}) \leq w(G(\Omega \cap \overline{U^w})) \leq w(G(\Omega)) < w(\Omega).$$

Thus $I : C \to C$ is a w–condensing map. Emmanuele's fixed point theorem (Theorem 7.1.1) implies that I has a fixed point, say, y in C. In addition since $I(x) = u_0 \in U$ for $x \in C \setminus \overline{U^w}$ we have $y \in \overline{U^w}$. Thus $y = I(y) = G(y)$ and since $G|_{\partial_C U} = u_0$ we have $y \in U$. Hence G has a fixed point $y \in U$ so F is essential. □

Theorem's 7.2.2 and 7.2.3 immediately yields the following nonlinear alternative of Leray–Schauder type.

Theorem 7.2.4. *Let Q and C be closed, bounded, convex subsets of a Banach space E with $Q \subseteq C$. In addition let U be a weakly open subset of Q, with $p \in U$ and $F : \overline{U^w} \to C$ a w–condensing map. Then either*

(A1). F has a fixed point

or

(A2). there is a point $u \in \partial_C U$ and $\lambda \in (0,1)$ with $u = \lambda F u + (1-\lambda)p$.

PROOF: We assume $F|_{\partial_C U}$ is fixed point free for otherwise $(A1)$ is satisfied. Let $G : \overline{U^w} \to C$ be the constant map $u \mapsto p$ and consider the homotopy $N : \overline{U^w} \times [0,1] \to C$ joining G and F given by

$$N(u,t) = tF(u) + (1-t)p.$$

Now N is a w–condensing map since if Ω is a subset of $\overline{U^w} \times [0,1]$ with $w(\pi\,\Omega) \neq 0$ then

$$N(\Omega) \subseteq co\,(F(\pi\,\Omega) \cup \{p\})$$

implies

$$\begin{aligned} w(N(\Omega)) &\leq w(\,co\,(F(\pi\,\Omega) \cup \{p\})) = w(F(\pi\,\Omega) \cup \{p\}) \\ &= w(F(\pi\,\Omega)) < w(\pi\,\Omega). \end{aligned}$$

Now either N_t is fixed point free on $\partial_C U$ for each $t \in [0,1]$ or it is not. If N_t is fixed point free on $\partial_C U$ for each $t \in [0,1]$ then Theorem's 7.2.2 and 7.2.3 imply that F must have a fixed point so $(A1)$ occurs. If N_t is not fixed point free on $\partial_C U$ for each $t \in [0,1]$ then there exists $u \in \partial_C U$ with $u = \lambda F u + (1-\lambda)p$ for some λ, $0 \leq \lambda \leq 1$. Now $\lambda \neq 0$ since $p \in U$ and $\lambda \neq 1$ since $F|_{\partial_C U}$ was assumed to be fixed point free. Hence $(A2)$ occurs, and we are finished. □

Next we present a very general and applicable fixed point result for weakly continuous maps in separable, reflexive Banach spaces.

Theorem 7.2.5. *Let $E = (E, \|\,.\,\|)$ be a separable and reflexive Banach space. Q and C are closed, bounded, convex subsets of E with $Q \subseteq C$ and $0 \in Q$. Assume $F : Q \to C$ is a weakly continuous map and suppose there exists a $\delta > 0$ with*

$$\Omega_\delta = \{x \in E : \, d(x,Q) \leq \delta\} \subseteq C;$$

here $d(x,y) = \|x-y\|$. In addition assume the following condition is satisfied:

$$(7.2.1)\quad \begin{cases} \textit{for any } \epsilon \leq \delta, \textit{ if } \{(x_j,\lambda_j)\}_{j=1}^{\infty} \textit{ is a sequence in } Q \times [0,1] \\ \textit{with } x_j \rightharpoonup x \in \partial_{\Omega_\epsilon} Q \textit{ and } \lambda_j \to \lambda \textit{ and if } x = \lambda F(x) \\ \textit{for } 0 \leq \lambda < 1, \textit{ then } \lambda_j F(x_j) \in Q \textit{ for } j \textit{ sufficiently} \\ \textit{large; here } \partial_{\Omega_\epsilon} Q \textit{ denotes the weak boundary of } Q \\ \textit{relative to } \Omega_\epsilon \textit{ (i.e. } \overline{(\Omega_\epsilon \setminus Q)^w} \cap Q). \end{cases}$$

Then F has a fixed point in Q.

Remark. A special case of (7.2.1) is the condition (this is all we need for the applications in Section 7.3):

$$(7.2.1)^\star \quad \begin{cases} \text{if } \{(x_j, \lambda_j)\}_{j=1}^{\infty} \text{ is a sequence in } Q \times [0,1] \text{ with } x_j \rightharpoonup x \\ \text{and } \lambda_j \to \lambda \text{ and if } x = \lambda F(x) \text{ for } 0 \le \lambda < 1, \text{ then} \\ \lambda_j F(x_j) \in Q \text{ for } j \text{ sufficiently large.} \end{cases}$$

PROOF: Let $r : E \to Q$ be the weakly continuous retraction guaranteed by Theorem 7.1.4 (notice (i). $E^\star$ is separable since E is separable and reflexive and (ii). Q is weakly compact by Theorem's 7.1.5 and 7.1.6). Consider

$$B = \{x \in E : x = Fr(x)\}.$$

We claim $B \neq \emptyset$. To see this look at rF. Notice $rF : Q \to Q$ is weakly continuous. Also rF is a w–condensing map. To see this first notice $F(Q)$ is relatively weakly compact by Theorem 7.1.6 since $F(Q) \subseteq C$ and C is bounded. This together with the fact that $r : E \to Q$ is weakly continuous implies $r(\overline{F(Q)^w})$ is a weakly compact set so $r(F(Q))$ is relatively weakly compact. Thus $w(r(F(Q))) = 0$ so $rF : Q \to Q$ is a w–condensing map. Emmanuele's fixed point theorem (Theorem 7.1.1) implies that there exists $y \in Q$ with $y = rF(y)$. Let $z = F(y)$ and we have

$$Fr(z) = Fr(F(y)) = F(y) = z,$$

so $z \in B$ and $B \neq \emptyset$. In addition the weak continuity of Fr implies that B is weakly closed. Also B is weakly compact since $B \subseteq Fr(B) \subseteq F(Q)$ and so $w(B) \le w(F(Q)) = 0$.

We now show $B \cap Q \neq \emptyset$. To do this we argue by contradiction. Suppose $B \cap Q = \emptyset$. Then since Q is a weakly closed set (since Q is closed and convex) and B is a weakly compact set we have from Theorem 7.1.7 that

$$d(B, Q) = \inf\{\, \|x - y\| : x \in B,\, y \in Q\,\} > 0.$$

Thus there exists $\epsilon > 0$, $\epsilon \le \delta$, with $\Omega_\epsilon \cap B = \emptyset$; here $\Omega_\epsilon = \{x \in E : d(x, Q) \le \epsilon\}$ and Ω_ϵ is weakly compact from Theorem 7.1.6 (since Ω_ϵ is a closed, convex, bounded subset of E). Also the weak topology on Ω_ϵ is metrizable [44]; let $d^\star$ denote the metric. For $i \in \{1, 2, ...\}$ let

$$U_i = \left\{ x \in \Omega_\epsilon : d^\star(x, Q) < \frac{\epsilon}{i} \right\}.$$

Fix $i \in \{1, 2, ...\}$. Now U_i is $d^\star$ open in Ω_ϵ so U_i is weakly open in Ω_ϵ. Also

$$\overline{U_i^w} = \overline{U_i^{d^\star}} = \left\{ x \in \Omega_\epsilon : d^\star(x, Q) \le \frac{\epsilon}{i} \right\}$$

and

$$\partial_{\Omega_\epsilon} U_i = \left\{ x \in \Omega_\epsilon : \, d^\star(x, Q) = \frac{\epsilon}{i} \right\}.$$

Now Theorem 7.2.4 (applied with $U = U_i$, $Q = \overline{U_i^w}$, $C = \Omega_\epsilon$, $F = Fr$ and $p = 0$) implies (since $\Omega_\epsilon \cap B = \emptyset$) that there exists $y_i \in \partial_{\Omega_\epsilon} U_i$ and $\lambda_i \in (0,1)$ with $y_i = \lambda_i Fr(y_i)$.

Remark. Notice $Fr : \overline{U_i^w} \to C$ (here $\overline{U_i^w}$ is the weak closure of U_i in Ω_ϵ) is a w–condensing map since $F(Q)$ is relatively weakly compact so $w(Fr(\overline{U_i^w})) = 0$.

We can do this argument for each $i \in \{1, 2, ...\}$. Consequently for each $j \in \{1, 2, ...\}$ there exists $(y_j, \lambda_j) \in \partial_{\Omega_\epsilon} U_j \times (0,1)$ with $y_j = \lambda_j \, Fr(y_j)$. Notice in particular since $y_j \in \partial_{\Omega_\epsilon} U_j$ that

$$\lambda_j \, Fr(y_j) \notin Q \quad \text{for} \quad j \in \{1, 2, ...\}. \tag{7.2.2}$$

We now claim that

$$D = \{ x \in E : \, x = \lambda Fr(x) \;\; \text{for some} \;\; \lambda \in [0,1] \}$$

is weakly compact. Clearly D is weakly closed since $Fr : E \to C$ is weakly continuous. Also since $D \subseteq co\,(F(Q) \cup \{0\})$ we have

$$w(D) \le w(co\,(F(Q) \cup \{0\})) = w(F(Q)) = 0.$$

Thus D is weakly compact (so weakly sequentially compact by the Eberlein Šmulian theorem (Theorem 7.1.8)).

This together with

$$d^\star(y_j, Q) = \frac{\epsilon}{j}, \;\; |\lambda_j| \le 1 \;\; \text{for} \;\; j \in \{1, 2, ...\},$$

implies that we may assume without loss of generality that

$$\lambda_j \to \lambda^\star \quad \text{and} \quad y_j \rightharpoonup y^\star \in \overline{(\Omega_\epsilon \setminus Q)^w} \cap \overline{Q^w} = \partial_{\Omega_\epsilon} Q.$$

Also since $y_j = \lambda_j Fr(y_j)$ we have that

$$y^\star = \lambda^\star Fr(y^\star) = \lambda^\star F(y^\star).$$

If $\lambda^\star = 1$ then $y^\star = F(y^\star)$ which contradicts $B \cap Q = \emptyset$. Hence we may assume $0 \le \lambda^\star < 1$. But in this case (7.2.1) with

$$x_j = r(y_j) \in Q \quad \text{and} \quad x = y^\star = r(y^\star) \in \partial_{\Omega_\epsilon} Q$$

implies $\lambda_j Fr(y_j) \in Q$ for j sufficiently large. This contradicts (7.2.2). Thus $B \cap Q \neq \emptyset$ so there exists $x \in Q$ with $x = Fr(x)$ i.e. $x = F(x)$. $\square$

7.3. Applications to operator equations.

In this section we present existence principles for nonlinear operators. We motivate our study by first considering the second order boundary value problem

$$\begin{cases} y'' + f(t, y, y') = 0 \text{ a.e. on } [0,1] \\ y(0) = y(1) = 0 \end{cases} \tag{7.3.1}$$

where $f : [0,1] \times \mathbf{R}^2 \to \mathbf{R}$ is a L^p–Carathéodory function, $p > 1$; by this we mean:
(a). $t \mapsto f(t, u, v)$ is measurable for all $(u, v) \in \mathbf{R}^2$
(b). $(u, v) \mapsto f(t, u, v)$ is continuous for a.e. $t \in [0,1]$
(c). for any $r > 0$ there exists $h_r \in L^p[0,1]$ such that $|f(t, u, v)| \leq h_r(t)$ for a.e. $t \in [0,1]$ and all $|u| \leq r$, $|v| \leq r$.

By a solution to (7.3.1) we mean a function $y \in W^{2,p}[0,1]$ (i.e. $y' \in AC[0,1]$ and $y'' \in L^p[0,1]$) which satisfies the differential equation in (7.3.1) a.e. and $y(0) = y(1) = 0$.

Define operators

$$H_1,\ H_2 : L^p[0,1] \to C[0,1] \subseteq L^p[0,1]$$

by

$$H_1 u(t) = \int_0^1 G(t, s) u(s)\, ds \quad \text{and} \quad H_2 u(t) = \int_0^1 G_t(t, s) u(s)\, ds$$

where

$$G(t, s) = \begin{cases} (t-1)s, & 0 \leq s \leq t \leq 1 \\ (s-1)t, & 0 \leq t \leq s \leq 1. \end{cases}$$

It is easy to check that solving (7.3.1) is equivalent to finding a solution $u \in L^p[0,1]$ to

$$u = -f(t, H_1(u), H_2(u)). \tag{7.3.2}$$

Remark. If u is a solution of (7.3.2) then $y(t) = \int_0^1 G(t, s) u(s)\, ds$ is a solution of (7.3.1) whereas if w is a solution of (7.3.1) then $v = w''$ is a solution of (7.3.2).

Define an operator $N : L^p[0,1] \to L^p[0,1]$ by

$$Nu(t) = -f(t, H_1(u(t)), H_2(u(t))). \tag{7.3.3}$$

Consequently solving (7.3.1) is equivalent to finding a fixed point $u \in L^p[0,1]$ to

$$u = N\,u. \tag{7.3.4}$$

Let Q be a bounded, closed, convex subset of $L^p[0,1]$. We claim that $N : Q \to L^p[0,1]$ is weakly continuous (we need only check N is weakly sequentially continuous [55, page 93]). Suppose $y_n, w \in Q$ with $y_n \rightharpoonup w$ in $L^p[0,1]$ (i.e. $\int_0^1 y_n\, g\, dt \to \int_0^1 w\, g\, dt$ for all $g \in L^q[0,1]$; here $\frac{1}{p} + \frac{1}{q} = 1$). We must show $Ny_n \rightharpoonup Nw$ in $L^p[0,1]$. Let $g \in L^q[0,1]$. Notice

$$\begin{aligned}
&\left|\int_0^1 (Ny_n - Nw) g\, dt\right| \\
= &\left|\int_0^1 [f(t, H_1(y_n), H_2(y_n)) - f(t, H_1(w), H_2(w))] g\, dt\right| \\
\le &\left(\int_0^1 |f(t, H_1(y_n), H_2(y_n)) - f(t, H_1(w), H_2(w))|^p\, dt\right)^{\frac{1}{p}} \\
&\times \left(\int_0^1 |g|^q\, dt\right)^{\frac{1}{q}}.
\end{aligned}$$

If we show

$$\int_0^1 |f(t, H_1(y_n), H_2(y_n)) - f(t, H_1(w), H_2(w))|^p\, dt \to 0 \tag{7.3.5}$$

as $y_n \rightharpoonup w$, then $\int_0^1 (Ny_n)\, g\, dt \to \int_0^1 (Nw)\, g\, dt$. As a result $Ny_n \rightharpoonup Nw$ in $L^p[0,1]$ and so $N : Q \to L^p[0,1]$ is weakly continuous. It remains to prove (7.3.5). First we show, for each $t \in [0,1]$,

$$\text{if } y_n \rightharpoonup w \text{ then } H_i(y_n(t)) \to H_i(w(t)),\ i = 1, 2. \tag{7.3.6}$$

We prove (7.3.6) with $i = 1$; the case $i = 2$ is similar. Fix $t \in [0,1]$. Then

$$|H_1(y_n(t)) - H_1(w(t))| = \left|\int_0^1 G(t,s)[y_n(s) - w(s)]\, ds\right| \to 0 \text{ as } y_n \rightharpoonup w$$

since $G(t,s) \in L^q[0,1]$. Now (7.3.6) together with the fact that f is L^p–Carathéodory implies if $y_n \rightharpoonup w$ then

$$f(t, H_1(y_n(t)), H_2(y_n(t))) \to f(t, H_1(w(t)), H_2(w(t))) \tag{7.3.7}$$

a.e. on $[0,1]$. Also for $u \in Q$ we have

$$|H_1 u(t)| = |\int_0^1 G(t,s)u(s)\,ds| \le \left(\int_0^1 |u|^p\,dt\right)^{\frac{1}{p}} \sup_{t\in[0,1]} \left(\int_0^1 |G(t,s)|^q\,ds\right)^{\frac{1}{q}}$$

for all $t \in [0,1]$. Since $Q \subseteq L^p[0,1]$ is bounded there exists $r > 0$ with

$$|H_i u(t)| \le r \text{ for all } t \in [0,1] \text{ and } u \in Q,\ i = 1,2. \tag{7.3.8}$$

Now (7.3.5) follows immediately from (7.3.7), (7.3.8), the fact that f is L^p–Carathéodory, and the Lebesgue dominated convergence theorem [57].

Theorem 7.3.1. *Let f be a L^p–Carathéodory function, $p > 1$. In addition suppose there is a constant M_0, independent of λ, with*

$$\|y''\|_{L^p} = \left(\int_0^1 |y''|^p\,dt\right)^{\frac{1}{p}} \le M_0$$

for any solution y to

$$\begin{cases} y'' + \lambda f(t,y,y') = 0 & a.e. \text{ on } [0,1] \\ y(0) = y(1) = 0 \end{cases} \tag{7.3.9$_\lambda$}$$

for each $\lambda \in (0,1)$. Then (7.3.1) has at least one solution.

PROOF: Let

$$Q = \{u \in L^p[0,1] : \|u\|_{L^p} \le M_0 + 1\}$$

and notice Q is a closed, bounded, convex subset of $L^p[0,1]$. Solving $(7.3.9)_\lambda$ is equivalent to finding a fixed point $u \in L^p[0,1]$ of

$$u = \lambda N u; \tag{7.3.10$_\lambda$}$$

here N is as defined in (7.3.3).

Remark. If y is a solution of $(7.3.9)_\lambda$ then $u = y''$ is a solution of $(7.3.10)_\lambda$ whereas if w a solution of $(7.3.10)_\lambda$ then $v(t) = \int_0^1 G(t,s)w(s)\,ds$ is a solution of $(7.3.9)_\lambda$.

We know $N : Q \to L^p[0,1]$ is weakly continuous. Also since $\overline{N(Q)}^w$ and $\Omega_1 = \{x \in L^p[0,1] : d(x,Q) \le 1\}$ are weakly compact then the Krein Šmulian theorem (Theorem 7.1.9) implies that

$$C = \overline{co}\,(\overline{N(Q)}^w \cup \Omega_1)$$

is a closed, bounded, convex subset of $L^p[0,1]$. The result follows immediately from Theorem 7.2.5 once we show that condition $(7.2.1)^\star$ is satisfied. Take a sequence $\{(x_j, \lambda_j)\}_{j=1}^{\infty}$ in $Q \times [0,1]$ with $\lambda_j \to \lambda$ and $x_j \rightharpoonup x$ with $x = \lambda N(x)$, $0 \le \lambda < 1$. We need to show $\lambda_j N(x_j) \in Q$ for j sufficiently large.

First notice since $x_j \rightharpoonup x$ we have (as in (7.3.5)) that

$$\begin{aligned} & \int_0^1 |N(x_j(t)) - N(x(t))|^p \, dt \\ = & \int_0^1 |f(t, H_1(x_j(t)), H_2(x_j(t))) - f(t, H_1(x(t)), H_2(x(t)))|^p \, dt \\ \to & \ 0. \end{aligned}$$

Then given $\epsilon > 0$ (say $\epsilon < \frac{1}{3}$) there exists $j_0 \in \{1, 2, ...\}$ with

$$\|N(x_j)\|_{L^p} \le \|N(x)\|_{L^p} + \epsilon \ \text{ for } \ j \ge j_0. \tag{7.3.11}$$

Also $x = \lambda N(x)$ together with the fact that $\|u\|_{L^p} \le M_0$ for any solution u to $(7.3.10)_\lambda$ implies

$$\|\lambda N(x)\|_{L^p} \le M_0. \tag{7.3.12}$$

Consequently (7.3.11), (7.3.12) and $\lambda_j \to \lambda$ implies that there exists $j_1 \in \{j_0, j_0 + 1, ...\}$ with

$$\|\lambda_j N(x_j)\|_{L^p} \le M_0 + \frac{1}{2}, \ j \ge j_1.$$

Thus $\lambda_j N(x_j) \in Q$ for j sufficiently large so $(7.2.1)^\star$ is satisfied. Theorem 7.2.5 now guarantees that $(7.3.10)_1$ has a solution (and consequently (7.3.1) has a solution). □

Essentially the same reasoning as in Theorem 7.3.1 now establishes immediately a general existence principle for the operator equation

$$u = T(u) \tag{7.3.13}$$

where $T : L^p([0,1], \mathbf{R}^n) \to L^p([0,1], \mathbf{R}^n)$, $p > 1$.

Theorem 7.3.2. *Let Q be a closed bounded convex subset of $L^p([0,1], \mathbf{R}^n)$, $p > 1$. Also assume $T : Q \to L^p([0,1], \mathbf{R}^n)$ is weakly continuous and that*

$$\begin{cases} \text{if } \{x_j\}_{j=1}^{\infty} \text{ is a sequence in } Q \text{ with } x_j \rightharpoonup x \\ \text{then } \int_0^1 |T(x_j(t)) - T(x(t))|^p \, dt \to 0 \end{cases} \tag{7.3.14}$$

holds. In addition suppose there is a constant M_0, independent of λ, with $\|u\|_{L^p} \leq M_0$ for any solution u to

$$u = \lambda T(u) \tag{7.3.15$_\lambda$}$$

for each $\lambda \in (0,1)$. Then (7.3.13) has at least one solution in Q.

Remarks. (i). In Theorem 7.3.2 to show $T : Q \to L^p([0,1], \mathbf{R}^n)$ is weakly continuous we need only show that $T : Q \to L^p([0,1], \mathbf{R}^n)$ is weakly sequentially continuous (this follows immediately from [55, pg. 93]).
(ii). In Theorem 7.3.2, $L^p([0,1], \mathbf{R}^n)$ may be replaced by $L^p([0,1], B)$ where B is a separable and reflexive Banach space (notice $L^p([0,1], B)$ is a separable and reflexive Banach space [57]).

CHAPTER 8

INTEGRAL EQUATIONS IN BANACH SPACES

8.1. Introduction.

This chapter studies Volterra and Urysohn integral equations in a ball of a Banach space E when the nonlinear kernel satisfies Carathéodory type conditions. Such integral equations are usually studied when the kernel $f(t, s, x)$ satisfies a global set-Lipschitz condition of the form

$$\alpha(f([0, T] \times [0, T] \times M)) \leq c\,\alpha(M) \tag{8.1.1}$$

for each bounded set $M \subset E$, with a suitable small positive constant c, α being the Kuratowski measure of noncompactness. In this chapter we consider the more general situation when the global condition (8.1.1) is replaced by a local one (see [140]), namely

$$\alpha(f(t, s, M)) \leq \omega(t, s, \alpha(M)) \tag{8.1.2}$$

for all $t \in [0, T]$, a.e. $s \in [0, T]$ and any bounded set $M \subset E$.

The existence principles do not require uniform continuity of f and are based upon the continuation theorem of Mönch [112] and a result by Heinz [79] concerning the interchanging of α and integral for countable sets of Bochner integrable functions. More applicable existence results are derived from the general principles by means of differential and integral inequalities.

Throughout this chapter E will be a real Banach space with norm $|\,.\,|$. For every $x \in E$ and $R > 0$, let $B_R(x)$ and $\overline{B}_R(x)$ be the open and closed balls

$$B_R(x) = \{y \in E;\ |x - y| < R\}, \quad \overline{B}_R(x) = \{y \in E;\ |x - y| \leq R\}.$$

We denote by $C([a, b]; E)$ the space of continuous functions $u : [a, b] \to E$ and by $|\,.\,|_\infty$ its max-norm $|u|_\infty = \max\limits_{t \in [a,b]} |u(t)|$. For any subset $M \subset E$, we denote by $C([a, b]; M)$ the set of all functions in $C([a, b]; E)$ which take values in M.

A function $u : [a,b] \to E$ is said to be *finitely-valued* if it is constant $\neq 0$ on each of a finite number of disjoint measurable sets I_j and $u(t) = 0$ on $[a,b] \setminus \bigcup_j I_j$. Let the value of u on I_j be x_j and let $\chi(I_j)$ be the characteristic function of I_j, $\chi(I_j)(t) = 1$ if $t \in I_j$, $\chi(I_j)(t) = 0$ if $t \in [a,b] \setminus I_j$. Then u can be represented as a finite sum

$$u = \sum_j x_j \chi(I_j)$$

and the element

$$\sum_j x_j \text{ mes } (I_j) \in E$$

is defined as the *Bochner integral* of u over $[a,b]$ and is denoted by $\int_a^b u(s)\,ds$. More generally, a function $u : [a,b] \to E$ is said to be *Bochner integrable* on $[a,b]$ if there exists a sequence of finitely-valued functions u_n with

(8.1.3) $$u_n(t) \to u(t) \quad \text{as } n \to \infty, \text{ a.e. } t \in [a,b]$$

(i.e. u is *strongly measurable*) and

(8.1.4) $$\int_a^b |u_n(s) - u(s)|\,ds \to 0 \quad \text{as } n \to \infty.$$

In this case the *Bochner integral* of u is defined by

$$\int_a^b u(s)\,ds = \lim_{n\to\infty} \int_a^b u_n(s)\,ds.$$

Recall that a strongly measurable function u is Bochner integrable if and only if $|u|$ is Lebesgue integrable (see [166, Theorem 5.5.1]).

For any real $p \in [1,\infty]$, we consider the space $L^p([a,b];E)$ of all strongly measurable functions $u : [a,b] \to E$ such that $|u|^p$ is Lebesgue integrable on $[a,b]$. $L^p([a,b];E)$ is a Banach space under the norm

$$|u|_p = \left(\int_a^b |u(s)|^p\,ds \right)^{1/p}$$

for $p < \infty$ and

$$|u|_\infty = \text{ess} \sup_{t\in[a,b]} |u(t)| = \inf\{c \geq 0;\ |u(t)| \leq c \text{ a.e. } t \in [a,b]\}.$$

When this will be important, we shall denote $|u|_p$ also by $|u|_{L^p([a,b];E)}$. In particular, $L^1([a,b];E)$ is the space of Bochner integrable functions on $[a,b]$. When $E = \mathbf{R}$, the space $L^p([a,b];\mathbf{R})$ is simply denoted by $L^p[a,b]$.

Recall that a function $\phi : [a,b] \times D \to E$, $D \subset E$, is said to be *L^p-Carathéodory* $(1 \leq p \leq \infty)$ if $\phi(\,.\,,x)$ is strongly measurable for each $x \in D$, $\phi(t,\,.\,)$ is continuous for a.e. $t \in [a,b]$ and for each $r > 0$ there exists $h_r \in L^p[a,b]$ with $|\phi(t,x)| \leq h_r(t)$ for all $x \in D$ satisfying $|x| \leq r$ and a.e. $t \in [a,b]$.

Now we recall the definition of the *Kuratowski measure of noncompactness* and the *Hausdorff ball measure of noncompactness*. Let $M \subset E$ be bounded. Then

$$\alpha(M) = \inf\left\{\varepsilon > 0;\ M \subset \cup_{j=1}^m M_j \text{ and diam } (M_j) \leq \varepsilon\right\}$$

and

$$\beta(M) = \inf\left\{\varepsilon > 0;\ M \subset \cup_{j=1}^m B_\varepsilon(x_j) \text{ where } x_j \in E\right\}.$$

If F is a linear subspace of E and $M \subset F$ is bounded, then we define

$$\beta_F(M) = \inf\left\{\varepsilon > 0;\ M \subset \cup_{j=1}^m B_\varepsilon(x_j) \text{ where } x_j \in F\right\}.$$

We have

$$\beta(M) \leq \alpha(M) \leq 2\beta(M) \text{ for } M \subset E \text{ bounded} \tag{8.1.5}$$

and

$$\beta(M) \leq \beta_F(M) \leq \alpha(M) \text{ for } M \subset F \text{ bounded.} \tag{8.1.6}$$

For a separable Banach space E, the ball measure of noncompactness β has the following representation on countable sets.

Proposition 8.1.1. *([112]) Let E be a separable Banach space and (E_n) an increasing sequence of finite dimensional subspaces with $E = \overline{\cup_{n\in\mathbf{N}} E_n}$. Then for every bounded countable set $M = \{x_m;\ m \in \mathbf{N}\} \subset E$, we have*

$$\beta(M) = \lim_{n\to\infty} \overline{\lim_{m\to\infty}}\, d(x_m, E_n)$$

(here $d(x, E_n) = \inf_{y\in E_n} |x - y|$).

Let γ be α or β. The next proposition gives the representation of γ on bounded equicontinuous sets of $C([a,b];E)$ and its property of interchanging with integral.

Proposition 8.1.2. *([30]) Let E be a Banach space and $M \subset C([a,b];E)$ bounded and equicontinuous. Then the function $\mu : [a,b] \to \mathbf{R}$ given by $\mu(t) = \gamma(M(t))$ is continuous on $[a,b]$,*

$$\gamma(M) = \max_{t\in[a,b]} \gamma(M(t))$$

and

$$\gamma\left(\int_a^b M(s)\,ds\right) \le \int_a^b \gamma(M(s))\,ds \tag{8.1.7}$$

(here $\int_a^b M(s)\,ds$ stands for the set $\left\{\int_a^b u(s)\,ds;\ u \in M\right\} \subset E$).

A result of type (8.1.7) holds without assuming the equicontinuity of M.

Proposition 8.1.3. *(Mönch - von Harten [113]) Let E be a separable Banach space and $M \subset C([a,b];E)$ countable with $|u(t)| \le h(t)$ on $[a,b]$ for every $u \in M$, where $h \in L^1[a,b]$. Then the function $\psi : [a,b] \to \mathbf{R}$ given by $\psi(t) = \beta(M(t))$ belongs to $L^1[a,b]$ and*

$$\beta\left(\int_a^b M(s)\,ds\right) \le \int_a^b \beta(M(s))\,ds. \tag{8.1.8}$$

Now we state the extension of Proposition 8.1.3 for countable sets of Bochner integrable functions.

Proposition 8.1.4. *(Heinz [79]) (a) If E is a separable Banach space and $M \subset L^1([a,b];E)$ countable with $|u(t)| \le h(t)$ for a.e. $t \in [a,b]$ and every $u \in M$, where $h \in L^1[a,b]$, then the function $\psi(t) = \beta(M(t))$ belongs to $L^1[a,b]$ and satisfies* (8.1.8).

(b) If E is a Banach space (which is not necessarily separable) and $M \subset L^1([a,b];E)$ countable with $|u(t)| \le h(t)$ for a.e. $t \in [a,b]$ and every $u \in M$, where $h \in L^1[a,b]$, then the function $\varphi(t) = \alpha(M(t))$ belongs to $L^1[a,b]$ and satisfies

$$\alpha\left(\int_a^b M(s)\,ds\right) \le 2\int_a^b \alpha(M(s))\,ds. \tag{8.1.9}$$

Remark. A proof of Proposition 8.1.4 can be found in Heinz [79] in a more general setting. However in our setting an easier proof can be presented for (a) and we include it here for the convenience of the reader.

PROOF: Let $M = \{u_m;\ m \in \mathbf{N}\}$.

(a) The fact that $\psi \in L^1[a,b]$ is a direct consequence of Proposition 8.1.1. Now, let (E_n) be any increasing sequence of finite dimensional subspaces of E with $E = \overline{\bigcup_{n\in\mathbf{N}} E_n}$. Let us fix $n, m \in \mathbf{N}$ and take any $\varepsilon > 0$. Since u_m is Bochner integrable, there is a finitely-valued function $\widetilde{u}_m$, say

$$\widetilde{u}_m = \sum_{j=1}^{k_m} x_{mj}\chi(I_{mj}),$$

where $\chi(I_{mj})$ is the characteristic function of a measurable set $I_{mj} \subset [a,b]$, such that

$$\int_a^b |\widetilde{u}_m(s) - u_m(s)|\,ds \le \varepsilon. \tag{8.1.10}$$

The sublinearity of $d(\,.\,, E_n)$ gives

$$d\left(\sum_{j=1}^{k_m} x_{mj}\ \mathrm{mes}\,(I_{mj}),\ E_n\right) \le \sum_{j-1}^{k_m} d(x_{mj}, E_n)\ \mathrm{mes}\,(I_{mj}). \tag{8.1.11}$$

It is clear that

$$\sum_{j=1}^{k_m} x_{mj}\ \mathrm{mes}\,(I_{mj}) = \int_a^b \widetilde{u}_m(s)\,ds$$

and

$$\sum_{j=1}^{k_m} d(x_{mj}, E_n)\ \mathrm{mes}\,(I_{mj}) = \int_a^b d(\widetilde{u}_m(s), E_n)\,ds.$$

Thus (8.1.11) can be rewritten as

$$d\left(\int_a^b \widetilde{u}_m(s)\,ds,\ E_n\right) \le \int_a^b d(\widetilde{u}_m(s), E_n)\,ds. \tag{8.1.12}$$

On the other hand, using (8.1.10), we have

$$\begin{aligned} d\left(\textstyle\int_a^b \widetilde{u}_m(s)\,ds,\ E_n\right) &\ge d\left(\textstyle\int_a^b u_m(s)\,ds,\ E_n\right) \\ &\quad -d\left(\textstyle\int_a^b u_m(s)\,ds,\ \int_a^b \widetilde{u}_m(s)\,ds\right) \\ &\ge d\left(\textstyle\int_a^b u_m(s)\,ds,\ E_n\right) - \varepsilon \end{aligned} \tag{8.1.13}$$

and

$$
\begin{aligned}
(8.1.14) \qquad \int_a^b d(\widetilde{u}_m(s), E_n)\, ds &\leq \int_a^b d(u_m(s), E_n)\, ds \\
&\quad + \int_a^b d(u_m(s), \widetilde{u}_m(s))\, ds \\
&\leq \int_a^b d(u_m(s), E_n)\, ds + \varepsilon.
\end{aligned}
$$

Now (8.1.12) – (8.1.14) imply

$$d\left(\int_a^b u_m(s)\, ds,\, E_n\right) \leq \int_a^b d(u_m(s), E_n)\, ds + 2\varepsilon.$$

Letting $\varepsilon \searrow 0$, we obtain

$$d\left(\int_a^b u_m(s)\, ds,\, E_n\right) \leq \int_a^b d(u_m(s), E_n)\, ds.$$

By means of Fatou's lemma, we have

$$\overline{\lim_{m\to\infty}}\, d\left(\int_a^b u_m(s)\, ds,\, E_n\right) \leq \int_a^b \overline{\lim_{m\to\infty}}\, d(u_m(s), E_n)\, ds.$$

Finally, since

$$\overline{\lim_{m\to\infty}}\, d(u_m(s), E_n) \leq h(t),$$

using the Lebesgue dominated convergence theorem, we find

$$\lim_{n\to\infty} \overline{\lim_{m\to\infty}}\, d\left(\int_a^b u_m(s)\, ds,\, E_n\right) \leq \int_a^b \lim_{n\to\infty} \overline{\lim_{m\to\infty}}\, d(u_m(s), E_n)\, ds$$

which is exactly (8.1.8).

(b) Let F be a separable closed linear subspace of E such that $u_m(t) \in F$ for every $m \in \mathbf{N}$ and for every $t \in [a,b]$ outside a fixed set of measure zero (the subspace F can be obtained as follows: for any m, u_m being Bochner integrable can be approximated in the sense of (8.1.3) – (8.1.4) by a sequence $(u_{mn})_{n\in\mathbf{N}}$ of finitely-valued functions,

$$u_{mn} = \sum_{j=1}^{k_{mn}} x_{mnj}\, \chi(I_{mnj}).$$

Then we take as F the closure in E of the subspace generated by the countable set of elements x_{mnj}, $1 \leq j \leq k_{mn}$, $m \in \mathbf{N}$, $n \in \mathbf{N}$). From (a), (8.1.5) and (8.1.6) we have

$$\alpha(M) \leq 2\beta_F(M) \leq 2\int_a^b \beta_F(M(t))\, dt \leq 2\int_a^b \alpha(M(t))\, dt. \ \square$$

We finish this section with two fixed point results due to Mönch [112] (see also [55, Ch. 5.18]). The first one contains as particular cases the fixed point theorems of Schauder, Darbo and Sadovskii. The proofs of Theorem 8.1.5 and Theorem 8.1.6 can be found in Chapter 10.

Theorem 8.1.5. *(Mönch [112]) Let K be a closed convex subset of a Banach space X and $N : K \to K$ continuous with the further property that for some $x_0 \in K$, we have*

$$\text{(8.1.15)} \quad M \subset K \ \textit{countable}, \ \overline{M} = \overline{co}\left(\{x_0\} \cup N(M)\right) \Longrightarrow \overline{M} \ \textit{compact}.$$

Then N has a fixed point.

The second result is the continuation analogue of Theorem 8.1.5.

Theorem 8.1.6. *(Mönch [112]) Let K be a closed convex subset of a Banach space X, U a relatively open subset of K and $N : \overline{U} \to K$ continuous with the further property that for some $x_0 \in U$, we have*

$$\text{(8.1.16)} \quad M \subset \overline{U} \ \textit{countable}, \ M \subset \overline{co}\left(\{x_0\} \cup N(M)\right) \Longrightarrow \overline{M} \ \textit{compact}.$$

In addition, assume

$$\text{(8.1.17)} \quad x \neq (1-\lambda)\,x_0 + \lambda N(x) \ \textit{ for all } x \in \overline{U} \setminus U \ \textit{ and } \lambda \in (0,1).$$

Then N has a fixed point in $\overline{U}$.

8.2. Existence criteria for Volterra integral equations.

In this section we establish existence criteria for the Volterra integral equation

$$\text{(8.2.1)} \qquad u(t) = \int_0^t f(t,s,u(s))\,ds, \quad t \in [0,T]$$

in a ball of the Banach space $(E, |.|)$, under Carathéodory conditions on f.

Let $R > 0$ and $T > 0$. We denote by B the closed ball $\overline{B}_R(0)$ of E and we consider

$$\Lambda = \{(t,s)\,;\ t \in [0,T], \ \ s \in [0,t]\}$$

and $D = \Lambda \times B$. Hence

$$D = \{(t,s,x)\,;\ t \in [0,T], \ \ s \in [0,t], \ x \in E \ \text{ and } \ |x| \le R\}.$$

We assume that $f : D \to E$ and we look for solutions u in $C([0,T];E)$ with $|u(t)| \leq R$ for all $t \in [0,T]$.

For a fixed $t \in [0,T]$, let $f_t : [0,t] \times B \to E$ be the map given by

$$f_t(s,x) = f(t,s,x).$$

Theorem 8.2.1. *Suppose*

(A). for each $t \in [0,T]$, the map f_t is L^1-Carathéodory uniformly in t, in the sense that there exists a bounded function $\eta : \Lambda \to \mathbf{R}_+$ with

$$\eta(t,t') \to 0 \quad \text{as } t - t' \searrow 0$$

and

$$\int_{t'}^{t} \sup_{|x| \leq R} |f_t(s,x)|\, ds \leq \eta(t,t')$$

for $0 \leq t' < t \leq T$;

(B). for each $t \in [0,T]$,

$$\int_0^{t_0} \sup_{|x| \leq R} |f_t(s,x) - f_{t'}(s,x)|\, ds \to 0 \quad \text{as } t' \to t$$

where $t_0 = \min\{t, t'\}$;

(C). there exists $\omega : \Lambda \times [0,2R] \to \mathbf{R}$ such that for each $t \in [0,T]$,

$$\omega_t = \omega(t,.,.) \quad \text{is } L^1\text{-Carathéodory,}$$

$$\alpha(f(t,s,M)) \leq \omega(t,s,\alpha(M)) \tag{8.2.2}$$

for a.e. $s \in [0,t]$ and every $M \subset B$, and the unique solution $\varphi \in C([0,T];[0,2R])$ of the inequality

$$\varphi(t) \leq 2\int_0^t \omega(t,s,\varphi(s))\, ds, \quad t \in [0,T]$$

is $\varphi \equiv 0$;

(D). $|u|_\infty < R$ for any solution $u \in C([0,T];B)$ to

$$u(t) = \lambda \int_0^t f(t,s,u(s))\, ds, \quad t \in [0,T] \tag{8.2.3}$$

for each $\lambda \in (0,1)$.

Then (8.2.1) *has a solution in* $C([0,T];B)$.

PROOF: We shall apply Theorem 8.1.6 to $K = X = C([0,T];E)$ with norm $|\,.\,|_\infty$, $U = \{u \in C([0,T];E);\ |u|_\infty < R\}$, x_0 the null function and $N : \overline{U} \to C([0,T];E)$ given by

$$N(u)(t) = \int_0^t f(t,s,u(s))\,ds.$$

Since f_t is L^1-Carathéodory, standard arguments yield the conclusion that for each $u \in \overline{U}$, the function $f_t(\,.\,,u(\,.\,))$ is Bochner integrable on $[0,t]$. In addition

$$\text{(8.2.4)}\qquad \begin{aligned}|N(u)(t)| &\le \textstyle\int_0^t |f(t,s,u(s))|\,ds \le \int_0^t \sup_{|x|\le R} |f_t(s,x)|\,ds\\ &\le \eta(t,0) \le \sup_\Lambda \eta < \infty.\end{aligned}$$

Also, for $u \in \overline{U}$ and every $t,\ t' \in [0,T]$, we have: if $t < t'$, then

$$\begin{aligned}|N(u)(t) - N(u)(t')| &\le \int_0^t |f(t,s,u(s)) - f(t',s,u(s))|\,ds\\ &+ \int_t^{t'} |f(t',s,u(s))|\,ds\\ &\le \int_0^t \sup_{|x|\le R} |f_t(s,x) - f_{t'}(s,x)|\,ds + \eta(t',t),\end{aligned}$$

while if $t' < t$, then

$$|N(u)(t) - N(u)(t')| \le \int_0^{t'} \sup_{|x|\le R} |f_t(s,x) - f_{t'}(s,x)|\,ds + \eta(t,t').$$

Hence, in both cases

$$\text{(8.2.5)}\qquad \begin{aligned}&|N(u)(t) - N(u)(t')|\\ &\quad\le \textstyle\int_0^{t_0} \sup_{|x|\le R} |f_t(s,x) - f_{t'}(s,x)|\,ds + \eta(t+t'-t_0,\ t_0)\end{aligned}$$

where $t_0 = \min\{t,\ t'\}$. Now (8.2.5) shows that $N(u) \in C([0,T];E)$ and $N(\overline{U})$ is equicontinuous on $[0,T]$. In addition, (8.2.4) shows that $N(\overline{U})$ is bounded.

Next we show that N is continuous. To see this, let $u_n \to u$ in $C([0,T];E)$, where $u_n,\ u \in \overline{U}$. Since f_t is L^1-Carathéodory, $f_t(s,\,.\,)$ is continuous for a.e. $s \in [0,t]$ and there exists $h_t \in L^1[0,t]$ with $|f_t(s,x)| \le h_t(s)$ for a.e. $s \in [0,t]$ and all $x \in B$. It follows that

$$f(t,s,u_n(s)) \to f(t,s,u(s)) \quad \text{as } n \to \infty$$

and

$$|f(t,s,u_n(s))| \le h_t(s)$$

for a.e. $s \in [0,t]$ and all $t \in [0,T]$. These together with the Lebesgue dominated convergence theorem yield

$$N(u_n)(t) \to N(u)(t) \quad \text{as } n \to \infty$$

for any $t \in [0,T]$. Now (8.2.5) guarantees that the convergence is uniform in t. Hence $N(u_n) \to N(u)$ in $C([0,T];E)$ as desired.

To check (8.1.16), let $M \subset \overline{U}$ be countable with

$$M \subset \overline{\text{co}}(\{0\} \cup N(M)). \tag{8.2.6}$$

Since $N(M) \subset N\left(\overline{U}\right)$ and $N\left(\overline{U}\right)$ is bounded and equicontinuous, from (8.2.6) we have that M is bounded and equicontinuous. To deduce that $\overline{M}$ is compact, that is $\alpha(M) = 0$ (in $C([0,T];E)$), by Proposition 8.1.2 we have to prove that $\alpha(M(t)) = 0$ (in E) for any $t \in [0,T]$. For this, let $\varphi :$ $[0,T] \to \mathbf{R}$ be given by $\varphi(t) = \alpha(M(t))$. Clearly $\varphi \in C([0,T];[0,2R])$. Now, using Proposition 8.1.4 (b), (8.2.2) and (8.2.6), we obtain

$$\begin{aligned}
\varphi(t) &= \alpha(M(t)) \le \alpha(N(M)(t)) = \alpha\left(\int_0^t f_t(s,M(s))\,ds\right) \\
&\le 2\int_0^t \alpha(f_t(s,M(s)))\,ds \le 2\int_0^t \omega(t,s,\alpha(M(s)))\,ds \\
&= 2\int_0^t \omega(t,s,\varphi(s))\,ds.
\end{aligned}$$

Now (C) guarantees $\varphi \equiv 0$ as desired.

Finally (D) guarantees (8.1.17). Thus Theorem 8.1.6 applies. □

A special case of (8.2.1) is

$$u(t) = \int_0^t k(t,s)\,g(s,u(s))\,ds, \quad t \in [0,T] \tag{8.2.7}$$

where $k : \Lambda \to \mathbf{R}$.

Theorem 8.2.2. *Let $k : \Lambda \to \mathbf{R}$ and $g : [0,T] \times B \to E$. Suppose*

(a). g is L^q-Carathéodory for some $q \ge 1$ and $k_t = k(t,.) \in L^p[0,t]$ for any $t \in [0,T]$, where $1/p + 1/q = 1$;

(b). for each $t \in [0,T]$, we have

$$|k_t - k_{t'}|_{L^p[0,t_0]} \to 0 \quad \text{as } t' \to t$$

where $t_0 = \min\{t, t'\}$;

(c). there exists $\omega_0 : [0,T] \times [0,2R] \to \mathbf{R}$ L^q*-Carathéodory with*

$$\alpha(g(s,M)) \le \omega_0(s, \alpha(M)) \tag{8.2.8}$$

for a.e. $s \in [0,T]$ *and* $M \subset B$, *such that the unique solution* $\varphi \in C([0,T];[0,2R])$ *of the inequality*

$$\varphi(t) \le 2\int_0^t |k(t,s)|\, \omega_0(s, \varphi(s))\, ds, \quad t \in [0,T]$$

is $\varphi \equiv 0$;

(d). $|u|_\infty < R$ *for any solution* $u \in C([0,T];B)$ *to*

$$u(t) = \lambda \int_0^t k(t,s)\, g(s, u(s))\, ds, \quad t \in [0,T] \tag{8.2.9}$$

for each $\lambda \in (0,1)$.

Then (8.2.7) *has a solution in* $C([0,T];B)$.

PROOF: The result follows from Theorem 8.2.1. Here we take $f(t,s,x) = k(t,s)\, g(s,x)$ and

$$\eta(t,t') = \left(\int_{t'}^t h(s)^q\, ds\right)^{1/q} \sup_{\tau\in[0,T]} |k_\tau|_{L^p[0,\tau]}, \quad 0 \le t' \le t \le T \tag{8.2.10}$$

where $h \in L^q[0,T]$ is such that $|g(s,x)| \le h(s)$ for all $x \in B$ and a.e. $s \in [0,T]$. Also $\omega(t,s,\tau) = |k(t,s)|\, \omega_0(s,\tau)$. Note that the supremum in (8.2.10) is finite because of (b). □

The next result contains a sufficient condition for (d).

Theorem 8.2.3. *Let* $k : \Lambda \to \mathbf{R}$ *and* $g : [0,T] \times B \to E$. *Assume (a)-(c) hold. Also suppose that*

(d'). there exists $\delta \in L^1[0,T]$ *and* $w : (0,R] \to (0,\infty)$ *continuous and nondecreasing such that*

$$|k(t,s)\, g(s,x)| \le \delta(s)\, w(|x|)$$

for a.e. $s \in [0,t]$ *and all* $t \in [0,T]$, $x \in B \setminus \{0\}$, *and*

$$\int_0^T \delta(s)\, ds \le \int_0^R \frac{dr}{w(r)}. \tag{8.2.11}$$

Then (8.2.7) *has a solution in* $C([0,T];B)$.

PROOF: The result follows from Theorem 8.2.2 once we show (d) is true. Let $u \in C([0,T];B)$ be any solution to (8.2.9) for some $\lambda \in (0,1)$. Then

$$|u(t)| \leq \lambda \int_0^t |k(t,s)\, g(s,u(s))|\, ds \leq \lambda \int_0^t \delta(s)\, w(|u(s)|)\, ds$$

for all $t \in [0,T]$ (we put $w(0) = \lim_{t \searrow 0} w(t)$). Let

$$c(t) = \min\left\{R,\ \lambda \int_0^t \delta(s)\ w(|u(s)|)\, ds\right\}.$$

Clearly c is nondecreasing. We claim that $c(T) < R$. Suppose the contrary. Then since $c(0) = 0$, there exists a subinterval $[a,b] \subset [0,T]$ with

$$c(a) = 0,\ c(b) = R \text{ and } c(t) \in (0,R) \quad \text{for } t \in (a,b).$$

Since $|u(t)| \leq c(t) \leq R$ on $[a,b]$ and w is nondecreasing on $[0,R]$, we have

$$c'(s) = \lambda\, \delta(s)\ w(|u(s)|) \leq \lambda\, \delta(s)\ w(c(s)) \quad \text{a.e. } s \in [a,b].$$

Now integration from a to b yields

$$\begin{aligned} \int_a^b \frac{c'(s)}{w(c(s))} ds &= \int_0^R \frac{dr}{w(r)} \leq \lambda \int_a^b \delta(s)\, ds \\ &\leq \lambda \int_0^T \delta(s)\, ds < \int_0^T \delta(s)\, ds, \end{aligned}$$

a contradiction. Notice we may assume $|\delta|_{L^1[0,T]} > 0$ since otherwise we have nothing to prove. □

Observe that Theorem 8.2.3 can be derived directly from Theorem 8.1.5 if we take

$$K = \{u \in C([0,T];E);\ |u(t)| \leq b(t) \quad \text{for } t \in [0,T]\},$$

where

$$b(t) = I^{-1}\left(\int_0^t \delta(s)\, ds\right) \quad \text{and} \quad I(\tau) = \int_0^\tau \frac{dr}{w(r)}$$

(see the proof of Theorem 2.3 in [125]). Notice $b(t) \leq R$ for all $t \in [0,T]$ because of (8.2.11).

Corollary 8.2.4. *Let $k : \Lambda \to \mathbf{R}$ and $g : [0,T] \times B \to E$ with $g = g_1 + g_2$, where $g_1(\,.\,,0) = 0$ and g_2 is completely continuous. Assume (a)-(b) hold with $q = 1$, $p = \infty$ and*

$$|k_t|_{L^\infty[0,t]} \leq 1 \quad \text{for all } t \in [0,T]. \tag{8.2.12}$$

Also suppose that

(c). there exists $\delta \in L^1[0,T]$ and $w_1 : (0,2R] \to (0,\infty)$ continuous and nondecreasing with*

$$\int_0^{2R} \frac{dr}{w_1(r)} = \infty \tag{8.2.13}$$

and

$$|g_1(s,x) - g_1(s,y)| \leq \delta(s)\, w_1(|x-y|) \tag{8.2.14}$$

for a.e. $s \in [0,T]$ and all $x, y \in B$, $x \neq y$;

(d). there exists $w_2 : [0,R] \to \mathbf{R}_+$ continuous and nondecreasing such that*

$$|g_2(s,x)| \leq \delta(s)\, w_2(|x|) \quad \text{for a.e. } s \in [0,T] \text{ and all } x \in B \tag{8.2.15}$$

and (8.2.11) holds with $w = w_1 + w_2$.

Then (8.2.7) has a solution in $C([0,T];B)$.

PROOF: First we check (c). Using (8.2.14) we see that (8.2.8) holds for $\omega_0(s,r) = \delta(s)\, w_1(r)$. Now let $\varphi \in C([0,T];\,[0,2R])$ satisfies

$$\varphi(t) \leq 2\int_0^t |k(t,s)|\,\delta(s)\, w_1(\varphi(s))\, ds.$$

Then, by (8.2.12), we have

$$\varphi(t) \leq 2\int_0^t \delta(s)\, w_1(\varphi(s))\, ds.$$

Let

$$c(t) = 2\int_0^t \delta(s)\, w_1(\varphi(s))\, ds.$$

It is clear that c is nondecreasing. Then $\varphi \equiv 0$ once we show $c(T) = 0$. Suppose the contrary, i.e. $c(T) > 0$. Then, since $c(0) = 0$, for each $\varepsilon \in$

$(0, A)$, where $A = \min\{c(T),\ 2R\}$, there is a subinterval $[a, b] \subset [0, T]$ with

$$c(a) = \varepsilon,\ \ c(b) = A \ \text{ and } \ c(t) \in (\varepsilon,\ A) \quad \text{for all } \ t \in (a, b).$$

Now

$$\varphi(t) \le c(t) \le 2R$$

on $[a, b]$ and w_1 nondecreasing on $[0, 2R]$ guarantee that

$$c'(s) = 2\delta(s)\, w_1(\varphi(s)) \le 2\delta(s)\, w_1(c(s))$$

a.e. $s \in [a, b]$. Consequently

$$\int_{\varepsilon}^{A} \frac{dr}{w_1(r)} \le 2 \int_a^b \delta(s)\, ds \le 2\, |\delta|_{L^1[0,T]}\,.$$

This, for $\varepsilon \searrow 0$, yields a contradiction to (8.2.13). Thus $c(T) = 0$ and so $\varphi \equiv 0$.

Finally notice that (d') follows from (8.2.12), (8.2.14), $g_1(\,.\,, 0) = 0$ and (8.2.15). □

Example 8.2.1. We give an example of a function g_1 which satisfies (c*). Let $E = C(I; \mathbf{R})$, $I \subset \mathbf{R}$ compact interval, and let $g_1 : [0, T] \times B \to E$ given by

$$g_1(s, x)(\tau) = \delta(s)\, w_1(|x(\tau)|)\,, \quad \tau \in I \ \text{ and } \ x \in B,$$

where $\delta \in L^1([0, T]; \mathbf{R}_+)$ and $w_1 : [0, 2R] \to \mathbf{R}_+$ is continuous, nondecreasing and satisfies the following conditions:

$$w_1(0) = 0,\ w_1(r) > 0 \quad \text{on } \ (0, 2R]$$

(8.2.16) $$|w_1(r) - w_1(r')| \le w_1(|r - r'|) \quad \text{for } r,\, r' \in [0, 2R]$$

and

(8.2.17) $$\int_0^{2R} \frac{dr}{w_1(r)} = \infty.$$

Notice (8.2.14) can easily be deduced from (8.2.16). Examples of w_1 with the above properties are: $w_1(r) = r$ for any $R > 0$ and $w_1(r) = -r \ln r$ for $R = 1/(2e)$ (see [154]).

8.3. Existence criteria for Urysohn integral equations.

In this section we discuss the Urysohn integral equation

$$u(t) = \int_0^T f(t, s, u(s))\, ds, \quad t \in [0, T] \tag{8.3.1}$$

in a ball $B = \{x \in E;\ |x| \le R\}$ of the Banach space $(E, |\,.\,|)$.

Essentially the same reasoning as in Section 8.2 establish the following existence principles for (8.3.1).

Theorem 8.3.1. *Let* $f : [0, T]^2 \times B \to E$. *Suppose*
(A). for each $t \in [0, T]$, f_t *is* L^1*-Carathéodory and*

$$\sup_{t \in [0,T]} \int_0^T \sup_{|x| \le R} |f_t(s, x)|\, ds < \infty;$$

(B). one has

$$\int_0^T \sup_{|x| \le R} |f_t(s, x) - f_{t'}(s, x)|\, ds \to 0 \quad \text{as } t' \to t;$$

(C). there exists $\omega : [0, T]^2 \times [0, 2R] \to \mathbf{R}$ *such that for each* $t \in [0, T]$,

$$\omega_t \quad \text{is } L^1\text{-Carathéodory},$$

$$\alpha(f(t, s, M)) \le \omega(t, s, \alpha(M))$$

for a.e. $s \in [0, T]$, $M \subset B$, *and the unique* $\varphi \in C([0, T]; [0, 2R])$ *satisfying*

$$\varphi(t) \le 2 \int_0^T \omega(t, s, \varphi(s))\, ds, \quad t \in [0, T]$$

is $\varphi \equiv 0$;
(D). $|u|_\infty < R$ *for any solution* $u \in C([0, T]; B)$ *to*

$$u(t) = \lambda \int_0^T f(t, s, u(s))\, ds, \quad t \in [0, T]$$

for each $\lambda \in (0, 1)$.

Then (8.3.1) *has a solution in* $C([0, T]; B)$.

An immediate consequence of Theorem 8.3.1 is the following result for the Hammerstein integral equation

$$u(t) = \int_0^T k(t, s)\, g(s, u(s))\, ds, \quad t \in [0, T]. \tag{8.3.2}$$

Theorem 8.3.2. *Let* $k : [0,T]^2 \to \mathbf{R}$ *and* $g : [0,T] \times B \to E$. *Suppose*
(a). g is L^q-Carathéodory for some $q \geq 1$ and for each $t \in [0,T]$, $k_t \in L^p[0,T]$ where $1/p + 1/q = 1$;
(b). the map $t \longmapsto k_t$ is continuous from $[0,T]$ to $L^p[0,T]$;
(c). there exists $\omega_0 : [0,T] \times [0,2R] \to \mathbf{R}$ L^q-Carathéodory with

$$\alpha(g(s,M)) \leq \omega_0(s, \alpha(M)) \tag{8.3.3}$$

for a.e. $s \in [0,T]$ and $M \subset B$, such that the unique solution $\varphi \in C([0,T];[0,2R])$ of the inequality

$$\varphi(t) \leq 2\int_0^T |k(t,s)|\,\omega_0(s,\varphi(s))\,ds, \quad t \in [0,T] \tag{8.3.4}$$

is $\varphi \equiv 0$;
(d). $|u|_\infty < R$ for any solution $u \in C([0,T];B)$ to

$$u(t) = \lambda \int_0^T k(t,s)\, g(s,u(s))\,ds, \quad t \in [0,T] \tag{8.3.5}$$

for each $\lambda \in (0,1)$.
Then (8.3.2) has a solution in $C([0,T];B)$.

Theorem 8.3.2 is now used to obtain an applicable result for (8.3.2).

Corollary 8.3.3. *Let $k : [0,T]^2 \to \mathbf{R}$ and $g : [0,T] \times B \to E$ with $g = g_1 + g_2$, where $g_1(\,.\,,0) = 0$ and g_2 is completely continuous. Assume (a)-(b) hold with $q = 1$, $p = \infty$ and*

$$|k_t|_{L^\infty[0,T]} \leq 1 \quad \text{for all } t \in [0,T].$$

In addition suppose
(c). there exists $\delta \in L^1[0,T]$ and $w_1 : (0,2R] \to (0,\infty)$ continuous and nondecreasing with*

$$\inf_{r \in (0,2R]} \frac{r}{w_1(r)} > 2\,|\delta|_{L^1[0,T]} \tag{8.3.6}$$

and

$$|g_1(s,x) - g_1(s,y)| \leq \delta(s)\, w_1(|x-y|) \tag{8.3.7}$$

for a.e. $s \in [0,T]$ and all $x, y \in B$, $x \neq y$;

(d). there exists* $w_2 : [0, R] \to \mathbf{R}_+$ *continuous and nondecreasing such that*

$$|g_2(s, x)| \le \delta(s)\, w_2(|x|)$$

for a.e. $s \in [0, T]$, *all* $x \in B$, *and*

$$\frac{R}{w(R)} \ge |\delta|_{L^1[0,T]} \tag{8.3.8}$$

where $w = w_1 + w_2$.

Then (8.3.2) *has a solution in* $C([0, T]; B)$.

PROOF: First we claim that (c) holds with $\omega_0(s, r) = \delta(s)\, w_1(r)$. It is clear that (8.3.3) follows from (8.3.7) and the complete continuity of g_2. Now let $\varphi \in C([0, T]; [0, 2R])$ be any solution to (8.3.4) and suppose that $\varphi \not\equiv 0$. Then $r_0 = \max_{t\in[0,T]} \varphi(t) \in (0, 2R]$. Let $t_0 \in [0, T]$ be such that $\varphi(t_0) = r_0$. From (8.3.4), since $|k_t|_\infty \le 1$ and w_1 is nondecreasing, we deduce that

$$r_0 = \varphi(t_0) \le 2\int_0^T |k(t_0, s)|\, \delta(s)\, w_1(\varphi(s))\, ds \le 2w_1(r_0)\, |\delta|_{L^1[0,T]}.$$

Hence $r_0/w_1(r_0) \le 2|\delta|_{L^1[0,T]}$, which contradicts (8.3.6). Thus $r_0 = 0$.

Next we check (d). Suppose $|u|_\infty = R$ for some $u \in C([0, T]; B)$ solution to (8.3.5). Let $t_1 \in [0, T]$ with $|u(t_1)| = R$. From (8.3.5) we obtain

$$R = |u(t_1)| \le \lambda \int_0^T \delta(s)\, w(|u(s)|)\, ds \le \lambda\, w(R)\, |\delta|_{L^1[0,T]}.$$

Since $\lambda \in (0, 1)$, $w(R) > 0$ and we may assume $|\delta|_{L^1[0,T]} > 0$, we find $R < w(R)\, |\delta|_{L^1[0,T]}$, a contradiction to (8.3.8). Thus (d) holds and Theorem 8.3.2 applies. □

Example 8.3.1. Let $E = C(I; \mathbf{R})$, $I \subset \mathbf{R}$ compact interval. Then an example of a function g_1 satisfying (c*) is given by Example 8.2.1, where (8.2.17) is now replaced by (8.3.6). For instance, we may take

$$w_1(r) = Cr \quad \text{for an arbitrary } R > 0 \text{ and } C < 1/(2|\delta|_1)$$

or

$$w_1(r) = C \sin r \quad \text{for } 0 < R \le \pi/4 \text{ and } C < 1/(2|\delta|_1).$$

CHAPTER 9

EQUATIONS IN BANACH SPACES RELATIVE TO THE WEAK TOPOLOGY

9.1. Introduction.

In this chapter we present fixed point theorems for weakly sequentially continuous maps between Banach spaces (or more generally metrizable locally convex spaces). In particular we extend Emmanuele's and other fixed point theory in the literature [61, 115, 123]. Also we present an analogue of Mönch's fixed point theorem [113, 114] in the weak topology setting for weakly sequentially continuous maps. In addition an existence result is established for operator equations in Banach spaces relative to the weak topology.

9.2. Fixed point theory.

We establish two fixed point results in this section. However before we do so we present a well known result from the literature which will be used throughout this section (for completness we include its proof).

Theorem 9.2.1. *Let E be a metrizable locally convex linear topological space with Q a weakly compact subset of E. Suppose $F: Q \to E$ is weakly sequentially continuous. Then $F: Q \to E$ is weakly continuous.*

PROOF: Let A be a weakly closed subset of E. We first show $F^{-1}(A)$ is weakly sequentially closed in Q. To see this let $y_n \in F^{-1}(A)$ and $y_n \rightharpoonup y$ (here $\rightharpoonup$ denotes weak convergence). Then $F(y_n) \rightharpoonup F(y)$. Also $F(y_n) \in A$ and A weakly closed implies $F(y) \in A$ i.e. $y \in F^{-1}(A)$. Thus $F^{-1}(A)$ is weakly sequentially closed. Now since Q is weakly compact we have $\overline{F^{-1}(A)}^w$ weakly compact. Let $x \in \overline{F^{-1}(A)}^w$. The Eberlein–Smulian Theorem [58 p. 549] guarantees that there is a sequence $x_n \in F^{-1}(A)$ with $x_n \rightharpoonup x$. Since $F^{-1}(A)$ is weakly sequentially closed we have $x \in F^{-1}(A)$ i.e. $\overline{F^{-1}(A)}^w = F^{-1}(A)$. Thus $F^{-1}(A)$ is weakly closed. □

Our first result is the analogue of Mönch's fixed point theorem in the weak topology setting (see [114]).

Theorem 9.2.2. *Let E be a Banach space (or more generally a quasicomplete metrizable locally convex linear topological space), Q a closed, convex subset of E and $x_0 \in Q$. Suppose there is a weakly sequentially continuous map $F : Q \to Q$ and assume the following properties hold:*

$$\text{(9.2.1)} \qquad \begin{cases} C \subseteq Q \ \textit{ is countable and } \ \overline{C^w} \subseteq \overline{co}\,(\{x_0\} \cup F(\overline{C^w})) \\ \textit{implies } \ \overline{C^w} \ \textit{ is weakly compact} \end{cases}$$

and

$$\text{(9.2.2)} \qquad \begin{cases} \textit{for any relatively weakly compact subset } A \textit{ of } E \textit{ there} \\ \textit{exists a countable set } B \textit{ of } E \textit{ with } \overline{B^w} = \overline{A^w} \end{cases}$$

Then F has a fixed point in Q.

Remark. If E is a Banach space and $E^\star$ (the dual of E) is separable then (9.2.2) is true. To see this recall if K is a weakly compact subset of E then K with the relative weak topology is metrizable. This together with the fact that compact metric spaces are separable yields (9.2.2).

PROOF: Let

$$D_0 = \{x_0\} \quad \text{and} \quad D_n = co\,(\{x_0\} \cup F(D_{n-1})) \quad \text{for} \quad n = 1, 2, \ldots\ldots .$$

We claim D_n is relatively weakly compact for $n = 0, 1, \ldots..$ Certainly it is true if $n = 0$. Now suppose D_k is relatively weakly compact for some $k \in \{1, 2, \ldots.\}$. Notice from Theorem 9.2.1 that

$$F : \overline{D_k^w} \to E \quad \text{is weakly continuous,}$$

and as a result

$$F(\overline{D_k^w}) \quad \text{is weakly compact.}$$

The Krein–Smulian Theorem [57 p. 434; 6 p. 82] guarantees that D_{k+1} is relatively weakly compact.

From (9.2.2) there exists a sequence of countable sets $\{C_n\}_0^\infty$ with $\overline{C_n^w} = \overline{D_n^w}$ for $n = 0, 1, \ldots\ldots$. Let

$$D = \bigcup_{n=0}^{\infty} D_n \quad \text{and} \quad C = \bigcup_{n=0}^{\infty} C_n.$$

Note since $D_{n-1} \subseteq D_n$ for $n = 1, 2, \ldots$ that D is convex. It is immediate since (D_n) is increasing that

$$D = \bigcup_{n=0}^{\infty} D_n = \bigcup_{n=1}^{\infty} co\,(\{x_0\} \cup F(D_{n-1})) = co\,(\{x_0\} \cup F(D))$$

and so [147 p. 66] guarantees that

$$\overline{D^w}\,(=\overline{D}) \;=\; \overline{co}\,(\{x_0\}\cup F(D)). \tag{9.2.3}$$

Also since

$$\bigcup_{n=0}^{\infty} D_n \subseteq \bigcup_{n=0}^{\infty} \overline{D_n^w} \subseteq \overline{\bigcup_{n=0}^{\infty} D_n}^{\,w}$$

we have

$$\overline{\bigcup_{n=0}^{\infty} \overline{D_n^w}}^{\,w} = \overline{\bigcup_{n=0}^{\infty} D_n}^{\,w} = \overline{D^w}$$

and

$$\overline{\bigcup_{n=0}^{\infty} \overline{D_n^w}}^{\,w} = \overline{\bigcup_{n=0}^{\infty} \overline{C_n^w}}^{\,w} = \overline{\bigcup_{n=0}^{\infty} C_n}^{\,w} = \overline{C^w}. \tag{9.2.4}$$

Now (9.2.3) and (9.2.4) imply

$$\overline{C^w} = \overline{D^w} = \overline{co}\,(F(D)\cup\{x_0\}) \subseteq \overline{co}\,(F(\overline{D^w})\cup\{x_0\}) = \overline{co}\,(F(\overline{C^w})\cup\{x_0\}).$$

This together with (9.2.1) guarantees that $\overline{C^w}$ is weakly compact. Hence $\overline{D^w}$ is weakly compact. From Theorem 9.2.1 we have that

$$F:\overline{D^w}\to E \quad \text{is weakly continuous.} \tag{9.2.5}$$

Also (9.2.3) implies $F(D)\subseteq \overline{D^w}$ and so

$$\overline{F(D)}^{\,w} \subseteq \overline{D^w}. \tag{9.2.6}$$

Now (9.2.5) and (9.2.6) (note the weak closure of $F(D)$ (respectively D) in $\overline{D^w}$ equals the weak closure of $F(D)$ (respectively D) in E) gives

$$F(\overline{D^w}) \subseteq \overline{F(D)}^{\,w} \subseteq \overline{D^w}.$$

We may apply the Schauder–Tychonoff Theorem (consider E with the weak topology and note $F:\overline{D^w}\to\overline{D^w}$ is continuous with $\overline{D^w}$ compact) to deduce that F has a fixed point in $\overline{D^w}$. □

We now present a fixed point result when (9.2.2) is not assumed. As one would expect (9.2.1) needs to be adjusted also.

Theorem 9.2.3. *Let E be a Banach space (or more generally a metrizable locally convex linear topological space), Q a closed, convex subset of E and*

$x_0 \in Q$. *Suppose there is a weakly sequentially continuous map* $F : Q \to Q$ *with the following property holding:*

$$\text{(9.2.7)} \qquad \begin{cases} A \subseteq Q \ \textit{and} \ \overline{A} = \overline{co}\,(\{x_0\} \cup F(A)) \\ \textit{implies} \ \overline{A^w} \ \textit{is weakly compact.} \end{cases}$$

Then F *has a fixed point in* Q.

PROOF: Let

$$D_0 = \{x_0\}, \quad D_n = co\,(\{x_0\} \cup F(D_{n-1})) \quad \text{for} \quad n = 1, 2, \ldots\ldots,$$

and

$$D = \bigcup_{n=0}^{\infty} D_n.$$

As in Theorem 9.2.2 we have

$$\overline{D} = \overline{D^w} = \overline{co}\,(F(D) \cup \{x_0\}).$$

Now (9.2.7) implies $\overline{D^w}$ is weakly compact. Note also that $F(D) \subseteq \overline{D^w}$ and so $\overline{F(D)}^w \subseteq \overline{D^w}$. In addition Theorem 9.2.1 guarantees that $F : \overline{D^w} \to E$ is weakly continuous, so

$$F(\overline{D^w}) \subseteq \overline{F(D)}^w \subseteq \overline{D^w}.$$

Apply the Schauder–Tychonoff Theorem to deduce the result . □

9.3. Application.

We begin with a discussion of the operator equation

$$\text{(9.3.1)} \qquad x(t) = F\,x(t) \quad \text{on} \quad [0, T].$$

Solutions to (9.3.1) will be sought in $C([0, T], E)$.

Theorem 9.3.1. *Let* E *be a Banach space with* Q *a nonempty, bounded, closed, convex, equicontinuous subset of* $C([0, T], E)$. *Suppose* $F : Q \to Q$ *is wk–sequentially continuous (i.e. if for any sequence* (x_n) *in* Q *with* $x_n(t) \to x(t)$ *in* (E, w) *for each* $t \in [0, T]$, *then* $F\,x_n(t) \to F\,x(t)$ *in* (E, w) *for each* $t \in [0, T]$*) and assume* (9.2.7) *holds. Then* (9.3.1) *has a solution in* Q.

PROOF: We show $F : Q \to Q$ is weakly sequentially continuous. Let (x_n) be a sequence in Q. Recall [111] since Q is equicontinuous, then

$x_n \rightharpoonup x$ iff $x_n(t) \to x(t)$ in (E, w) for each $t \in [0,T]$; here $x_n \rightharpoonup x$ means x_n converges weakly to $x \in C([0,T], E)$. Consequently $F : Q \to Q$ wk–sequentially continuous is equivalent to saying $F : Q \to Q$ is weakly sequentially continuous. To see this suppose F is wk–sequentially continuous and suppose (x_n) be a sequence in Q with $x_n \rightharpoonup x \in C([0,T], E)$. Then $x_n(t) \to x(t)$ in (E, w) for each $t \in [0,T]$. Now since F is wk–sequentially continuous we have $F x_n(t) \to F x(t)$ in (E, w) for each $t \in [0,T]$. Thus $F x_n \rightharpoonup F x$. The other way is similar. The result now follows from Theorem 9.2.3. □

We next gather together some facts that will be needed in this section. Let Ω_E be the bounded subsets of a Banach space E and let K^w be the family of all weakly compact subsets of E. Also let B_E be the closed unit ball of E. The De Blasi measure of weak noncompactness is the map $\beta : \Omega_E \to [0, \infty)$ defined by

$$\beta(X) = \inf \{t > 0 : \ \exists Y \in K^w \ \text{ with } \ X \subseteq Y + t\, B_E\} : \quad \text{here } \ X \in \Omega_E.$$

We now state the following well known result [111].

Theorem 9.3.2. *Let $H \subseteq C([0,T], E)$ be bounded and equicontinuous. Then*

$$\beta(H) = \sup_{t \in [0,T]} \beta(H(t)) = \beta(H[0,T])$$

and the function $t \mapsto \beta(H(t))$ is continuous; here

$$H(t) = \{\phi(t) : \ \phi \in H\} \quad \text{and} \quad H[0,T] = \cup_{t \in [0,T]} \{\phi(t) : \ \phi \in H\}.$$

We now discuss a special case of (9.3.1), namely

$$(9.3.2) \qquad y(t) = x_0 + \int_0^t f(s, y(s))\, ds \quad \text{for } \ t \in [0,T];$$

here $x_0 \in E$ and $E = (E, |\,.\,|)$ is a real Banach space. Assume that the following conditions hold:

$$(9.3.3) \qquad \begin{cases} \text{for each } \ t \in [0,T], \ f_t = f(t, \,.\,) \ \text{ is weakly sequentially} \\ \text{continuous (i.e. for each } \ t \in [0,T], \ \text{ and for each} \\ \text{convergent sequence } \ (x_n), \ \text{ the sequence } \ f_t(x_n) \\ \text{is weakly convergent)} \end{cases}$$

$$(9.3.4) \qquad \begin{cases} \text{for each continuous } \ y : [0,T] \to E, \ f(\,.\,, y(\,.\,)) \\ \text{is Pettis integrable on } \ [0,T] \end{cases}$$

and

$$(9.3.5) \qquad \begin{cases} \text{for any } r > 0 \text{ there exists } h_r \in L^1[0,T] \text{ with} \\ |f(t,y)| \leq h_r(t) \text{ for a.e. } t \in [0,T] \text{ and all} \\ y \in E \text{ with } |y| \leq r. \end{cases}$$

Define the operator F by

$$(9.3.6) \qquad F\,x(t) = x_0 + \int_0^t f(s, x(s)).$$

We claim

$$F : C([0,T], E) \to C([0,T], E).$$

To see this let $y \in C([0,T], E)$. By (9.3.4) we know that F is well defined. To see that $F\,y \in C([0,T], E)$ first notice that there exists $r > 0$ with $|y|_0 = \sup_{[0,T]} |y(t)| \leq r$. Now (9.3.5) implies that there exists $h_r \in L^1[0,T]$ with

$$|f(t, y(t))| \leq h_r(t) \quad \text{for a.e. } t \in [0,T].$$

Let $t,\, x \in [0,T]$ with $t > x$. Without loss of generality assume $F\,y(t) - F\,y(x) \neq 0$. Then there exists (consequence of the Hahn–Banach theorem) $\phi \in E^\star$ with $|\phi| = 1$ and $|F\,y(t) - F\,y(x)| = \phi(F\,y(t) - F\,y(x))$. Thus

$$\begin{aligned} |F\,y(t) - F\,y(x)| &= \phi\Big(\int_x^t f(s, y(s))\, ds\Big) \\ &\leq \int_x^t |\phi(f(s, y(s)))|\, ds \\ &\leq \int_x^t h_r(s)\, ds, \end{aligned}$$

and so $F\,y \in C([0,T], E)$.

We say $w : [0,\infty) \to [0,\infty)$ is a Kamke function if the unique solution to the integral inequality

$$u(t) \leq \int_0^t w(u(\tau))\, d\tau, \quad t \in [0,T]$$

which satisfies $u(0) = 0$ is $u \equiv 0$.

Theorem 9.3.3. *Let E be a Banach space with Q a nonempty, bounded, closed, convex, equicontinuous subset of $C([0,T], E)$. In addition suppose (9.3.3), (9.3.4) and (9.3.5) hold. Also assume*

$$(9.3.7) \qquad \begin{cases} \beta(\, f([0,T] \times X)\,) \leq w\,(\,\beta(X)\,) \\ \textit{for all bounded subsets } X \textit{ of } E \end{cases}$$

is satisfied. Let F be as defined in (9.3.6) *and assume $F : Q \to Q$. Then* (9.3.2) *has a solution in Q.*

PROOF: We will apply Theorem 9.3.1. First we show $F : Q \to Q$ is wk–sequentially continuous. Let (x_n) be a sequence in Q and let $x_n(t) \to x(t)$ in (E, w) for each $t \in [0, T]$. Fix $t \in (0, T]$. Since f_t is weakly sequentially continuous, the Lebesgue dominated convergence theorem for the Pettis integral [69 Corollary 4] implies for each $\phi \in E^\star$ that

$$\phi\,(\,F\,x_n(t)\,) \to \phi\,(\,F\,x(t)\,).$$

We can do this for each $t \in [0, T]$ and so $F : Q \to Q$ is wk–sequentially continuous. It remains to show (9.2.7). Let $x^\star \in Q$ and

$$\overline{C} = \overline{co}\,(\{x^\star\} \cup F(C))$$

for some $C \subseteq Q$. We must show C is relatively weakly compact. Notice from Theorem 9.3.2 that the function

$$v : t \to \beta\,(C(t)\,) \quad \text{is continuous on} \quad [0, T].$$

For fixed $t \in (0, T]$ divide $[0, t]$ into m parts:

$$0 = t_0 < t_1 < \ldots\ldots < t_m = t,$$

where $t_i = i\,\frac{t}{m}$ for $i = 0, 1, \ldots, m$. Let

$$C[t_{i-1}, t_i] = \{u(s) :\ u \in C \ \text{ and } \ t_{i-1} \le s \le t_i\}\,.$$

There exists (Theorem 9.3.2 and the continuity of v) $s_i \in [t_{i-1}, t_i]$ with

$$(9.3.8) \qquad \beta\,(C[t_{i-1}, t_i]) = \sup\,\{\beta\,(C(s)\,) :\ t_{i-1} \le s \le t_i\} \;=\; v(s_i).$$

Also by the Pettis integral mean value theorem we obtain for $u \in C$,

$$\begin{aligned} F\,u(t) \;&=\; x_0 + \sum_{i=0}^{m-1} \int_{t_i}^{t_{i+1}} f(s, u(s))\,ds \\ &\in\; x_0 + \sum_{i=0}^{m-1} (t_{i+1} - t_i)\,\overline{co}\,(f([0, T] \times C[t_i, t_{i+1}])\,)\,, \end{aligned}$$

and so

$$F\,C(t) \subseteq x_0 + \sum_{i=0}^{m-1} (t_{i+1} - t_i)\,\overline{co}\,(f([0, T] \times C[t_i, t_{i+1}])\,)\,.$$

Now

$$\begin{aligned}
\beta\,(F\,C(t)) &\le \sum_{i=0}^{m-1} (t_{i+1}-t_i)\,\beta\,(\overline{co}\,(f([0,T]\times C[t_i,t_{i+1}])\,)\,) \\
&= \sum_{i=0}^{m-1} (t_{i+1}-t_i)\,\beta\,(f([0,T]\times C[t_i,t_{i+1}])\,) \\
&\le \sum_{i=0}^{m-1} (t_{i+1}-t_i)\,w\,(\beta\,(C[t_i,t_{i+1}]\,)\,) \\
&= \sum_{i=0}^{m-1} (t_{i+1}-t_i)\,w\,(\,v(s_i)\,)
\end{aligned}$$

using (9.3.7) and (9.3.8). Let $m \to \infty$ and we have

$$\sum_{i=0}^{m-1} (t_{i+1}-t_i)\,w\,(\,v(s_i)\,) \to \int_0^t w\,(\,v(s)\,)\,ds$$

so

(9.3.9) $$\beta\,(F\,C(t)) \le \int_0^t w\,(\,v(s)\,)\,ds \quad \text{for} \quad t\in[0,T].$$

Since $\overline{C} = \overline{co}\,(\{x^\star\}\cup F(C))$ we have for $t\in[0,T]$ that

$$\beta\,(\,C(t)\,) \le \beta\,(F\,C(t))$$

and this together with (9.3.9) yields

$$v(t) = \beta\,(\,C(t)\,) \le \int_0^t w\,(\,v(s)\,)\,ds.$$

Now since w is a Kamke function we get $v(t)=0$ for each $t\in[0,T]$ i.e. $\beta\,(C(t)\,)=0$ for each $t\in[0,T]$. This together with Theorem 9.3.2 implies C is relatively weakly compact. Thus (9.2.7) holds. Our result follows from Theorem 9.3.1. □

We now use Theorem 9.3.3 to establish a general existence result for (9.3.2).

Theorem 9.3.4. *Let E be a Banach space and suppose* (9.3.3), (9.3.4) *and* (9.3.7) *hold. In addition assume*

(9.3.10) $$\begin{cases} \textit{there exists } \alpha\in L^1[0,T] \textit{ and } \psi:[0,\infty)\to(0,\infty) \\ \textit{a nondecreasing continuous function with} \\ |f(s,u)| \le \alpha(s)\,\psi(|u|) \textit{ for a.e. } s\in[0,T] \\ \textit{and all } u\in E \end{cases}$$

and

$$\int_0^T \alpha(s)\,ds < \int_{|x_0|}^{\infty} \frac{dx}{\psi(x)} \tag{9.3.11}$$

are satisfied. Then (9.3.2) *has a solution in* $C([0,T],E)$.

PROOF: Let

$$Q = \{\, y \in C([0,T],E) : \quad \begin{array}{l} |y(t)| \le b(t) \ \text{ for } \ t \in [0,T] \ \text{ and} \\ |y(t) - y(s)| \le b(t) - b(s) \ \text{ for } \ t,\ s \in [0,T] \,\} \end{array}$$

where

$$b(t) = I^{-1}\left(\int_0^t \alpha(s)\,ds\right) \quad \text{and} \quad I(z) = \int_{|x_0|}^{z} \frac{dx}{\psi(x)}.$$

First we notice that Q is a closed, convex, bounded, equicontinuous subset of $C([0,T],E)$. Let F be as defined in (9.3.6). The result follows immediately from Theorem 9.3.3 once we show F maps Q to Q. To see this take $y \in Q$. Without loss of generality assume $F\,y\,(s) \neq 0$ for all $s \in [0,T]$. Then there exists $\phi_s \in E^\star$ with

$$|\phi_s| = 1 \quad \text{and} \quad \phi_s(F\,y\,(s)) = |F\,y\,(s)|.$$

Consequently for each fixed $t \in [0,T]$ we have

$$\begin{aligned} |F\,y\,(t)| &= \phi_t\left(x_0 + \int_0^t f(s,y(s))\,ds\right) \\ &\le |x_0| + \int_0^t \alpha(s)\,\psi(|y(s)|)\,ds \\ &\le |x_0| + \int_0^t \alpha(s)\,\psi(b(s))\,ds \\ &= |x_0| + \int_0^t b'(s)\,ds = b(t) \end{aligned}$$

since

$$\int_{|x_0|}^{b(s)} \frac{dx}{\psi(x)} = \int_0^s \alpha(x)\,dx.$$

Next suppose $t,\ s \in [0,T]$ with $t > s$. Without loss of generality assume $F\,y\,(t) - F\,y\,(s) \neq 0$. Then there exists $\phi \in E^\star$ with

$$|\phi| = 1 \quad \text{and} \quad \phi(F\,y\,(t) - F\,y\,(s)) = |F\,y\,(t) - F\,y\,(s)|.$$

Consequently

$$
\begin{aligned}
|F\,y\,(t) - F\,y\,(s)| \;&=\; \phi\left(\int_s^t f(x, y(x))\,dx\right) \\
&\le\; \int_s^t \alpha(x)\,\psi(|y(x)|)\,dx \\
&\le\; \int_s^t \alpha(x)\,\psi(b(x))\,dx \\
&=\; \int_s^t b'(x)\,dx \;=\; b(t) - b(s).
\end{aligned}
$$

Thus $F\,y \in Q$ and we are finished. □

CHAPTER 10

EXISTENCE PRINCIPLES FOR INTEGRAL INCLUSIONS

10.1. Introduction.

In Chapter 8, we stated that Mönch [112] obtained the following common generalization of the fixed point theorems of Schauder, Krasnoselskii, Darbo and Sadovskii:

Theorem 10.1.1. *Let D be a closed, convex subset of a Banach space X and $N : D \to D$ continuous with the further property that for some $x_0 \in D$ one has*

$$(10.1.1) \quad C \subset D \text{ countable, } \overline{C} = \overline{conv}\left(\{x_0\} \cup N(C)\right) \implies \overline{C} \text{ compact.}$$

Then N has a fixed point.

The continuation theorem of Leray-Schauder type accompanying Theorem 10.1.1 is also due to Mönch [112]:

Theorem 10.1.2. *Let K be a closed, convex subset of a Banach space X, U a relatively open subset of K and $N : \overline{U} \to K$ continuous with the further property that for some $x_0 \in U$ one has*

$$(10.1.2) \quad C \subset \overline{U} \text{ countable, } C \subset \overline{conv}\left(\{x_0\} \cup N(C)\right) \implies \overline{C} \text{ compact.}$$

In addition, assume

$$x \neq (1-\lambda)\,x_0 + \lambda N(x) \text{ for all } x \in \overline{U} \setminus U \text{ and } \lambda \in (0,1).$$

Then N has a fixed point in $\overline{U}$.

Applications of Theorem 10.1.1 and Theorem 10.1.2 to differential and integral equations in abstract spaces can be found in Chapter 8. To check (10.1.1) or (10.1.2) in applications we make use of the following result of Heinz [79].

Theorem 10.1.3. *Let E be a Banach space and $C \subset L^1([a,b];E)$ countable with $|u(t)| \leq h(t)$ for a.e. $t \in [a,b]$ and every $u \in C$, where*

$h \in L^1[a,b]$. *Then the function* $\varphi(t) = \alpha(C(t))$ *belongs to* $L^1[a,b]$ *and satisfies*

$$\alpha\left(\left\{\int_a^b u(t)\,dt : \ u \in C\right\}\right) \leq 2\int_a^b \alpha(C(t))\,dt$$

(here α *is the Kuratowski measure of noncompactness).*

We note that the factor 2 in the above inequality (and in all inequalities based on it) can be dropped if instead of α we use β, the ball measure of noncompactness provided some extra assumptions are placed on E.

In Chapter 8 we discussed integral equations of the form

$$u(t) = \int_0^t k(t,s)\,g(s,u(s))\,ds, \quad t \in [0,T]$$

and

$$u(t) = \int_0^T k(t,s)\,g(s,u(s))\,ds, \quad t \in [0,T]$$

in a ball of a Banach space. There we assumed a condition of the form

$$\alpha(g(s,M)) \leq \omega(s,\alpha(M))$$

for a.e. $s \in [0,T]$ and any bounded set M, and we established some existence criteria using Theorems 10.1.1-10.1.3.

The goal of this chapter is to extend our results from Chapter 8 to Hammerstein and Volterra integral inclusions. Such inclusions arise naturally in mathematical physics, control theory and critical point theory for nonsmooth energy functionals (see [24, 39]).

In this chapter we first extend the Mönch theorems to set-valued maps. The extension is possible if we replace (10.1.1), (10.1.2) by some more general conditions which are expressed in terms of a pair (M,C) instead of a single set C. Next, the fixed point theorems are applied to establish existence principles for Hammmerstein integral inclusions in general Banach spaces.

10.2. Preliminaries.

For any nonempty set X we let 2^X be the collection of all subsets of X and $P(X) = 2^X \setminus \{\emptyset\}$. In case that X is a Hausdorff topological space, we let

$$P_f(X) = \{A \subset X : \ A \text{ is nonempty, closed}\},$$

$$P_k(X) = \{A \subset X : \ A \text{ is nonempty, compact}\}.$$

If X is a closed, convex subset of a normed space $(E, |\,.\,|)$, then we define

$$P_c(X) = \{A \subset X : \ A \text{ is nonempty, convex}\},$$

$$P_{fc}(X) = \{A \subset X : \ A \text{ is nonempty, closed, convex}\},$$

$$P_{kc}(X) = \{A \subset X : \ A \text{ is nonempty, compact, convex}\}$$

and for any $A \in P(E)$ we let $|A| = \sup\{|x| : \ x \in A\}$ and by $\operatorname{conv}(A)$ we mean the convex hull of A.

Let X, Y be two sets, $N : X \to 2^Y$ a set-valued map and $A \subset Y$. We define

$$\operatorname{graph}(N) = \{(x,y) : \ x \in X, \ y \in N(x)\} \text{ (the } \textit{graph} \text{ of } N),$$

$$N(A) = \cup\{N(x) : \ x \in A\} \text{ (the } \textit{image} \text{ of } A),$$

$$N^-(A) = \{x \in X : \ N(x) \cap A \neq \emptyset\} \text{ (the } \textit{weak inverse} \text{ of } A).$$

Suppose now that X, Y are Hausdorff topological spaces. A map $N : X \to P(Y)$ is said to be *upper semicontinuous* if $N^-(A)$ is closed for every closed set $A \in P(Y)$. The map N ie said to be *lower semicontinuous* if $N^-(A)$ is open for every open set $A \in P(Y)$.

We now state some well-known results of set-valued analysis (for proofs see [26, 54, 81]).

(P1) Let X, Y be Hausdorff topological spaces and $N : X \to P_f(Y)$. If N is upper semicontinuous, then graph (N) is closed in $X \times Y$. On the other hand, if graph (N) is closed and $\overline{N(X)}$ is compact, then N is upper semicontinuous.

(P2) Let X, Y be Hausdorff topological spaces and $N : X \to P_k(Y)$ upper semicontinuous. Then $N(A)$ is compact for each compact $A \subset X$.

(P3) (Bohnenblust-Karlin's fixed point theorem) If X is a Banach space, $K \in P_{kc}(X)$ and $N : K \to P_{fc}(K)$ is upper semicontinuous, then N has a fixed point (i.e. there exists $x \in K$ with $x \in N(x)$).

(P4) (Michael's selection theorem) If X is a metric space, Y a Banach space and $N : X \to P_{fc}(Y)$ is lower semicontinuous, then there exists a continuous map $N_0 : X \to Y$ with $N_0(x) \in N(x)$ for all $x \in X$.

Throughout this chapter E will be a real Banach space with norm $|\,.\,|$. We denote by $C([a,b]; E)$ the space of continuous functions $u : [a,b] \to E$ and by $|\,.\,|_\infty$ its max-norm $|u|_\infty = \max_{t\in[a,b]} |u(t)|$. For any subset $M \subset$

E, we denote by $C([a,b];M)$ the set of all functions in $C([a,b];E)$ which take values in M.

A function $u : [a,b] \to E$ is said to be *strongly measurable* on $[a,b]$ if there exists a sequence of finitely-valued functions u_n with

$$u_n(t) \to u(t) \quad \text{as } n \to \infty, \text{ a.e. } t \in [a,b]$$

By $\int_a^b u(t)\,dt$ we mean the Bochner integral of u, assuming it exists. Recall that a strongly measurable function u is Bochner integrable if and only if $|u|$ is Lebesgue integrable.

For any real $p \in [1,\infty]$, we consider the space $L^p([a,b];E)$ of all strongly measurable functions $u : [a,b] \to E$ such that $|u|^p$ is Lebesgue integrable on $[a,b]$. $L^p([a,b];E)$ is a Banach space under the norm

$$|u|_p = \left(\int_a^b |u(s)|^p\,ds\right)^{1/p}$$

for $p < \infty$ and

$$|u|_\infty = \operatorname{ess}\sup_{t\in[a,b]} |u(t)| = \inf\{c \geq 0 : |u(t)| \leq c \text{ a.e. } t \in [a,b]\}.$$

In particular, $L^1([a,b];E)$ is the space of Bochner integrable functions on $[a,b]$. When $E = \mathbf{R}$, the space $L^p([a,b];\mathbf{R})$ is simply denoted by $L^p[a,b]$.

We say that a map $\phi : [a,b] \times D \to P(E)$, $D \subset E$, is *L^p-Carathéodory* $(1 \leq p \leq \infty)$ if

(C1) $\phi(.,x)$ has a strongly measurable selection for each $x \in D$;

(C2) $\phi(t,.)$ is upper semicontinuous for a.e. $t \in [a,b]$;

(C3) for each $r > 0$ there exists $h_r \in L^p[a,b]$ with $|\phi(t,x)| \leq h_r(t)$ for all $x \in D$ satisfying $|x| \leq r$ and a.e. $t \in [a,b]$.

Recall the definition of the *Kuratowski measure of noncompactness.* Let $M \subset E$ be bounded. Then

$$\alpha(M) = \inf\left\{\varepsilon > 0 : \ M \subset \cup_{j=1}^m M_j \text{ and } \operatorname{diam}(M_j) \leq \varepsilon\right\}.$$

10.3. Fixed Point Theorems for Set-Valued Maps.

Our first result is the set-valued analog of the Mönch's fixed point theorem for a self-map of a closed, convex subset of a Banach space (Theorem 10.1.1).

Theorem 10.3.1. *Let D be a closed, convex subset of a Banach space X and $N : D \to P_c(D)$. Assume graph(N) is closed, N maps compact sets into relatively compact sets and that for some $x_0 \in D$ one has*

$$\left.\begin{array}{l} M \subset D,\ M = conv\,(\{x_0\} \cup N(M)) \\ and\ \overline{M} = \overline{C}\ with\ C \subset M\ countable \end{array}\right\} \Longrightarrow \overline{M}\ compact. \tag{10.3.1}$$

Then there exists $x \in D$ with $x \in N(x)$.

Remark. (a) Under the assumptions of Theorem 10.3.1, $N : D \to P_{kc}(D)$.

(b) According to (P1)-(P2), all the assumptions on N in Theorem 10.3.1 are fulfilled if $N : D \to P_{kc}(D)$ is upper semicontinuous and satisfies (10.3.1).

PROOF: (a) Define

$$M_0 = \{x_0\}, \quad M_n = \text{conv}\,(\{x_0\} \cup N(M_{n-1})) \quad \text{for}\, n \geq 1.$$

It is clear that $M_n \subset D$ and M_n is convex. Also, it is immediately seen by induction, that $M_{n-1} \subset M_n$. Consequently, the set

$$M = \cup\{M_n :\ n \geq 0\}$$

is convex and $M \subset D$. In addition, it is easy to show that $M = \text{conv}\,(\{x_0\} \cup N(M))$.

(b) M_n is relatively compact. This follows by induction, using the property of N of sending compact sets into relatively compact sets and Mazur's lemma. In consequence, there exists a countable set $C_n \subset M_n$ with $\overline{C_n} = \overline{M_n}$. Let

$$C = \cup\{C_n :\ n \geq 0\}.$$

It is clear that $C \subset M$, is countable and $\overline{M} = \overline{C}$. Now (10.3.1) guarantees that the set $K = \overline{M}$ is compact (nonempty and convex, obviously). Hence $K \in P_{kc}(X)$.

(c) We also have $K \subset N^-(K)$. Indeed, if $x \in K$, then $x_n \to x$ as $n \to \infty$, for some $x_n \in M$. Now take any $y_n \in N(x_n)$. Since $N(M) \subset M$, we have $y_n \in K$. Due to the compactness of K, we may suppose that $y_n \to y$ as $n \to \infty$, for some $y \in K$. Since $(x_n, y_n) \in \text{graph}\,(N)$ and graph(N) is closed, we also have $(x, y) \in \text{graph}\,(N)$. Thus, $y \in N(x) \cap K$, that is $x \in N^-(K)$ as claimed.

(d) Define the map $\widehat{N} : K \to P_{fc}(K)$ by

$$\widehat{N}(x) = N(x) \cap K.$$

Notice $\widehat{N}(x) \neq \emptyset$ because of $K \subset N^{-}(K)$.

It is clear from the definition, that graph $\left(\widehat{N}\right)$ is closed like graph (N). Then, from (P1), we find that $\widehat{N}$ is upper semicontinuous. The Bohnenblust Karlin's fixed point theorem for $\widehat{N}$ guarantees the existence of an $x \in K$ with $x \in \widehat{N}(x)$. Since $K \subset D$ and $\widehat{N}(x) \subset N(x)$, we also have $x \in D$ and $x \in N(x)$. $\square$

Remark. In case that N also satisfies

$$N\left(\overline{M}\right) \subset \overline{N(M)} \text{ for every } M \subset D \tag{10.3.2}$$

(which always holds if N is lower semicontinuous), a sufficient condition for (10.3.1) is the Mönch's original condition:

$$C \subset D \text{ countable, } \overline{C} = \overline{\text{conv}}\left(\{x_0\} \cup N(C)\right) \Longrightarrow \overline{C} \text{ compact.} \tag{10.3.3}$$

Indeed, if $M \subset D$ satisfies $M = \text{conv}\left(\{x_0\} \cup N(M)\right)$ and $\overline{M} = \overline{C}$ for some countable set $C \subset M$, then from (10.3.2), we obtain

$$N(M) \subset N\left(\overline{M}\right) = N\left(\overline{C}\right) \subset \overline{N(C)} \subset \overline{\text{conv}}\left(\{x_0\} \cup N(C)\right).$$

This implies

$$\overline{\text{conv}}\left(\{x_0\} \cup N(M)\right) \subset \overline{\text{conv}}\left(\{x_0\} \cup N(C)\right).$$

The converse inclusion being obvious, we find that

$$\overline{\text{conv}}\left(\{x_0\} \cup N(M)\right) = \overline{\text{conv}}\left(\{x_0\} \cup N(C)\right).$$

Consequently,

$$\overline{C} = \overline{\text{conv}}\left(\{x_0\} \cup N(C)\right).$$

Now (10.3.3) guarantees that $\overline{C} = \overline{M}$ is compact.

The next result is the continuation principle accompanying Theorem 10.3.1.

Theorem 10.3.2. *Let K be a closed, convex subset of a Banach space X, U a relatively open subset of K and $N : \overline{U} \to P_c(K)$. Assume graph (N) is closed, N maps compact sets into relatively compact sets and that for some $x_0 \in U$, the following two conditions are satisfied:*

$$\left.\begin{array}{c} M \subset \overline{U},\ M \subset \textit{conv}\left(\{x_0\} \cup N(M)\right) \\ \textit{and } \overline{M} = \overline{C} \textit{ with } C \subset M \textit{ countable} \end{array}\right\} \Longrightarrow \overline{M} \textit{ compact;} \tag{10.3.4}$$

(10.3.5) $$x \notin (1-\lambda)\, x_0 + \lambda N(x) \quad \textit{for all } x \in \overline{U} \setminus U,\ \lambda \in (0,1).$$

Then there exists $x \in \overline{U}$ *with* $x \in N(x)$.

PROOF: We may assume that (10.3.5) holds for all $\lambda \in [0,1]$ (the conclusion being trivial if (10.3.5) does not hold for $\lambda = 1$). Also, we may suppose that $U \neq K$ (otherwise the conclusion follows immediately from Theorem 10.3.1), so $\overline{U} \setminus U \neq \emptyset$.

Define $H : \overline{U} \times [0,1] \to P(K)$ by

$$H(x,\lambda) = (1-\lambda)\, x_0 + \lambda N(x).$$

Let

$$\Sigma = \left\{ x \in \overline{U} : \ x \in H(x,\lambda) \text{ for some } \lambda \in [0,1] \right\}.$$

The sets Σ and $\overline{U} \setminus U$ are nonempty ($x_0 \in \Sigma$) and disjoint by (10.3.5) extended to all $\lambda \in [0,1]$. We claim that Σ is closed like is $\overline{U} \setminus U$. Indeed, consider $x_n \in \Sigma$ with $x_n \to x$ as $n \to \infty$. Then there exist $\lambda_n \in [0,1]$ and $y_n \in N(x_n)$ with

(10.3.6) $$x_n - (1-\lambda_n)\, x_0 = \lambda_n y_n.$$

Clearly, we may suppose that $\lambda_n \to \lambda$ as $n \to \infty$ for some $\lambda \in [0,1]$. Two cases are possible:

(a) $\lambda \neq 0$. Then

$$y_n = \lambda_n^{-1} (x_n - (1-\lambda_n)\, x_0) \to \lambda^{-1} (x - (1-\lambda)\, x_0)$$

and, since $\operatorname{graph}(N)$ is closed, we deduce that

$$\lambda^{-1} (x - (1-\lambda)\, x_0) \in N(x),$$

i.e. $x \in (1-\lambda)\, x_0 + \lambda N(x)$. This shows that $x \in \Sigma$.

(b) $\lambda = 0$. Since

$$\overline{\{x_n : \ n \geq 1\}}$$

is compact, its image by N is relatively compact and in consequence, bounded. It follows that

$$\{y_n : \ n \geq 1\}$$

is bounded. Now letting $n \to \infty$ in (10.3.6), we obtain $x - x_0 = 0$. Thus $x = x_0 \in \Sigma$.

Therefore Σ is closed as claimed. The Urysohn's lemma now guarantees the existence of a continuous function $\theta : \overline{U} \to [0,1]$ with $\theta(x) = 0$ on $\overline{U} \setminus U$ and $\theta(x) = 1$ on Σ.

Now we define

$$D = \overline{\text{conv}}\left(\{x_0\} \cup N\left(\overline{U}\right)\right)$$

and $\widetilde{N} : D \to P_c(D)$ by

$$\widetilde{N}(x) = \begin{cases} (1 - \theta(x))\, x_0 + \theta(x)\, N(x) & \text{for } x \in D \cap U \\ x_0 & \text{for } x \in D \setminus U. \end{cases}$$

It is not difficult to prove that graph $\left(\widetilde{N}\right)$ is closed and $\widetilde{N}$ maps compact sets into relatively compact sets. Next we check (10.3.1) for $\widetilde{N}$. Let $M \subset D$ be such that

$$M = \text{conv}\left(\{x_0\} \cup \widetilde{N}(M)\right)$$

and $\overline{M} = \overline{C}$ for some countable set $C \subset M$. Since

$$(10.3.7) \quad M \cap U \subset \text{conv}\left(\{x_0\} \cup \widetilde{N}(M)\right) = \text{conv}\left(\{x_0\} \cup N(M \cap U)\right),$$

$$\overline{M \cap U} = \overline{C \cap U} \text{ and } C \cap U \subset M \cap U \text{ is countable,}$$

from (10.3.4) we deduce that $M \cap U$ is relatively compact. Then

$$N(M \cap U)$$

is relatively compact and Mazur's lemma and (10.3.7) imply that $\overline{M}$ is compact. Thus (10.3.1) holds for $\widetilde{N}$.

Now we apply Theorem 10.3.1 to deduce that there exists an $x \in D \subset \overline{U}$ with $x \in \widetilde{N}(x)$. The case $x \notin U$ is impossible because, otherwise, $x \in \widetilde{N}(x) = \{x_0\}$, that is $x = x_0$, which contradicts $x_0 \in U$. Hence $x \in U$ and in consequence,

$$(10.3.8) \qquad x \in (1 - \theta(x))\, x_0 + \theta(x)\, N(x).$$

Thus $x \in \Sigma$ and so $\theta(x) = 1$. Then (10.3.8) shows that $x \in N(x)$. □

The analogs of Theorems 10.3.1 and 10.3.2 for lower semicontinuous maps are as follows:

Theorem 10.3.3. *Let D be a closed, convex subset of a Banach space X and $N : D \to P_{fc}(D)$. Assume N is lower semicontinuous and that for some $x_0 \in D$ one has*

$$(10.3.9) \qquad \left.\begin{array}{l} M \subset D,\ M \subset \textit{conv}\left(\{x_0\} \cup N(M)\right) \\ \textit{and } \overline{M} = \overline{C} \textit{ with } C \subset M \textit{ countable} \end{array}\right\} \Longrightarrow \overline{M} \textit{ compact.}$$

Then there exists $x \in D$ *with* $x \in N(x)$.

PROOF: Michael's selection theorem guarantees the existence of a continuous selection $N_0 : D \to D$ of N. Now (10.3.9) implies that N_0 satisfies (10.3.1). Thus, Theorem 10.3.1 applies to N_0. □

Theorem 10.3.4. *Let* K *be a closed, convex subset of a Banach space* X, U *a relatively open subset of* K *and* $N : \overline{U} \to P_{fc}(K)$. *Assume* N *is lower semicontinuous and that for some* $x_0 \in U$, (10.3.4) *and* (10.3.5) *hold. Then there exists* $x \in \overline{U}$ *with* $x \in N(x)$.

PROOF: Use a continuous selection of N and apply Theorem 10.3.2. □

Remark. (the use of two norms on X). Let $|\,.\,|$ be the norm of the Banach space X and let $|\,.\,|'$ be another norm on X not necessarily complete. Assume that there is a constant $c > 0$ such that

$$|x|' \leq c\,|x| \quad \text{for all } x \in X.$$

Then the results in Theorems 10.3.1-10.3.4 remain true if the conclusion in the implications (10.3.1), (10.3.4) and (10.3.9) is that M is relatively compact with respect to the topology induced by $|\,.\,|'$, provided that all the topological properties of $N : D(N) \to P(X)$ are understood with respect to the $|\,.\,|'$-topology on $D(N)$ and the $|\,.\,|$-topology on X. Indeed, if the graph of N is $\left(|\,.\,|', |\,.\,|\right)$-closed, then it is also $(|\,.\,|, |\,.\,|)$-closed; if N maps $|\,.\,|'$-compact sets into $|\,.\,|$-relatively compact maps, then it also maps $|\,.\,|$-compact sets into $|\,.\,|$-relatively compact sets; if N is $\left(|\,.\,|', |\,.\,|\right)$-upper (lower) semicontinuous, then N is also $(|\,.\,|, |\,.\,|)$-upper (respectively, lower) semicontinuous.

10.4. Hammerstein Integral Inclusions.

In this section we study the Hammerstein integral inclusion

$$u(t) \in \int_0^T k(t,s)\, g(s, u(s))\, ds, \quad t \in [0,T] \tag{10.4.1}$$

in a ball of a Banach space $(E, |\,.\,|)$; here k is a real single-valued function and g is a set-valued map with compact, convex values in E. In particular, if

$$k(t,s) = 0 \ \text{ for } \ 0 \leq t < s \leq T$$

(the *Volterra case*), (10.4.1) becomes the Volterra integral inclusion

$$u(t) \in \int_0^t k(t,s)\, g(s, u(s))\, ds, \quad t \in [0,T]. \tag{10.4.2}$$

Thus, the existence theorems for (10.4.1) yield automatically existence results for (10.4.2). However, some of the assumptions made for (10.4.1) can be stated in a different way or improved for (10.4.2).

Let $0 < T < \infty$, $0 < R < \infty$ and let B be the closed ball

$$\{x \in E : \ |x| \leq R\}.$$

Suppose that for some $p \in (1, \infty]$ we have

(10.4.3) $$k : [0,T]^2 \to \mathbf{R},\ k(t, .) \in L^p[0,T] \text{ for each } t \in [0,T]$$

and let $g : [0,T] \times B \to P(E)$. By a solution of (10.4.1) we mean:

$$\begin{cases} u \in C([0,T];B) \text{ with } u(t) = \int_0^T k(t,s)\,w(s)\,ds \text{ for } t \in [0,T], \\ \text{where} \\ w \in L^q([0,T];E) \text{ and } w(s) \in g(s,u(s)) \text{ for a.e. } s \in [0,T]. \end{cases}$$

Throughout this section it is assumed that $1/p + 1/q = 1$.

Let

$$U = \{u \in C([0,T];E) : \ |u|_\infty < R\}.$$

Clearly, $\overline{U} = C([0,T];B)$. Assign to g and $q \in [1,\infty)$, a set-valued operator (the *Nemitsky operator*)

$$G_q : \overline{U} \to 2^{L^q([0,T];E)}$$

by letting

$$G_q(u) = \{w \in L^q([0,T];E) : \ w(s) \in g(s,u(s)) \text{ for a.e. } s \in [0,T]\}.$$

If in addition to (10.4.3), we assume that

(10.4.4) $$|k(t,.) - k(t',.)|_p \to 0 \text{ as } t' \to t,$$

then we may define the linear operator

$$J_q : L^q([0,T];E) \to C([0,T];E),$$

$$J_q(w)(t) = \int_0^T k(t,s)\,w(s)\,ds, \quad t \in [0,T].$$

Now, the solutions of (10.4.1) appear as fixed points of the set-valued operator

$$N : \overline{U} \to 2^{C([0,T];E)}, \quad N = J_q \circ G_q.$$

Theorem 10.3.2 yields the following very general existence principle for Hammerstein integral inclusions of upper semicontinuous type.

Theorem 10.4.1. *Assume that* (10.4.3), (10.4.4) *hold. In addition suppose*

(H1) $N : \overline{U} \to P_c\left(C\left([0,T];E\right)\right)$, *has closed graph and maps compact sets into relatively compact sets;*

(H2) $\left.\begin{array}{c} M \subset \overline{U},\ M \subset conv\left(\{0\} \cup N(M)\right) \\ \text{and } \overline{M} = \overline{C} \text{ with } C \subset M \text{ countable} \end{array}\right\} \Longrightarrow \overline{M}$ *compact;*

(H3) $u \notin \lambda N(u)$ *for all* $u \in \overline{U}$ *with* $|u|_\infty = R$ *and all* $\lambda \in (0,1)$.

Then (10.4.1) *has a solution in* $C\left([0,T];B\right)$.

PROOF: Apply Theorem 10.3.2 with $K = X = C\left([0,T];E\right)$, X endowed with the norm $|\,.\,|_\infty$, and $x_0 = 0$. □

Remark. If $N\left(\overline{U}\right) \subset \overline{U}$, then (H3) trivially holds and the conclusion of Theorem 10.4.1 follows directly from Theorem 10.3.1.

To obtain applicable existence criteria we have to find sufficient conditions for (H1)-(H3).

Lemma 10.4.2. *Assume that* (10.4.3) *and* (10.4.4) *hold. If* $g : [0,T] \times B \to P_{kc}(E)$ *is* L^q*-Carathéodory, then* N *satisfies (H1).*

PROOF: (a) $N(u) \neq \emptyset$ for each $u \in \overline{U}$. Indeed, since g takes nonempty, compact values and satisfies (C1)-(C2), for each $u \in C\left([0,T];B\right)$ there exists a strongly measurable selection w of $g\left(.,u(.)\right)$ (see [55, Proposition 3.5 (a)]). Next, (C3) guarantees $w \in L^q\left([0,T];E\right)$. Then $v = J_q(w) \in N(u)$.

(b) $N(u)$ is convex. This follows immediately from the convexity of the values of g.

(c) graph (N) is closed. To show this, let $(u_n, v_n) \in \text{graph}(N)$, $n \geq 1$, with

$$|u_n - u|_\infty,\ |v_n - v|_\infty \to 0 \quad \text{as} \quad n \to \infty.$$

Let

$$v_n = J_q(w_n), \tag{10.4.5}$$

$$w_n \in L^q\left([0,T];E\right),\ w_n(s) \in g\left(s, u_n(s)\right) \text{ a.e. } s \in [0,T].$$

The sequence (w_n) has an L^q-weakly convergent subsequence; see [55, Proposition 9.4 and Problem 9.6]. Let w be its limit. As in [54, Proof of

Lemma 2.3], we can show that $w(s) \in g(s, u(s))$ for a.e. $s \in [0, T]$. Also, from (10.4.5), we deduce $v = J_q(w)$. Therefore, $(u, v) \in \text{graph}(N)$.

(d) $N(M)$ is relatively compact for each compact $M \subset \overline{U}$. To prove this, let $M \subset \overline{U}$ be a compact set and (v_n) be any sequence of elements of $N(M)$. We show that (v_n) has a convergent subsequence by using the Arzèla-Ascoli criterion of compactness in $C([0, T]; E)$ (see [52, Proposition 7.3 (b)]). Since $v_n \in N(M)$, there exist $u_n \in M$ and $w_n \in L^q([0, T]; E)$ with

$$v_n = J_q(w_n) \text{ and } w_n(s) \in g(s, u_n(s)) \text{ for a.e. } s \in [0, T].$$

Using Theorem 10.1.3, we obtain that

$$\text{(10.4.6)} \qquad \alpha(\{v_n(t) : n \geq 1\}) \leq 2 \int_0^T \alpha(\{k(t, s) w_n(s) : n \geq 1\}) \, ds.$$

On the other hand, since $g(s, .)$ is upper semicontinuous, with compact values and $M(s)$ is compact in E, the set $g(s, M(s))$ is compact. Consequently,

$$\alpha(\{w_n(s) : n \geq 1\}) = 0 \quad \text{for a.e.} \quad s \in [0, T].$$

Furthermore,

$$\alpha(\{k(t, s) w_n(s) : n \geq 1\}) = |k(t, s)| \, \alpha(\{w_n(s) : n \geq 1\}) = 0$$

for a.e. $s \in [0, T]$. Now (10.4.6) implies that $\{v_n(t) : n \geq 1\}$ is relatively compact in E, for each $t \in [0, T]$. In addition, we have

$$\text{(10.4.7)} \qquad \begin{aligned} |v_n(t) - v_n(t')| &\leq |k(t, .) - k(t', .)|_p \, |w_n|_q \\ &\leq |k(t, .) - k(t', .)|_p \, |h|_q, \end{aligned}$$

where $h \in L^q[0, T]$ is such that $|g(s, x)| \leq h(s)$ for a.e. $s \in [0, T]$ and all $x \in B$. This, and (10.4.4), shows that $\{v_n : n \geq 1\}$ is equicontinuous. Consequently, $\{v_n : n \geq 1\}$ is relatively compact in $C([0, T]; E)$.

Thus, N satisfies (H1). □

Remark. Another way to assure (H1) is to put conditions so that the set-valued Nemitsky operator G_q is upper semicontinuous from the L^p-topology to the L^q-topology and then use the Remark after Theorem 10.3.4; here $|\,.\,|_p \leq c\,|\,.\,|_\infty$. For a result concerning the upper semicontinuity of the Nemitsky set-valued operator from L^p to L^q, we refer the reader to Cellina-Fryszkowski-Rzezuchowski [41]. See also Couchouron-Kamenski [52] for a direct result about the upper semicontinuity of N.

Lemma 10.4.3. *Assume that* (10.4.3), (10.4.4) *hold and that* $g : [0,T] \times B \to P(E)$ *satisfies*

$$|g(s,x)| \leq h(s) \tag{10.4.8}$$

for a.e. $s \in [0,T]$, *all* $x \in B$ *and some* $h \in L^q[0,T]$. *In addition, suppose that there exists an* L^q*-Carathéodory function* $\omega : [0,T] \times [0,2R] \to \mathbf{R}_+$ *such that*

$$\alpha(g(s,M)) \leq \omega(s,\alpha(M)) \tag{10.4.9}$$

for a.e. $s \in [0,T]$, *every* $M \subset B$, *and that the unique solution* $\varphi \in C([0,T];[0,2R])$ *of the inequality*

$$\varphi(t) \leq 2\int_0^T |k(t,s)|\,\omega(s,\varphi(s))\,ds, \quad t \in [0,T] \tag{10.4.10}$$

is $\varphi \equiv 0$. *Then* N *satisfies (H2).*

PROOF: Suppose $M \subset \overline{U}$,

$$M \subset \operatorname{conv}(\{0\} \cup N(M))$$

and $\overline{M} = \overline{C}$ for some countable set $C \subset M$. Using (10.4.8) and an estimation of type (10.4.7), we see that $N(M)$ is equicontinuous. Then, from

$$M \subset \operatorname{conv}(\{0\} \cup N(M)),$$

we deduce that M is equicontinuous too. In order to apply the Arzèla-Ascoli theorem, it remains to show that $M(t)$ is relatively compact in E for each $t \in [0,T]$. Since

$$C \subset M \subset \operatorname{conv}(\{0\} \cup N(M)) \text{ and } C \text{ is countable,}$$

we can find a countable set $V = \{v_n : n \geq 1\} \subset N(M)$ with

$$C \subset \operatorname{conv}(\{0\} \cup V).$$

Then, there exist $u_n \in M$ and $w_n \in L^q([0,T];E)$ with

$$v_n = J_q(w_n) \text{ and } w_n(s) \in g(s,u_n(s)) \text{ for a.e. } s \in [0,T].$$

From

$$M \subset \overline{C} \subset \overline{\operatorname{conv}}(\{0\} \cup V),$$

according to Theorem 10.1.3, we have

$$\begin{aligned}\alpha(M(t)) &\leq \alpha\left(\overline{C}(t)\right) \leq \alpha(V(t)) \\ &= \alpha(\{J_q(w_n)(t) : n \geq 1\}) \\ &\leq 2\int_0^T \alpha(\{k(t,s)\, w_n(s) : n \geq 1\})\, ds.\end{aligned}$$

Now, since $w_n(s) \in g(s, u_n(s))$ and $u_n \in M$, (10.4.9) guarantees

$$\begin{aligned}\alpha(\{k(t,s)\, w_n(s) : n \geq 1\}) &\leq |k(t,s)|\, \alpha(g(s, M(s))) \\ &\leq |k(t,s)|\, \omega(s, \alpha(M(s))).\end{aligned}$$

It follows that

$$\alpha(M(t)) \leq 2\int_0^T |k(t,s)|\, \omega(s, \alpha(M(s)))\, ds, \ \ t \in [0,T].$$

Also, the function φ given by $\varphi(t) = \alpha(M(t))$ belongs to $C([0,T];[0,2R])$. Consequently, $\varphi \equiv 0$, that is $\alpha(M(t)) = 0$ for all $t \in [0,T]$.

Now, by the Arzèla-Ascoli theorem, M is relatively compact in $C([0,T];E)$. □

Remark. In the Volterra case, the inequality (10.4.10) is

$$\varphi(t) \leq 2\int_0^t |k(t,s)|\, \omega(s, \varphi(s))\, ds, \ \ t \in [0,T].$$

Lemma 10.4.4. *Assume* (10.4.3), (10.4.4). *In addition, suppose that*

$$|k(t,.)|_p \leq 1$$

for all $t \in [0,T]$, $g : [0,T] \times B \to P(E)$ *and that there exist* $\delta \in L^q[0,T]$ *and* $\psi : (0,R] \to (0,\infty)$ *continuous and nondecreasing such that*

$$|g(s,x)| \leq \delta(s)\, \psi(|x|)$$

for a.e. $s \in [0,T]$ *and all* $x \in B \setminus \{0\}$.

(*a*) *If*

$$|\delta|_q \leq R/\psi(R), \tag{10.4.11}$$

then N *satisfies (H3) with* $N\left(\overline{U}\right) \subset \overline{U}$.

(*b*) *In the Volterra case, if*

$$|\delta|_q^q \leq q \int_0^R \left(r^{q-1} / \psi(r)^q \right) dr, \tag{10.4.12}$$

then N *satisfies (H3).*

PROOF: (a) Let $v \in N(u)$ with $u \in \overline{U}$. Then $v = J_q(w)$ for some $w \in L^p([0,T]; E)$ with $w(s) \in g(s, u(s))$ a.e. on $[0,T]$. Since $|u(s)| \leq R$, ψ is nondecreasing and $|k(t,.)|_p \leq 1$, we have

$$\begin{aligned} |v(t)| &\leq \int_0^T |k(t,s)| \, |w(s)| \, ds \\ &\leq \int_0^T |k(t,s)| \, \delta(s) \, \psi(|u(s)|) \, ds \\ &\leq \psi(R) \int_0^T |k(t,s)| \, \delta(s) \, ds \\ &\leq \psi(R) \, |k(t,.)|_p \, |\delta|_q \\ &\leq \psi(R) \, |\delta|_q \end{aligned}$$

(we put $\psi(0) = \lim_{t \to 0+} \psi(t)$). Now this together with (10.4.11) implies

$$|v(t)| \leq R$$

for all $t \in [0,T]$. Hence $v \in \overline{U}$.

(b) Assume the Volterra case. Let $u \in \lambda N(u)$ for some $\lambda \in (0,1)$. Then

$$\begin{aligned} |u(t)| &\leq \lambda \textstyle\int_0^t |k(t,s)| \, \delta(s) \, \psi(|u(s)|) \, ds \\ &\leq \lambda \left(\textstyle\int_0^t [\delta(s) \, \psi(|u(s)|)]^q \, ds \right)^{1/q} \end{aligned} \tag{10.4.13}$$

for all $t \in [0,T]$. Let

$$c(t) = \min \left\{ R, \ \lambda \left(\int_0^t [\delta(s) \, \psi(|u(s)|)]^q \, ds \right)^{1/q} \right\}.$$

Clearly c is nondecreasing. We claim that $c(T) < R$. Suppose the contrary. Then, since $c(0) = 0$, there exists a subinterval $[a,b] \subset [0,T]$ with

$$c(a) = 0, \ c(b) = R \text{ and } 0 < c(t) < R \text{ on } (a,b).$$

Since by (10.4.3), $|u(t)| \leq c(t) \leq R$ on $[a,b]$ and ψ is nondecreasing, we have

$$(c(s)^q)' = \lambda^q [\delta(s) \, \psi(|u(s)|)]^q \leq \lambda^q \delta(s)^q \, \psi(c(s))^q \quad \text{a.e. } s \in [a,b].$$

This yields

$$\begin{aligned}\int_a^b \frac{(c(s)^q)'}{\psi(c(s))^q} ds &= q\int_0^R \frac{r^{q-1}}{\psi(r)^q} dr \\ &\leq \lambda^q \int_a^b \delta(s)^q ds \\ &\leq \lambda^q |\delta|_q^q < |\delta|_q^q,\end{aligned}$$

which contradicts (10.4.12). Notice we may assume $|\delta|_q > 0$ for the last inequality, since otherwise we have nothing to prove. □

Remark. Condition (10.4.12) is less restrictive than (10.4.11). Indeed, since ψ is nondecreasing, one has $\psi(r) \leq \psi(R)$ for all $r \in [0, R]$. Consequently,

$$q\int_0^R \left(r^{q-1}/\psi(r)^q\right) dr \geq q(1/\psi(R))^q \int_0^R r^{q-1} dr = (R/\psi(R))^q,$$

which proves our claim.

Now, from Theorem 10.4.1 and Lemmas 10.4.2-10.4.4, we can state the main existence result for (10.4.1).

Theorem 10.4.5. *Let* $k : [0,T]^2 \to \mathbf{R}$ *and* $g : [0,T] \times B \to P_{kc}(E)$ *be* L^q*-Carathéodory for some* $q \in [1, \infty)$. *Suppose*

(A) $k(t,.) \in L^p[0,T]$, $|k(t,.)|_p \leq 1$ *for each* $t \in [0,T]$ *and*

$$|k(t,.) - k(t',.)|_p \to 0 \text{ as } t' \to t \quad (1/p + 1/q = 1);$$

(B) there exists an L^q*-Carathéodory function* $\omega : [0,T] \times [0, 2R] \to \mathbf{R}_+$ *such that*

$$\alpha(g(s,M)) \leq \omega(s, \alpha(M)) \text{ for a.e. } s \in [0,T], \tag{10.4.14}$$

for any $M \subset B$, *and that the unique solution in* $C([0,T]; [0, 2R])$ *of the inequality*

$$\varphi(t) \leq 2\int_0^T |k(t,s)| \, \omega(s, \varphi(s)) \, ds, \quad t \in [0,T], \tag{10.4.15}$$

is $\varphi \equiv 0$;

(C) there exists $\delta \in L^q[0,T]$ *and* $\psi : (0, R] \to (0, \infty)$ *continuous and nondecreasing such that*

$$|g(s,x)| \leq \delta(s)\, \psi(|x|) \tag{10.4.16}$$

for a.e. $s \in [0,T]$, $x \in B \setminus \{0\}$, *and*

$$|\delta|_q \le R/\psi(R). \tag{10.4.17}$$

Then (10.4.1) *has a solution in* $C([0,T];B)$.

For the Volterra case, we have:

Theorem 10.4.6. *Assume the Volterra case. Suppose that all the assumptions of Theorem 10.4.5 are satisfied except* (10.4.17) *is relaxed to*

$$|\delta|_q^q \le q \int_0^R \left(r^{q-1}/\psi(r)^q \right) dr. \tag{10.4.18}$$

Then (10.4.2) *has a solution in* $C([0,T];B)$.

Many more applicable results can be derived from Theorems 10.4.5 and 10.4.6.

Corollary 10.4.7. *Let* $k : [0,T]^2 \to \mathbf{R}$ *and* $g : [0,T] \times B \to P_{fc}(E)$ *with*

$$g = g_1 + g_2,$$

where g_1 *is single-valued and* $g_1(.,0) = 0$ *and the values of* g_2 *are subsets of a compact set of* E. *Assume (A) holds. In addition suppose*

(B) there exists* $\delta \in L^q[0,T]$ *and* $\psi_1 : (0,2R] \to (0,\infty)$ *continuous and nondecreasing with*

$$2|\delta|_q < \inf_{r \in (0,2R]} (r/\psi_1(r)) \tag{10.4.19}$$

and

$$|g_1(s,x) - g_1(s,y)| \le \delta(s)\, \psi_1(|x-y|) \tag{10.4.20}$$

for a.e. $s \in [0,T]$ *and all* $x,\ y \in B$, $x \ne y$;

(C) there exists* $\psi_2 : [0,R] \to \mathbf{R}_+$ *continuous and nondecreasing such that*

$$|g_2(s,x)| \le \delta(s)\, \psi_2(|x|) \tag{10.4.21}$$

for a.e. $s \in [0,T]$, *all* $x \in B$, *and*

$$|\delta|_q \le R/\psi(R) \tag{10.4.22}$$

where $\psi = \psi_1 + \psi_2$.

Then (10.4.1) *has a solution in* $C([0,T];B)$.

PROOF: We show that (B) holds. Indeed, by (10.4.20), (10.4.14) holds with

$$\omega(s,r) = \delta(s)\,\psi_1(r).$$

Next, suppose that $\varphi \in C([0,T];[0,2R])$ solves (10.4.15) and $\varphi \not\equiv 0$. Then

$$r_0 = \max_{t\in[0,T]} \varphi(t) \in (0,2R].$$

Let $t_0 \in [0,T]$ be such that $\varphi(t_0) = r_0$. From (10.4.15), using Hölder's inequality, $|k(t,.)|_p \le 1$ and ψ_1 nondecreasing, we obtain

$$r_0 = \varphi(t_0) \le 2\left(\int_0^T [\delta(s)\,\psi_1(\varphi(s))]^q\, ds\right)^{1/q} \le 2\psi_1(r_0)\,|\delta|_q\,.$$

Hence $r_0/\psi_1(r_0) \le 2|\delta|_q$, which contradicts (10.4.19). Thus $r_0 = 0$.

Next we check (C). Indeed, from (10.4.20) and (10.4.21), since $g_1(.,0) = 0$, we have

$$\begin{aligned} |g(s,x)| &= |(g_1+g_2)(s,x)| \\ &\le |g_1(s,x) - g_1(s,0)| + |g_2(s,x)| \\ &\le \delta(s)\,\psi_1(|x|) + \delta(s)\,\psi_2(|x|) \\ &= \delta(s)\,\psi(|x|)\,. \end{aligned}$$

Thus (10.4.16) holds. Now (10.4.17) is just (10.4.22). □

In [145] it is shown that a condition like (10.4.19) can be relaxed in some particular cases.

The above result can be improved for the Volterra case as follows.

Corollary 10.4.8. *Assume the Volterra case. Suppose that all the assumptions of Corollary 10.4.7 are satisfied except* (10.4.19) *and* (10.4.22) *are relaxed to*

$$\int_0^{2R} \left(r^{q-1}/\psi_1(r)^q\right) dr = \infty \tag{10.4.23}$$

and (10.4.18), *respectively. Then* (10.4.2) *has a solution in* $C([0,T];B)$.

PROOF: We check the conditions of Theorem 10.4.6. First we check (B). From (10.4.20) we see that (10.4.14) holds with

$$\omega(s,r) = \delta(s)\,\psi_1(r)\,.$$

Now, let $\varphi \in C([0,T];[0,2R])$ be any solution of (10.4.15). By Hölder's inequality, we obtain

$$\varphi(t) \leq 2\left(\int_0^t [\delta(s)\,\psi_1(\varphi(s))]^q\, ds\right)^{1/q}.$$

Let

$$c(t) = 2\left(\int_0^t [\delta(s)\,\psi_1(\varphi(s))]^q\, ds\right)^{1/q}.$$

It is clear that c is nondecreasing. Then $\varphi \equiv 0$ once we show $c(T) = 0$. Suppose the contrary, i.e. $c(T) > 0$. Then, since $c(0) = 0$, for each $\varepsilon \in (0, A)$, where $A = \min\{c(T),\ 2R\}$, there is a subinterval $[a,b] \subset [0,T]$ with

$$c(a) = \varepsilon,\ \ c(b) = A \ \text{ and } \ c(t) \in (\varepsilon,\ A) \ \text{ for all } \ t \in (a,b).$$

Now

$$\varphi(t) \leq c(t) \leq 2R$$

on $[a,b]$ and ψ_1 nondecreasing on $[0,2R]$ guarantee that

$$(c(s)^q)' = 2^q\delta(s)^q\,\psi_1(\varphi(s))^q \leq 2^q\delta(s)^q\,\psi_1(c(s))^q$$

a.e. $s \in [a,b]$. Consequently

$$q\int_\varepsilon^A \frac{r^{q-1}}{\psi_1(r)^q}dr = \int_a^b \frac{(c(s)^q)'}{\psi_1(c(s))^q}ds \leq 2^q\,|\delta|_q^q.$$

This, for $\varepsilon \searrow 0$, yields a contradiction to (10.4.23). Thus $c(T) = 0$ and so $\varphi \equiv 0$.

Finally (C) with (10.4.18) instead of (10.4.17), follows from (10.4.20), $g_1(.,0) = 0$, and (10.4.21). $\square$

Remark. Condition (10.4.23) is less restrictive than (10.4.19). Indeed, if (10.4.19) holds, then

$$\begin{aligned}\int_0^{2R}\left(r^{q-1}/\psi_1(r)^q\right)dr &= \lim_{\varepsilon\to 0+}\int_\varepsilon^{2R}\left(r^{q-1}/\psi_1(r)^q\right)dr \\ &\geq \inf_{r\in(0,2R]}(r/\psi_1(r))^q \lim_{\varepsilon\to 0+}\int_\varepsilon^{2R} r^{-1}dr = \infty,\end{aligned}$$

that is (10.4.23) also holds.

We conclude with a general existence principle of lower semicontinuous type.

Theorem 10.4.9. *Assume that* (10.4.3), (10.4.4) *hold. In addition suppose*

(h1) $N : \overline{U} \to P_{fc}(C([0,T];E))$ *is lower semicontinuous;*

(h2) $\left.\begin{array}{c} M \subset \overline{U},\ M \subset conv(\{0\} \cup N(M)) \\ \text{and } \overline{M} = \overline{C} \text{ with } C \subset M \text{ countable} \end{array}\right\} \Longrightarrow \overline{M}$ *compact;*

(h3) $u \notin \lambda N(u)$ *for all* $u \in \overline{U}$ *with* $|u|_\infty = R$ *and* $\lambda \in (0,1)$.

Then (10.4.1) *has a solution in* $C([0,T];B)$.

Notice we can guarantee (h1) by using lower semicontinuity results for the Nemitsky set-valued operator from L^p into L^q. Such results can be found in Cardinali-Papageorgiou [39] and Hu-Papageorgiou [81].

CHAPTER 11

CAUCHY AND PERIODIC INTEGRODIFFERENTIAL INCLUSIONS

11.1. Introduction.

In this chapter our main goal is to present an existence principle for the periodic integrodifferential inclusion

$$(11.1.1) \qquad \begin{cases} y'(t) \in F(t, y(t)) + \int_0^t k(t,s)\, G(s, y(s))\, ds \quad \text{a.e. on} \quad [0,T] \\ y(0) = y(T); \end{cases}$$

here $0 < T < \infty$ is fixed. Throughout $k : [0,T] \times [0,t] \to \mathbf{R}$ and

$$F,\, G : [0,T] \times \mathbf{R}^n \to CK(\mathbf{R}^n);$$

here $CK(\mathbf{R}^n)$ denotes the family of nonempty, compact, convex subsets of $\mathbf{R}^n$. By a solution to (11.1.1) we mean a function $y \in AC[0,T]$ which satisfies the integrodifferential inclusion in (11.1.1) and also $y(0) = y(T)$. The chapter will be divided into two main sections. In Section 11.2 we present results for the Cauchy problem

$$(11.1.2) \qquad \begin{cases} y'(t) \in F(t, y(t)) + \int_0^t k(t,s)\, G(s, y(s))\, ds \quad \text{a.e. on} \quad [0,T] \\ y(0) = x_0 \in \mathbf{R}^n. \end{cases}$$

Conditions are put on the nonlinearities F and G which guarantee that the set of solutions of (11.1.2) is an R_δ set. In Section 11.3 an existence principle is established for (11.1.1). To establish our existence result we use the Poincaré operator together with results from Section 11.2.

For the remainder of this introduction we gather together some definitions and facts from the literature. Let X and Y be two metric spaces and $H : [0,T] \times X \to 2^Y$ (here 2^Y denotes the family of nonempty subsets of Y). H is **almost uppersemicontinuous** (almost u.s.c.) if for every $\epsilon > 0$ there exists a measurable set $A_\epsilon \subset [0,T]$ such that $m([0,T] \backslash A_\epsilon) < \epsilon$ and $H|_{A_\epsilon \times X}$ is u.s.c. (here m stands for the Lebesgue measure).

Theorem 11.1.1. [54, 70] *Let X be a separable metric space and suppose $H : [0,T] \times X \to CK(\mathbf{R}^n)$ is a Carathéodory map (By this we mean: (i). $t \mapsto H(t,y)$ is measurable for all $y \in \mathbf{R}^n$ and (ii). $y \mapsto H(t,y)$ is u.s.c. for all $t \in [0,T]$). Then there exists an almost u.s.c. map $\Phi : [0,T] \times X \to CK(\mathbf{R}^n)$ such that*

(a). $\Phi(t,x) \subset H(t,x)$ for every $(t,x) \in [0,T] \times X$
and

(b). if $\triangle \subset [0,T]$ is measurable, $u : \triangle \to \mathbf{R}^n$ and $v : \triangle \to X$ are measurable maps and $u(t) \in H(t,v(t))$ for almost all $t \in \triangle$, then $u(t) \in \Phi(t,v(t))$ for almost all $t \in \triangle$.

A single valued map $h : [0,T] \times X \to Y$ is said to be **measurable–locally Lipschitz** if for every $x \in X$ there exists a neighborhood U_x of x in X and a $L_x \in L^1[0,T]$ (here $L_x : [0,T] \to [0,\infty)$) such that

$$|h(t,x_1) - h(t,x_2)| \leq L_x(t)\,|x_1 - x_2|$$

for every $t \in [0,T]$ and x_1, $x_2 \in U_x$, and also $h(\,.\,,x)$ is measurable for every $x \in X$. A map $S : [0,T] \times \mathbf{R}^n \to 2^{\mathbf{R}^n}$ is said to be **integrably bounded** if there exists a $\mu \in L^1[0,T]$ (here $\mu : [0,T] \to [0,\infty)$) such that $|y| \leq \mu(t)$ for every $x \in \mathbf{R}^n$, $t \in [0,T]$ and $y \in S(t,x)$ [and we write $|S(t,x)| \leq \mu(t)$ for every $x \in \mathbf{R}^n$, $t \in [0,T]$ for convenience].

Let X and Y be metric spaces with $H : X \to 2^Y$ and $h : X \to Y$. We say h is a **selection** of H (and we write $h \subset H$) if $h(x) \in H(x)$ for every $x \in X$.

Definition 11.1.1. $H : [0,T] \times X \to 2^Y$ is Ca–selectionable if there exists a Carathéodory selection of H.

Definition 11.1.2. $H : [0,T] \times X \to 2^Y$ is mLL–selectionable if there exists a measurable–locally Lipschitz selection of H.

Theorem 11.1.2. [70] *Let E_1 and E_2 be two separable Banach spaces and $\Phi : [0,T] \times E_1 \to CK(E_2)$ be an almost u.s.c. map. Then Φ is σ-Ca–selectionable i.e.*

$$\Phi(t,x) = \cap_{k=1}^{\infty} \Phi_k(t,x)$$

and $\Phi_{k+1}(t,x) \subset \Phi_k(t,x)$ for almost all $t \in [0,T]$ and all $x \in \mathbf{R}^n$ and Φ_k is Ca–selectionable for every $k = \{1, 2, ...\}$. The maps $\Phi_k : [0,T] \times E_1 \to 2^{E_2}$ are almost u.s.c. and

$$\Phi_k(t,x) \subset \overline{conv}\,(\cup_{x \in E_1} \Phi(t,x)) \quad \textit{for all} \;\; (t,y) \in [0,T] \times E_1.$$

In addition, if Φ is integrably bounded then Φ is σ–mLL–selectionable i.e. it is an intersection of a decreasing sequence of mLL–selectionable mappings.

Definition 11.1.3. A nonempty set A is **contractible** provided there exists $x_0 \in A$ and a homotopy $H : A \times [0,1] \to A$ such that $H(x,1) = x$ and $H(x,0) = x_0$ for every $x \in A$.

Definition 11.1.4. A set A is called an R_δ set provided there exists a decreasing sequence $\{A_n\}_1^\infty$ of nonempty compact, contractible sets such that $A = \cap \{A_n : \ n = 1, 2, ...\}$.

Next we state the following existence–uniqueness result from the literature (see [137 Chapter 2] for existence and the ideas in [2 Chapter 1] for uniqueness)for the integrodifferential equation

$$(11.1.3) \qquad \begin{cases} y(t) = f(t, y(t)) + \int_0^t k(t,s)\, g(s, y(s))\, ds \quad \text{a.e. on } [0,T] \\ y(0) = y_0. \end{cases}$$

Theorem 11.1.3. *Let $k : [0,T] \times [0,t] \to \mathbf{R}$ and $f,\, g : [0,T] \times \mathbf{R}^n \to \mathbf{R}^n$. Suppose the following conditions hold:*

$$(11.1.4) \qquad t \mapsto f(t,y) \ \text{ is measurable for all } \ y \in \mathbf{R}^n$$

$$(11.1.5) \qquad \begin{cases} \text{there exists } \ \mu \in L^1[0,T] \ \text{ with } \ |f(t,y)| \le \mu(t) \\ \text{for a.e. } \ t \in [0,T] \ \text{ and all } \ y \in \mathbf{R}^n \end{cases}$$

$$(11.1.6) \qquad t \mapsto g(t,y) \ \text{ is measurable for all } \ y \in \mathbf{R}^n$$

$$(11.1.7) \qquad \begin{cases} \text{there exists } \ \tau \in L^1[0,T] \ \text{ with } \ |g(t,y)| \le \tau(t) \\ \text{for a.e. } \ t \in [0,T] \ \text{ and all } \ y \in \mathbf{R}^n \end{cases}$$

$$(11.1.8) \qquad \begin{cases} k : [0,T] \times [0,t] \to \mathbf{R} \ \text{ is continuous and} \\ k(t) = \sup_{s \in [0,t]} |k(t,s)| \ \text{ is bounded on } \ [0,T] \end{cases}$$

$$(11.1.9) \qquad \begin{cases} \text{for every } \ x \in \mathbf{R}^n \ \text{ there exists a neighborhood} \\ U_x \ \text{ of } x \text{ in } \mathbf{R}^n \ \text{ and a } \ L_x \in L^1[0,T] \\ (\text{here } \ L_x : [0,T] \to [0,\infty)) \ \text{ with} \\ |f(t,x_1) - f(t,x_2)| \le L_x(t)\, |x_1 - x_2| \\ \text{for every } \ t \in [0,T] \ \text{ and } \ x_1,\, x_2 \in U_x \end{cases}$$

and

$$(11.1.10)\qquad \begin{cases} \text{for every } x \in \mathbf{R}^n \text{ there exists a neighborhood} \\ V_x \text{ of } x \text{ in } \mathbf{R}^n \text{ and a } N_x \in L^1[0,T] \\ (\text{here } N_x : [0,T] \to [0,\infty)) \text{ with} \\ |g(t,x_1) - g(t,x_2)| \le N_x(t)\,|x_1 - x_2| \\ \text{for every } t \in [0,T] \text{ and } x_1,\, x_2 \in V_x. \end{cases}$$

Then there exists a unique solution $y \in AC[0,T]$ *to* (11.1.3).

Let X and Y be metric spaces. A continuous single valued map $p : Y \to X$ is called a Vietoris map if the following two conditions hold:
(i). for each $x \in X$, the set $p^{-1}(x)$ is acyclic,
(ii). p is a proper map i.e. for every compact set $A \subseteq X$ we have that $p^{-1}(A)$ is compact.

Definition 11.1.5. A multifunction $\phi : X \to C(Y)$ (here $C(Y)$ denotes the family of nonempty compact subsets of Y) is **admissible** (strongly) if there exists a metric space Z and two continuous maps $p : Z \to X$ and $q : Z \to Y$ such that
(i). p is a Vietoris map
and
(ii). $\phi(x) = q(p^{-1}(x))$ for any $x \in X$.

For notational purposes, for $r > 0$, let

$$B_r = B(0,r) = \{x \in \mathbf{R}^n : \ |x| < r\} \quad \text{and} \quad \overline{B_r} = \{x \in \mathbf{R}^n : \ |x| \le r\}.$$

Theorem 11.1.4. *Let* $M > 0$ *and suppose* $G : \overline{B_M} \to \overline{B_M}$ *is admissible. Then* G *has a fixed point.*

PROOF: The result follows immediately from [71, Theorem 1.5]. □

Theorem 11.1.5. *Let* $M > 0$ *and suppose* $H : \overline{B_M} \to \mathbf{R}^n$ *is admissible. Then either*

(A1). H *has a fixed point in* $\overline{B_M}$*; or*

(A2). there exists $\lambda \in (0,1)$ *and* $u \in \partial\overline{B_M}$ *with* $u \in \lambda H(u)$.

PROOF: Let $r : \mathbf{R}^n \to \overline{B_M}$ be the continuous retraction given by

$$r(x) = \begin{cases} x, & x \in \overline{B_M} \\ M\,\frac{x}{\|x\|}, & x \notin \overline{B_M}. \end{cases}$$

Now [71] guarantees that $r\,H : \overline{B_M} \to \overline{B_M}$ is admissible. Consequently Theorem 11.1.4 implies that there exists $x \in \mathbf{R}^n$ with $x \in r\,H(x)$. Either

$H(x) \in \overline{B_M}$ or not. If $H(x) \in \overline{B_M}$ then $x \in H(x)$ so $(A1)$ occurs. On the other hand if $H(x) \notin \overline{B_M}$ then

$$x \in M \frac{H(x)}{\|H(x)\|},$$

so $x \in \lambda H(x)$ with $\lambda = \frac{M}{\|H(x)\|} < 1$ i.e. $(A2)$ occurs. □

11.2. Cauchy problem.

This section is concerned with the solution set of the Cauchy problem (11.1.2). By a solution to (11.1.2) we mean a function $y \in AC[0,T]$ which satisfies the integrodifferential inclusion in (11.1.2) and also $y(0) = x_0$.

We begin by establishing a result for the case when F and G are integrably bounded. Let $S(F,G;x_0)$ denote the solution set of (11.1.2).

Theorem 2.1. *Let $k : [0,T] \times [0,t] \to \mathbf{R}$ and F, $G : [0,T] \times \mathbf{R}^n \to 2^{\mathbf{R}^n}$ and suppose* (11.1.8) *holds. In addition assume the following conditions are satisfied:*

$$(11.2.1) \qquad \begin{cases} F \text{ is integrably bounded i.e. there exists a } \mu \in L^1[0,T] \\ \text{such that } |y| \le \mu(t) \text{ for every } x \in \mathbf{R}^n,\ t \in [0,T] \\ \text{and } y \in F(t,x) \end{cases}$$

$$(11.2.2) \qquad \begin{cases} G \text{ is integrably bounded i.e. there exists a } \tau \in L^1[0,T] \\ \text{such that } |y| \le \tau(t) \text{ for every } x \in \mathbf{R}^n,\ t \in [0,T] \\ \text{and } y \in G(t,x) \end{cases}$$

$$(11.2.3) \qquad \begin{cases} F : [0,T] \times \mathbf{R}^n \to 2^{\mathbf{R}^n} \text{ is } mLL - \text{selectionable} \\ \text{i.e. there exists } f : [0,T] \times \mathbf{R}^n \to \mathbf{R}^n \text{ with} \\ f \subset F \text{ such that for every } x \in \mathbf{R}^n \text{ there exists} \\ \text{a neighborhood } U_x \text{ of } x \text{ in } \mathbf{R}^n \text{ and a} \\ L_x \in L^1[0,T] \text{ (here } L_x : [0,T] \to [0,\infty)\text{) with} \\ |f(t,x_1) - f(t,x_2)| \le L_x(t)\,|x_1 - x_2| \\ \text{for every } t \in [0,T] \text{ and } x_1,\ x_2 \in U_x, \text{ and also} \\ f(\,.\,,x) \text{ is measurable for every } x \in \mathbf{R}^n \end{cases}$$

and

$$(11.2.4) \qquad G : [0,T] \times \mathbf{R}^n \to 2^{\mathbf{R}^n} \text{ is } mLL - \text{selectionable.}$$

Then $S(F,G;x_0)$ is contractible.

PROOF: Let f (respectively g) be a measurable–locally Lipschitz selection of F (respectively G). Now Theorem 11.1.3 implies

$$\begin{cases} y'(t) = f(t, y(t)) + \int_0^t k(t,s)\, g(s, y(s))\, ds \quad \text{a.e. } t \in [0,T] \\ y(t_0) = u_0 \end{cases}$$

has exactly one solution $y[t_0; u_0]$ for every $u_0 \in \mathbf{R}^n$ and $t_0 \in [0,T]$. Let

$$A = S(F, G; x_0).$$

If we define a homotopy $H : A \times [0,1] \to A$ such that

$$H(x,s) = \begin{cases} x \quad \text{when } s = 1 \text{ and } x \in A \\ y[0; x_0] \quad \text{when } s = 0 \text{ and } x \in A \end{cases}$$

then we are finished. Put

$$H(x,s)(t) = \begin{cases} x(t), \; 0 \le t \le sT, \; 0 \le s \le 1 \\ y[sT; x(sT)]\,(t), \; sT \le t \le T, \; 0 \le s \le 1. \end{cases}$$

Note $y[sT; x(sT)]$, for fixed $s \in [0,1]$ and fixed $x \in A$, is the unique solution of

$$\begin{cases} y'(t) = f(t, y(t)) + \int_0^t k(t,s)\, g(s, y(s))\, ds \quad \text{a.e. } t \in [0,T] \\ y(sT) = x(sT). \end{cases}$$

We have immediately that $H(x,1) = x$ for $x \in A$ and $H(x,0) = y[0; x_0]$ (note $H(x,0)(t) = y[0; x(0)]\,(t) = y[0; x_0]\,(t)$, $0 \le t \le T$ since $x \in A$). Notice also that $H : A \times [0,1] \to A$. To see this fix $x \in A$ and let $s \in [0,1]$ be fixed. Then with

$$\eta(t) = \begin{cases} x(t), \; 0 \le t \le sT \\ y[sT; x(xT)]\,(t), \; sT \le t \le T \end{cases}$$

it is easy to check that

$$\eta'(t) \in F(t, \eta(t)) + \int_0^t k(t,s)\, G(s, \eta(s))\, ds \quad \text{a.e.}$$

(consider $0 \le t \le sT$ and $sT \le t \le T$ seperately) since $x \in A$ and $\eta(sT) = x(sT)$. This H is a continuous homotopy contracting $S(F, G; x_0)$ to the point $y[0; x_0]$. $\square$

Suppose Φ (respectively Ψ) : $[0,T] \times \mathbf{R}^n \to 2^{\mathbf{R}^n}$ is the intersection of the decreasing sequence Φ_k (respectively Ψ_k) : $[0,T] \times \mathbf{R}^n \to 2^{\mathbf{R}^n}$ i.e.

$$\Phi(t,x) = \cap_{k=1}^{\infty} \Phi_k(t,x)$$

(respecively $\Psi(t,x) = \cap_{k=1}^{\infty} \Psi_k(t,x)$) and

$$\Phi_{k+1}(t,x) \subset \Phi_k(t,x)$$

(respectively $\Psi_{k+1}(t,x) \subset \Psi_k(t,x)$) for almost all $t \in [0,T]$ and all $x \in \mathbf{R}^n$. If $y \in \cap_{k=1}^{\infty} S(\Phi_k, \Psi_k; x_0)$ then for **each** $k \in \{1,2,...\}$ there exists a set $A_k \subseteq [0,T]$ of Lebesgue measure zero with

$$y'(t) \in \Phi_k(t, y(t)) + \int_0^t k(t,s)\, \Psi_k(s, y(s))\, ds \quad \text{for} \quad t \in [0,T] \backslash A_k.$$

Let $A = \cup_{k=1}^{\infty} A_k$. Then A has Lebesgue measure zero and

$$y'(t) \in \Phi_k(t, y(t)) + \int_0^t k(t,s)\, \Psi_k(s, y(s))\, ds \quad \text{for} \quad t \in [0,T] \backslash A.$$

Hence $y \in S(\Phi, \Psi; x_0)$ and so

(11.2.5) $$S(\Phi, \Psi; x_0) = \cap_{k=1}^{\infty} S(\Phi_k, \Psi_k; x_0).$$

Theorem 11.2.2. *Let* $k : [0,T] \times [0,t] \to \mathbf{R}$ *and* $F, G : [0,T] \times \mathbf{R}^n \to CK(\mathbf{R}^n)$ *and assume*

(11.2.6) $y \mapsto F(t,y)$ *is u.s.c. for a.e.* $t \in [0,T]$

(11.2.7) $y \mapsto G(t,y)$ *is u.s.c. for a.e.* $t \in [0,T]$

(11.2.8) $t \mapsto F(t,y)$ *is measurable for every* $y \in \mathbf{R}^n$

and

(11.2.9) $t \mapsto G(t,y)$ *is measurable for every* $y \in \mathbf{R}^n$

hold. In addition suppose (11.1.8), (11.2.1) *and* (11.2.2) *are satisfied. Then* $S(F, G; x_0)$ *is an* R_δ *set.*

PROOF: From Theorem 11.1.1 there exist almost u.s.c. maps

$$\Phi,\ \Psi : [0,T] \times \mathbf{R}^n \to CK(\mathbf{R}^n)$$

with

$$S(\Phi, \Psi; x_0) = S(F, G; x_0).$$

Also Theorem 11.1.2 implies Φ and Ψ are σ–mLL–selectionable so we can find sequences of maps Φ_k, Ψ_k, $k \in \{1, 2, ...\}$, as described in Theorem 11.1.2. Notice Theorem 11.2.1 and (11.2.5) implies that $S(\Phi, \Psi; x_0)$ is the interection of a decreasing sequence of contractible spaces (Of course Theorem 11.1.3 guarantees that $S(\Phi_k, \Psi_k; x_0)$ is nonempty for each $k \in \{1, 2, ...\}$). Finally it is easy to see from the Arzela–Ascoli Theorem that $S(\Phi_k, \Psi_k; x_0)$ is compact for every $k \in \{1, 2, ...\}$. Thus $S(\Phi, \Psi; x_0)$ is an R_δ set. □

We now discuss the more general situation. Assume for the remainder of this section that (11.1.8), (11.2.6), (11.2.7), (11.2.8) and (11.2.9) hold. In addition assume the following conditions are satisfied:

$$\text{(11.2.10)} \qquad \begin{cases} \text{for each } r > 0 \text{ there exists a function } h_r \in L^1[0, T] \\ \text{with } |F(t, x)| \le h_r(t) \text{ for } t \in [0, T] \text{ and every} \\ x \in \mathbf{R}^n \text{ with } |x| \le r \end{cases}$$

$$\text{(11.2.11)} \qquad \begin{cases} \text{for each } r > 0 \text{ there exists a function } m_r \in L^1[0, T] \\ \text{with } |G(t, x)| \le m_r(t) \text{ for } t \in [0, T] \text{ and every} \\ x \in \mathbf{R}^n \text{ with } |x| \le r \end{cases}$$

and

$$\text{(11.2.12)} \qquad \begin{cases} \text{there exists } M > 0 \text{ with } |y|_0 < M \text{ for any} \\ \text{possible solution } y \in AC[0, T] \text{ to (11.1.2);} \\ \text{here } |y|_0 = \sup_{t \in [0,T]} |y(t)|. \end{cases}$$

Let $\epsilon > 0$ be given and let $\tau_\epsilon : \mathbf{R}^n \to [0, 1]$ be the Urysohn function for

$$\left(\overline{B}(0, M)\, ,\, \mathbf{R}^n \backslash B(0, M + \epsilon)\right)$$

such that

$$\tau_\epsilon(x) = 1 \text{ if } |x| \le M \text{ and } \tau_\epsilon(x) = 0 \text{ if } |x| \ge M + \epsilon.$$

Let

$$F^\star(t, x) = \tau_\epsilon(x)\, F(t, x) \quad \text{and} \quad G^\star(t, x) = \tau_\epsilon(x)\, G(t, x)$$

and consider the problem

$$\text{(11.2.13)} \qquad \begin{cases} y'(t) \in F^\star(t, y(t)) + \int_0^t k(t, s)\, G^\star(s, y(s))\, ds \quad \text{a.e. on } [0, T] \\ y(0) = x_0 \in \mathbf{R}^n. \end{cases}$$

Let $S_\epsilon(F^\star, G^\star; x_0)$ denote the solution set of (11.2.13).

Theorem 11.2.3. *Let* $k : [0,T] \times [0,t] \to \mathbf{R}$ *and* $F, G : [0,T] \times \mathbf{R}^n \to CK(\mathbf{R}^n)$ *and suppose* (11.1.8), (11.2.6) $- -$(11.2.12) *hold. Let* $\epsilon > 0$ *be given and assume*

$$(11.2.14) \qquad \begin{cases} |w|_0 < M & \text{for any possible solution} \\ w \in AC[0,T] & \text{to (11.2.13).} \end{cases}$$

Then $S(F, G; x_0)$ *is an* R_δ *set.*

PROOF: Notice (11.2.12) and (11.2.14) imply $S(F, G; x_0) = S_\epsilon(F^\star, G^\star; x_0)$. It is easy to see that $F^\star : [0,T] \times \mathbf{R}^n \to CK(\mathbf{R}^n)$ satisfies (11.2.6) and (11.2.8) (with F replaced by $F^\star$) and $G^\star : [0,T] \times \mathbf{R}^n \to CK(\mathbf{R}^n)$ satisfies (11.2.7) and (11.2.9) (with G replaced by $G^\star$). Also (11.2.10) (respectively (11.2.11)) and the definition of τ_ϵ implies $F^\star$ (respectively $G^\star$) is integrably bounded. Now Theorem 11.2.2 implies that $S_\epsilon(F^\star, G^\star; x_0)$ is an R_δ set. □

11.3. Periodic problem.

This section is concerned with existence results for the periodic problem (11.1.1). Our analysis relies on the following well known result from the literature [70].

Theorem 11.3.1. *Let* E *be a normed space,* X *a metric space and* $H : E \times X \to E$ *a continuous (single valued) map such that for any compact subset* $A \subseteq X$, $\overline{H(E \times A)}$ *is a compact subset of* E. *Then the (multivalued) map* $J : X \to E$ *defined by*

$$J(x) = Fix\,(H(\,.\,, x))$$

is a u.s.c. mapping.

PROOF: Schauder's fixed point theorem guarantees for each $x \in X$, that $J(x)$ is nonempty (and also in fact compact).

Now let $x_0 \in X$ and let U be an open neighborhood of $J(x_0)$ in E. It remains for us to prove that there exists $r > 0$ such that

$$\text{for every} \quad x \in B(x_0, r) \quad \text{we have} \quad J(x) \subseteq U.$$

Suppose this is false. Then for every $n = 1, 2, \ldots$ there exists

$$x_n \in B\left(x_0, \frac{1}{n}\right) \quad \text{and} \quad y_n \in Fix\,(H(\,.\,, x_n)) \quad \text{with} \quad y_n \notin U.$$

Let

$$A = \overline{\{x_n\}}, \quad \text{so} \quad A = \{x_n\} \cup \{x_0\}.$$

From our assumptions, we can assume without loss of generality that $\lim_{n\to\infty} y_n = y_0$. Notice $y_0 \notin U$. However

$$y_0 \in Fix\,(H(\,.\,, x_0)) = J(x_0) \subseteq U,$$

a contradiction. □

Theorem 11.3.2. *Let* $k : [0,T] \times [0,t] \to \mathbf{R}$ *and* $F, G : [0,T] \times \mathbf{R}^n \to CK(\mathbf{R}^n)$ *and suppose* (11.1.8), (11.2.1), (11.2.2) *and* (11.2.6) − −(11.2.9) *hold. Let* $P : \mathbf{R}^n \to C[0,T]$ *be defined by*

$$P(x) = S(F, G; x);$$

here $S(F, G; x)$ *is the solution set of*

$$\text{(11.3.1)} \qquad \begin{cases} y'(t) \in F(t, y(t)) + \int_0^t k(t,s)\, G(s, y(s))\, ds & a.e.\ on\ [0,T] \\ y(0) = x. \end{cases}$$

Then P *is a u.s.c. map with* R_δ *values.*

PROOF: By Theorem 11.2.2, P has R_δ values. Define $H : C[0,T] \times \mathbf{R}^n \to C[0,T]$ by

$$\begin{aligned} H(u, x) = \{v : \quad & v(t) = x + \int_0^t w_1(s)\, ds + \int_0^t \int_0^z k(z,s)\, w_2(s)\, ds\, dz \\ & \text{for } t \in [0,T], \text{ where } w_1(\tau) \in F(\tau, u(\tau)) \text{ and} \\ & w_2(\tau) \in G(\tau, u(\tau)) \text{ for every } \tau \in [0,T]\,\}. \end{aligned}$$

Remark. Note w_1 (respectively w_2) is a measurable selection of W_1 (respectively W_2) which exist from the Kuratowski–Ryll–Nardzewski Selection Theorem; here $W_1(t) = F(t, u(t))$ and $W_2(t) = G(t, u(t))$.

Notice that

$$P(x) = Fix\,(H(\,.\,, x)).$$

We wish to apply Theorem 11.3.1. It is easy to see that $H : C[0,T] \times \mathbf{R}^n \to C[0,T]$ is continuous. Next let A be a compact subset of $\mathbf{R}^n$. Then there exists $r > 0$ with $|y| \le r$ for all $y \in A$. Let $v_1 \in H(C[0,T] \times A)$. Then there exists $u_1 \in C[0,T]$ (which gives the existence of w_1 and w_2 such that $w_1(z) \in F(z, u_1(z))$ and $w_2(z) \in G(z, u_1(z))$ for $z \in [0,T]$) and $x_1 \in A$ with

$$v_1(t) = x_1 + \int_0^t w_1(s)\, ds + \int_0^t \int_0^z k(z,s)\, w_2(s)\, ds\, dz \quad \text{for} \quad t \in [0,T].$$

For $t \in [0,T]$ we have (see (11.2.1) and (11.2.2)),

$$|v_1(t)| \leq |x_1| + \int_0^t \mu(s)\, ds + \int_0^t \int_0^z |k(z,s)|\, \tau(s)\, ds\, dz$$

and for $t,\ s \in [0,T]$, with $s < t$, we have

$$|v_1(t) - v_1(s)| \leq \int_s^t \mu(x)\, dx + \int_s^t \int_0^z |k(z,x)|\, \tau(x)\, dx\, dz.$$

The Arzela–Ascoli Theorem now guarantees that $\overline{H(C[0,T] \times A)}$ is a compact subset of $C[0,T]$. Theorem 11.3.1 now implies that P is u.s.c. □

Let $P : \mathbf{R}^n \to C[0,T]$ be as described in Theorem 11.3.1. Also define the evaluation map $e_t : C[0,T] \to \mathbf{R}^n$ by

$$e_t(y) = y(t).$$

The composition

$$P_T = e_T \circ P : \mathbf{R}^n \to \mathbf{R}^n$$

is called the Poincaré translation operator. Notice solving (11.1.1) is equivalent to finding a point $x \in \mathbf{R}^n$ with $x \in P_T(x)$.

Theorem 11.3.3. *Let* $k : [0,T] \times [0,t] \to \mathbf{R}$ *and* $F,\, G : [0,T] \times \mathbf{R}^n \to CK(\mathbf{R}^n)$ *and suppose* (11.1.8), (11.2.1), (11.2.2) *and* (11.2.6) – –(11.2.9) *hold. Assume there exists a constant* $M > 0$, *independent of* λ, *with*

$$|y(0)| \neq M$$

for any solution $y \in AC[0,T]$ *to*

$$(11.3.2)_\lambda \qquad \begin{cases} y'(t) \in F(t, y(t)) + \int_0^t k(t,s)\, G(s, y(s))\, ds & a.e. \text{ on } [0,T] \\ y(0) = \lambda\, y(T) \end{cases}$$

for each $\lambda \in (0,1)$. *Then* (11.1.1) *has a solution* $u \in AC[0,T]$ *with* $|u(0)| \leq M$.

PROOF: The result follows immediately from Theorem 11.1.5 with $H = P_T$. Theorem 11.3.2 guarantees that $P_T : \mathbf{R}^n \to \mathbf{R}^n$ is an admissible map. Notice condition $(A2)$ cannot occur. To see this suppose there exists $\lambda \in (0,1)$ and $u \in \partial\overline{B_M}$ (i.e. $|u| = M$) with $u \in \lambda P_T(u)$. Then there exists $y \in P(u)$ with $u = \lambda y(T)$ i.e.

$$y'(t) \in F(t, y(t)) + \int_0^t k(t,s)\, G(s, y(s))\, ds \quad \text{for a.e. } t \in [0,T]$$

and $y(0) = u = \lambda y(T)$. Thus y is a solution of $(11.3.2)_\lambda$ with (since $u \in \partial\overline{B_M}$) $|y(0)| = M$. This is a contradiction. Now Theorem 11.1.5 guarantees that P_T has a fixed point $z \in \overline{B_M}$. Thus there exists $u \in P(z)$ with $z = u(T)$ and of course $z \in \overline{B_M}$ i.e.

$$u'(t) \in F(t, u(t)) + \int_0^t k(t,s)\, G(s, u(s))\, ds \quad \text{for a.e. } t \in [0,T],$$

$u(0) = z = u(T)$ and $|u(0)| \le M$. □

We now use Theorem 11.3.3 to establish an existence result for (11.1.1). For simplicity in the statement of our next theorem we rewrite (11.1.1) as

$$\text{(11.3.3)} \qquad \begin{cases} y'(t) \in N\, y(t) \quad \text{a.e. on } [0,T] \\ y(0) = y(T) \end{cases}$$

where $N : C[0,T] \to C[0,T]$ is given by

$$N\, y(t) = F(t, y(t)) + \int_0^t k(t,s)\, G(s, y(s))\, ds.$$

We will assume for the remainder of this paper that (11.1.8) and (11.2.6) – –(11.2.9) hold. In addition suppose the following conditions are satisfied:

$$\text{(11.3.4)} \qquad \begin{cases} \text{for any constant } a \ge 0 \text{ there exists } \eta_a \in L^1[0,T] \\ \text{such that for every } y \in C[0,T] \text{ with } |y|_0 = \\ \sup_{t\in[0,T]} |y(t)| \le a, \text{ we have } |N\, y(t)| \le \eta_a(t) \\ \text{for a.e. } t \in [0,T] \end{cases}$$

and

$$\text{(11.3.5)} \qquad \begin{cases} \text{there exists a constant } r > 0 \text{ with} \\ y(t)\,.\,N\, y(t) \subseteq (-\infty, 0] \text{ a.e. on } [0,T] \\ \text{for all } y \in C[0,T] \text{ with } |y(t)| > r. \end{cases}$$

Theorem 11.3.4. *Let* $k : [0,T] \times [0,t] \to \mathbf{R}$ *and* $F, G : [0,T] \times \mathbf{R}^n \to CK(\mathbf{R}^n)$ *and suppose* (11.1.8), (11.2.6) – –(11.2.9), (11.3.4) *and* (11.3.5) *hold. Then* (11.3.3) *(i.e.* (11.1.1)*) has a solution in* $AC[0,T]$.

PROOF: We wish to apply Theorem 11.3.3. Notice (11.3.4) and (11.3.5) guarantee that there exists $\eta \in L^1[0,T]$ with

$$(11.3.6)\qquad \begin{cases} y(t)\,.\,N\,y(t) \subseteq (-\infty, \eta(t)] & \text{for a.e.}\\ t \in [0,T] \text{ and all } y \in C[0,T]. \end{cases}$$

Let

$$M = \sqrt{r^2 + 4|\eta|_1 + 1}$$

where $|\eta|_1 = \int_0^T |\eta(x)|\,dx$. Let $\tau : \mathbf{R}^n \to [0,1]$ be the Urysohn function for

$$\left(\overline{B_M}\,,\, \mathbf{R}^n \backslash \overline{B_{M+1}}\right)$$

such that

$$\tau(x) = 1 \text{ if } |x| \le M \text{ and } \tau(x) = 0 \text{ if } |x| \ge M + 1.$$

Let $V(x) = \tau(x)\,N(x)$ and we will consider the periodic problem

$$(11.3.7)\qquad \begin{cases} y'(t) \in V\,y(t) & \text{a.e. on } [0,T]\\ y(0) = y(T). \end{cases}$$

It is easy to see that $\tau\,F$ and $\tau\,G$ are integrably bounded and so to show (11.3.7) has a solution we will apply Theorem 11.3.3. Let y be any solution of

$$(11.3.8)_\lambda\qquad \begin{cases} y'(t) \in V\,y(t) & \text{a.e. on } [0,T]\\ y(0) = \lambda\,y(T) \end{cases}$$

for $\lambda \in (0,1)$. Notice we have immediately that

$$(11.3.9)\qquad \begin{cases} y(t)\,.\,V\,y(t) \subseteq (-\infty, 0] & \text{a.e. on } [0,T]\\ \text{for all } y \in C[0,T] \text{ with } |y(t)| > r \end{cases}$$

and

$$(11.3.10)\qquad \begin{cases} y(t)\,.\,V\,y(t) \subseteq (-\infty, \eta(t)] & \text{for a.e.}\\ t \in [0,T] \text{ and all } y \in C[0,T]. \end{cases}$$

We now show that there exists $t_1 \in [0,T]$ with $|y(t_1)| \le r$. If not, then $|y(t)| > r$ for all $t \in [0,T]$ and so

$$|y(T)|^2\,(1 - \lambda^2) = |y(T)|^2 - |y(0)|^2 = 2 \int_0^T y(t)\,.\,y'(t)\,dt$$

and this together with $(11.3.8)_\lambda$ and (11.3.9) yields $|y(T)| = 0$; this contradicts the fact that $|y(t)| > r$ for all $t \in [0,T]$. Thus there exists $t_1 \in [0,T]$ with $|y(t_1)| \le r$. Let $|y(t)|$ assume its maximum at t_2 and its minimum at t_3 in $[0,T]$. Notice $|y(t_3)| \le r$. Also notice (see (11.3.10)) that

$$y(t)\,.\,y'(t) \le \eta(t) \ \text{ a.e. on } [0,T] \tag{11.3.11}$$

and so integration from t_3 to T and also from 0 to t_2 yields

$$|y(T)|^2 \le |y(t_3)|^2 + 2\,|\eta|_1 \quad \text{and} \quad |y(t_2)|^2 \le |y(0)|^2 + 2\,|\eta|_1.$$

Thus

$$\begin{aligned} |y(t_2)|^2 &\le |y(0)|^2 + 2\,|\eta|_1 = \lambda^2\,|y(T)|^2 + 2\,|\eta|_1 \le |y(T)|^2 + 2\,|\eta|_1 \\ &\le |y(t_3)|^2 + 4\,|\eta|_1 \le r^2 + 4\,|\eta|_1 < M^2. \end{aligned}$$

That is

$$|y|_0 < M \ \text{ so } \ |y(0)| \ne M.$$

As a result Theorem 11.3.3 guarantees that (11.3.7) has a solution $u \in AC[0,T]$ with $|u(0)| < M$. Next we <u>claim</u> that

$$|u|_0 < M. \tag{11.3.12}$$

If (11.3.12) is true then $V\,u(t) = N\,u(t)$ so u is a solution of (11.3.3) and we are finished.

It remains to prove the claim. There are two cases to consider, either there exists $s_1 \in [0,T]$ with $|u(s_1)| \le r$ or $|u(s)| > r$ for all $s \in [0,T]$.

Case (1). $|u(s)| > r$ for all $s \in [0,T]$.

Then (11.3.7) and (11.3.9) imply

$$|u(t)|^2 - |u(0)|^2 = 2\int_0^t u(s)\,.\,u'(s)\,ds \le 0$$

and also

$$|u(T)|^2 - |u(t)|^2 \le 0$$

for all $t \in [0,T]$. Thus

$$|u(T)|^2 \le |u(t)|^2 \le |u(0)|^2 \ \text{ for all } \ t \in [0,T].$$

However since $u(T) = u(0)$ we have $|u(t)|^2 = |u(0)|^2$ for all $t \in [0,T]$. Consequently

$$|u|_0 = |u(0)| < M$$

and (11.3.12) is true in this case.

Case (2). $|u(s_1)| \leq r$ for some $s_1 \in [0,T]$.

Let $|u(t)|$ assume its maximum at s_2 and its minimum at s_3 in $[0,T]$. Notice

$$|u(s_3)| \leq r.$$

As before

$$|u(T)|^2 \leq |u(s_3)|^2 + 2\,|\eta|_1 \quad \text{and} \quad |u(s_2)|^2 \leq |u(0)|^2 + 2\,|\eta|_1.$$

Thus

$$\begin{aligned} |u(s_2)|^2 &\leq |u(0)|^2 + 2\,|\eta|_1 = |u(T)|^2 + 2\,|\eta|_1 \\ &\leq |u(s_3)|^2 + 4\,|\eta|_1 \leq r^2 + 4\,|\eta|_1 \; < \; M^2, \end{aligned}$$

so $|u|_0 < M$ and (11.3.12) is true in this case also. □

Remark. It is possible to use a translational approach [122] to discuss (11.1.1) also.

CHAPTER 12

POSITIVE L^p AND CONTINUOUS SOLUTIONS FOR FREDHOLM INTEGRAL INCLUSIONS

12.1. Introduction.

In this chapter we present results which guarantee that the Fredholm integral inclusion

$$y(t) \in \int_0^T k(t,s)\, f(s, y(s))\, ds \tag{12.1.1}$$

has a positive solution $y \in L^p[0,T]$, $1 \leq p < \infty$, or has a nonnegative solution $y \in C[0,T]$. Throughout this chapter $T > 0$ is fixed, $k : [0,T] \times [0,T] \to \mathbf{R}$ and $f : [0,T] \times \mathbf{R} \to 2^{\mathbf{R}}$; here $2^{\mathbf{R}}$ denotes the family of nonempty subsets of $\mathbf{R}$. The main idea in this chapter relies on the multivalued analogue [7] of Krasnoselski's fixed point theorem in a cone. This fixed point result will be established in Section 12.2. In Section 12.3 using the 1991 paper of Cellina, Fryszkowski and Rzezuchowski [41] we are able to establish criteria which guarantee that (12.1.1) has a positive solution $y \in L^p[0,T]$. Section 12.4 discusses $C[0,T]$ solutions to (12.1.1).

For the remainder of this section we gather together some definitions and known facts. Let E be a Banach space and $\mathcal{P}_B(E)$ the bounded subsets of E. The *Kuratowskii measure of noncompactness* is the map $\alpha : \mathcal{P}_B(E) \to [0,\infty]$ defined by

$$\alpha(X) = \inf\{\epsilon > 0 : X \subseteq \cup_{i=1}^n X_i \text{ and } diam\,(X_i) \leq \epsilon\};$$

here $X \in \mathcal{P}_B(E)$. Let Z be a nonempty subset of E and $F : Z \to 2^E$. F is called a k–set contraction if $\alpha(F(Y)) \leq k\,\alpha(Y)$ for all bounded sets Y of Z and $F(Z)$ is bounded.

Let C be a closed convex subset of E and U an open subset of C.

Definition 12.1.1. $K(\overline{U}, C)$ denotes the set of all upper semicontinuous (u.s.c.), k–set contractive (here $0 \leq k < 1$) maps $F : \overline{U} \to CK(C)$; here

$CK(C)$ denotes the family of nonempty, compact, convex subsets of C and $\overline{U}$ denotes the closure of U in C.

Remark. If $F \in K(\overline{U}, C)$ then by definition $F(\overline{U})$ is a subset of a bounded set in C.

Definition 12.1.2. We let $K_{\partial U}(\overline{U}, C)$ denote the maps $F \in K(\overline{U}, C)$ with $x \notin F(x)$ for $x \in \partial U$; here ∂U denotes the boundary of U in C.

Definition 12.1.3. A map $F \in K_{\partial U}(\overline{U}, C)$ is *essential* in $K_{\partial U}(\overline{U}, C)$ if for every map $G \in K_{\partial U}(\overline{U}, C)$ with $G|_{\partial U} = F|_{\partial U}$ we have that there exists $x \in \overline{U}$ with $x \in G(x)$. Otherwise F is *inessential* in $K_{\partial U}(\overline{U}, C)$ i.e. there exists a map $G \in K_{\partial U}(\overline{U}, C)$ with $G|_{\partial U} = F|_{\partial U}$ and $x \notin G(x)$ for $x \in \overline{U}$.

Definition 12.1.4. Two maps $F, G \in K_{\partial U}(\overline{U}, C)$ are homotopic in $K_{\partial U}(\overline{U}, C)$, written

$$F \cong G \quad \text{in} \quad K_{\partial U}(\overline{U}, C),$$

if there exists a u.s.c., k–set contractive (here $0 \le k < 1$) mapping $N : \overline{U} \times [0,1] \to CK(C)$ such that $N_t(u) = N(u,t) : \overline{U} \to CK(C)$ belongs to $K_{\partial U}(\overline{U}, C)$ for each $t \in [0,1]$ and $N_0 = F$ with $N_1 = G$.

Remark. We say $N : \overline{U} \times [0,1] \to CK(C)$ is k–set contractive if

$$\alpha(N\, W) \le k\, \alpha(\Pi\, W) \quad \text{for all bounded sets} \quad W \subseteq \overline{U} \times [0,1]$$

and $N(\overline{U})$ is a subset of a bounded set in C; here $\Pi : \overline{U} \times [0,1] \to \overline{U}$ is the natural projection.

Theorem 12.1.1. [119, 120, 121] *Let E, C and U be as above. Suppose F and G are two maps in $K_{\partial U}(\overline{U}, C)$ such that $F \cong G$ in $K_{\partial U}(\overline{U}, C)$. Then F is essential in $K_{\partial U}(\overline{U}, C)$ iff G is essential in $K_{\partial U}(\overline{U}, C)$.*

In this chapter we will also need the following nonlinear alternative of Leray–Schauder type (see Chapter 10).

Theorem 12.1.2. *Let E, C and U be as above and let $0 \in U$. Suppose $F \in K(\overline{U}, C)$. Then either*

(A1). there exists $x \in \overline{U}$ with $x \in F(x)$; or

(A2). there exists $u \in \partial U$ and $\lambda \in (0,1)$ with $u \in \lambda\, F(u)$.

12.2. Multiple fixed points.

In this section $E = (E, \|\cdot\|)$ will be a Banach space and C a closed convex subset of E with $\alpha\, u + \beta\, v \in C$ for all $\alpha \ge 0$, $\beta \ge 0$ and $u, v \in C$.

Let $\rho > 0$ with

$$B_\rho = \{x : x \in C \text{ and } \|x\| < \rho\},$$

$$S_\rho = \{x : x \in C \text{ and } \|x\| = \rho\},$$

and of course $\overline{B_\rho} = B_\rho \cup S_\rho$.

Theorem 12.2.1. *Let E and C be as above and let r, R be constants with $0 < r < R$. Suppose*

$$F : \overline{B_R} \to CK(C)$$

is a u.s.c., k–set contractive (here $0 \leq k < 1$) map and assume the following conditions hold:

$$x \notin F(x) \quad \textit{for} \quad x \in S_R \cup S_r \tag{12.2.1}$$

$$\left\{ \begin{array}{l} F : \overline{B_r} \to CK(C) \quad \textit{is inessential in} \quad K_{S_r}(\overline{B_r}, C) \\ (\textit{i.e.} \;\; F|_{\overline{B_r}} \quad \textit{is inessential}) \end{array} \right. \tag{12.2.2}$$

and

$$F : \overline{B_R} \to CK(C) \quad \textit{is essential in} \quad K_{S_R}(\overline{B_R}, C). \tag{12.2.3}$$

Then F has at least one fixed point in

$$V = \{x : x \in C \;\; \textit{and} \;\; r < \|x\| < R\}.$$

Remark. In Theorem 12.2.1, F k–set contractive could be replaced by F condensing.

PROOF: Suppose F has no fixed points in V. Now (12.2.2) implies that there exists a u.s.c., k–set contractive map $\theta : \overline{B_r} \to CK(C)$ with

$$\theta|_{S_r} = F|_{S_r} \text{ and with } x \notin \theta(x) \text{ for } x \in \overline{B_r}.$$

Define the mapping $\Phi : \overline{B_R} \to CK(C)$ by

$$\Phi(x) = \left\{ \begin{array}{l} F(x), \; r < \|x\| \leq R \\ \theta(x), \; \|x\| \leq r. \end{array} \right.$$

Clearly $\Phi : \overline{B_R} \to CK(C)$ is a u.s.c., k–set contractive map and Φ has no fixed points in $\overline{B_R}$ (since θ has no fixed points in $\overline{B_r}$ and F has no fixed points in V). This contradicts the fact that $F : \overline{B_R} \to CK(C)$ is essential in $K_{S_R}(\overline{B_R}, C)$. □

Theorem 12.2.2. *Let E, C be as above and let r, R be constants with $0 < r < R$. Suppose the following conditions are satisfied:*

$$(12.2.4)\quad \begin{cases} N : \overline{B_R} \times [0,1] \to CK(C) \text{ is a u.s.c., } k\text{-set contractive} \\ (\text{here } 0 \le k < 1) \text{ map with } N(x,0) = \{0\} \text{ for all } x \in \overline{B_R} \\ \text{and such that for each } t \in [0,1] \text{ we have } x \notin N(x,t) \\ \text{for all } x \in S_R \end{cases}$$

$$(12.2.5)\quad \begin{cases} H : \overline{B_r} \times [0,1] \to CK(C) \text{ is a u.s.c., } k\text{-set contractive} \\ (\text{here } 0 \le k < 1) \text{ map such that for each } t \in [0,1] \\ \text{we have } x \notin H(x,t) \text{ for all } x \in S_r \end{cases}$$

$$(12.2.6)\qquad H(\,.\,,1)|_{\overline{B_r}} = N(\,.\,,1)|_{\overline{B_r}}$$

and

$$(12.2.7)\qquad x \notin H(x,0) \quad \text{for all } x \in B_r.$$

Then $N(x,1)$ has a fixed point in

$$V = \{x : x \in C \text{ and } r < \|x\| < R\}.$$

PROOF: We know (see [119,120,121]) that the zero map is essential in $K_{S_R}(\overline{B_R}, C)$. Now (12.2.4) together with Theorem 12.1.1 implies

$$(12.2.8)\qquad N(x,1) : \overline{B_R} \to CK(C) \text{ is essential in } K_{S_R}(\overline{B_R}, C).$$

In addition (12.2.7) (and also (12.2.5)) implies

$$H(x,0) \text{ is inessential in } K_{S_r}(\overline{B_r}, C).$$

This together with (12.2.5), (12.2.6) and Theorem 12.1.1 implies

$$(12.2.9)\quad N(x,1) = H(x,1) : \overline{B_r} \to CK(C) \text{ is inessential in } K_{S_r}(\overline{B_r}, C).$$

Now (12.2.8), (12.2.9) (and also (12.2.4), (12.2.5) and (12.2.6)) imply that (12.2.1), (12.2.2) and (12.2.3) of Theorem 12.2.1 hold with $F(x) = N(x,1)$. The result now follows from Theorem 12.2.1. □

We now use Theorem 12.2.2 to establish two results which are useful from an application viewpoint (see Section 12.3 and 12.4).

Theorem 12.2.3. *Let E, C be as above and let r, R be constants with $0 < r < R$. Suppose*

$$F : \overline{B_R} \to CK(C)$$

is a u.s.c., k-set contractive (here $0 \le k < 1$) map and assume the following conditions hold:

$$x \notin \lambda F(x) \quad \text{for} \quad \lambda \in [0,1) \quad \text{and} \quad x \in S_R \tag{12.2.10}$$

and

$$\begin{cases} \text{there exists a } v \in C\backslash\{0\} \text{ with } x \notin F(x) + \delta v \\ \text{for any } \delta > 0 \text{ and } x \in S_r. \end{cases} \tag{12.2.11}$$

Then F has a fixed point in

$$\{x : x \in C \ \text{ and } \ r \le \|x\| \le R\}.$$

PROOF: Suppose

$$x \notin F(x) \quad \text{for} \quad x \in S_r \cup S_R$$

(otherwise we are finished). Choose $M > 0$ such that

$$\|y\| \le M \quad \text{for all} \quad y \in F(x) \quad \text{and} \quad x \in \overline{B_r}.$$

Now choose $\delta_0 > 0$ such that

$$\|\delta_0 v\| > M + r. \tag{12.2.12}$$

Let

$$N(\,.\,,t) = t\,F(\,.\,) \quad \text{and} \quad H(\,.\,,t) = F(\,.\,) + (1-t)\,\delta_0\, v.$$

Now (12.2.10) and (12.2.11) (with $\delta = (1-t)\,\delta_0$) imply that (12.2.4) and (12.2.5) are satisfied. In addition (12.2.6) is true since

$$N(x,1) = F(x) = H(x,1) \quad \text{for} \quad x \in \overline{B_r}$$

and finally (12.2.12) implies (12.2.7) is satisfied (note $H(x,0) = F(x) + \delta_0\, v$). The result now follows from Theorem 12.2.2. □

In our next theorem $C \subseteq E$ will be a cone. Let $\rho > 0$ with

$$\Omega_\rho = \{x \in E : \ \|x\| < \rho\},$$

$$\partial_E \Omega_\rho = \{x \in E : \ \|x\| = \rho\},$$
$$B_\rho = \{x : x \in C \ \text{ and } \ \|x\| < \rho\}$$

and

$$S_\rho = \{x : x \in C \ \text{ and } \ \|x\| = \rho\}.$$

Notice

$$B_\rho = \Omega_\rho \cap C \quad \text{and} \quad S_\rho = \partial_C(\Omega_\rho \cap C) = \partial_E \Omega_\rho \cap C.$$

Theorem 12.2.4. *Let $E = (E, \|\,.\,\|)$ be a Banach space, $C \subseteq E$ a cone and let $\|\,.\,\|$ be increasing with respect to C. Also r, R are constants with $0 < r < R$. Suppose*

$$F : \overline{\Omega_R} \cap C \to CK(C)$$

is a u.s.c., k–set contractive (here $0 \le k < 1$) map and assume the following conditions hold:

$$\|y\| \le \|x\| \ \text{ for all } \ y \in F(x) \ \text{ and } \ x \in \partial_E \Omega_R \cap C \tag{12.2.13}$$

and

$$\|y\| > \|x\| \ \text{ for all } \ y \in F(x) \ \text{ and } \ x \in \partial_E \Omega_r \cap C. \tag{12.2.14}$$

Then F has a fixed point in

$$C \cap \{x \in E : \ r \le \|x\| \le R\}.$$

PROOF: Notice (12.2.13) implies that (12.2.10) is true. To see this suppose there exists $x \in S_R$ and $\lambda \in [0, 1)$ with $x \in \lambda F(x)$. Then there exists $y \in F(x)$ with $x = \lambda y$ and so

$$R = \|x\| = |\lambda| \, \|y\| < \|y\| \le \|x\| = R,$$

a contradiction. Also (12.2.14) implies that (12.2.11) is true. To see this suppose there exists $v \in C \backslash \{0\}$ with $x \in F(x) + \delta v$ for some $\delta > 0$ and $x \in S_r$. Then there exists $y \in F(x)$ with $x = y + \delta v$. Now since $\|\,.\,\|$ is increasing with respect to C we have

$$\|x\| = \|y + \delta v\| \ge \|y\| > \|x\|,$$

a contradiction. The result now follows from Theorem 12.2.3. □

For our next two results we will assume $C \subseteq E$ is a closed, convex, nonempty set with $\alpha u + \beta v \in C$ for all scalars $\alpha \ge 0$, $\beta \ge 0$ and u, $v \in C$.

Theorem 12.2.5. *Let E be a Banach space and $C \subseteq E$ a closed, convex, nonempty set with $\alpha u + \beta v \in C$ for all $\alpha \geq 0$, $\beta \geq 0$ and $u, v \in C$. Also r, R are constants with $0 < r < R$. Suppose*

$$F : \overline{B_R} \to CK(C)$$

is a u.s.c., k–set contractive (here $0 \leq k < 1$) map and assume the following conditions hold:

$$x \notin F(x) \quad \text{for} \quad x \in S_R \cup S_r \tag{12.2.15}$$

$$\begin{cases} F : \overline{B_r} \to CK(C) \quad \text{is essential in} \quad K_{S_r}(\overline{B_r}, C) \\ (\text{i.e.} \quad F|_{\overline{B_r}} \quad \text{is essential}) \end{cases} \tag{12.2.16}$$

and

$$F : \overline{B_R} \to CK(C) \quad \text{is inessential in} \quad K_{S_R}(\overline{B_R}, C). \tag{12.2.17}$$

Then F has at least two fixed points x_0 and x_1 with

$$x_0 \in B_r \quad \text{and} \quad x_1 \in V = \{x : x \in C \quad \text{and} \quad r < \|x\| < R\}.$$

PROOF: We know from (12.2.16) that F has a fixed point in B_r.

Let $\Psi = F|_{\overline{V}}$ and suppose $\Psi : \overline{V} \to CK(C)$ has no fixed points. Now $F : \overline{B_R} \to CK(C)$ being inessential in $K_{S_R}(\overline{B_R}, C)$ means that there exists a u.s.c., k–set contraction $\theta : \overline{B_R} \to CK(C)$ with

$$\theta|_{S_R} = F|_{S_R} \quad \text{and} \quad x \notin \theta(x) \quad \text{for} \quad x \in \overline{B_R}.$$

Fix $\rho \in (0, r)$ and consider the map Φ given by

$$\Phi(x) = \begin{cases} \frac{\rho}{R}\,\theta\left(\frac{R}{\rho}\,x\right), \quad \|x\| \leq \rho \\ \frac{(r-\rho)\|x\|}{(R-\rho)r-(R-r)\|x\|}\,\Psi\left(\frac{(R-\rho)r-(R-r)\|x\|}{(r-\rho)\|x\|}\,x\right), \quad \rho \leq \|x\| \leq r \\ \Psi(x), \quad r \leq \|x\| \leq R. \end{cases}$$

Notice $\Phi : \overline{B_R} \to CK(C)$ is well defined since if $\rho \leq \|x\| \leq r$ then

$$r \leq \left\| \frac{(R-\rho)r - (R-r)\|x\|}{(r-\rho)\|x\|}\,x \right\| \leq R.$$

Also $\Phi : \overline{B_R} \to CK(C)$ is a u.s.c., k–set contractive (see [136]) map. In addition

$$\Phi|_{S_R} = \Psi|_{S_R} = F|_{S_R} = \theta|_{S_R} \quad \text{and} \quad \Phi|_{\overline{V}} = \Psi|_{\overline{V}} = F|_{\overline{V}}$$

and so Φ has no fixed points in $\overline{B_R}$ (since θ has no fixed points in $\overline{B_R}$ and F has no fixed points in $\overline{V}$).

Lets concentrate on $\Phi : \overline{B_r} \to CK(C)$ (i.e. $\Phi|_{\overline{B_r}}$). Now

$$\Phi|_{S_r} = \Psi|_{S_r} = F|_{S_r}$$

so $\Phi : \overline{B_r} \to CK(C)$ is a u.s.c., k–set contractive map with $\Phi|_{S_r} = F|_{S_r}$ and Φ has no fixed points in $\overline{B_r}$. This contradicts (12.2.16). $\square$

Before we use Theorem 12.2.5 to establish some applicable fixed point theorems, we first recall the following result from the literature [119,120, 121].

Theorem 12.2.6. *Let E be a Banach space and $C \subseteq E$ a closed, convex, subset of E with $\rho > 0$ and $v_0 \in C$. Define the mapping $F : \overline{B_\rho} \to CK(C)$ by*

$$F(x) = \{v_0\} \quad \textit{for } x \in \overline{B_\rho}.$$

(i). If $v_0 \in B_\rho$ then F is essential in $K_{S_\rho}(\overline{B_\rho}, C)$.
(ii). If $v_0 \in C \backslash \overline{B_\rho}$ then F is inessential in $K_{S_\rho}(\overline{B_\rho}, C)$.

Theorem 12.2.7. *Let E be a Banach space and $C \subseteq E$ a closed, convex, nonempty set with $\alpha u + \beta v \in C$ for all $\alpha \geq 0$, $\beta \geq 0$ and $u, v \in C$. Also r, R are constants with $0 < r < R$. Suppose*

$$F : \overline{B_R} \to CK(C)$$

is a u.s.c., k–set contractive (here $0 \leq k < 1$) map and assume the following conditions hold:

$$x \notin \lambda F(x) \quad \textit{for } \lambda \in [0,1) \textit{ and } x \in S_r \tag{12.2.18}$$

and

$$\begin{cases} \textit{there exists a } v \in C \backslash \{0\} \textit{ with } x \notin F(x) + \delta v \\ \textit{for any } \delta > 0 \textit{ and } x \in S_R. \end{cases} \tag{12.2.19}$$

Then F has a fixed point in

$$\{x : x \in C \textit{ and } r \leq \|x\| \leq R\}.$$

PROOF: Assume

$$x \notin F(x) \quad \text{for} \quad x \in S_r \cup S_R$$

(otherwise we are finished). The result follows immediately from Theorem 12.2.5 once we show conditions (12.2.16) and (12.2.17) are satisfied.

Consider the homotopy $H : \overline{B_r} \times [0,1] \to CK(C)$ defined by

$$H(x,\lambda) = \lambda\, F(x).$$

Notice $H_0 = 0$, $H_1 = F$ and so since $x \notin F(x)$ on S_r we have

$$H_0 \cong H_1 \quad \text{in} \quad K_{S_r}(\overline{B_r}, C)$$

(this follows since $H : \overline{B_r} \times [0,1] \to CK(C)$ is a u.s.c., k–set contractive map with $H_t \in K_{S_r}(\overline{B_r}, C)$ for each $t \in [0,1]$ since (12.2.18) holds). From Theorem 12.2.6 (i) we have that

$$H_0 : \overline{B_r} \to CK(C) \quad \text{is essential in} \quad K_{S_r}(\overline{B_r}, C)$$

and this together with Theorem 12.1.1 implies

$$F : \overline{B_r} \to CK(C) \quad \text{is essential in} \quad K_{S_r}(\overline{B_r}, C).$$

Thus (12.2.16) holds.

Let $\delta_0 > 0$ be such that

$$\|\delta_0\, v\| > \sup_{x \in S_R} \|y\| + R \quad \text{for all} \quad y \in F(x). \tag{12.2.20}$$

Consider the homotopy $N : \overline{B_R} \times [0,1] \to CK(C)$ defined by

$$N(x,\lambda) = F(x) + \lambda\, \delta_0\, v.$$

Notice N is a u.s.c., k–set contractive map with $N_0 = F$ and $N_1 = F + \delta_0\, v$. Then since (12.2.19) holds (for all $\delta \geq 0$ since we are assuming $x \notin F(x)$ on S_R), we have

$$N_0 \cong N_1 \quad \text{in} \quad K_{S_R}(\overline{B_R}, C). \tag{12.2.21}$$

Notice (12.2.20) implies for $\lambda \in [0,1]$ and $x \in S_R$ that

$$\|\delta_0\, v + \lambda\, y\| > R = \|x\| \quad \text{for all} \quad y \in F(x).$$

Thus

$$x \notin \lambda F(x) + \delta_0 v \ \text{ if } \ \lambda \in [0,1] \ \text{ and } \ x \in S_R. \tag{12.2.22}$$

Let $G : \overline{B_R} \to CK(C)$ be the constant map

$$G(x) = \{\delta_0 v\}.$$

Consider the homotopy $J : \overline{B_R} \times [0,1] \to CK(C)$ defined by

$$J(x, \lambda) = \delta_0 v + \lambda F(x).$$

Now J is a u.s.c., k–set contractive map with $J_0 = G$ and $J_1 = N_1$. From (12.2.22) we have that

$$N_1 \cong G \ \text{ in } \ K_{S_R}(\overline{B_R}, C). \tag{12.2.23}$$

Now (12.2.21) and (12.2.23) imply

$$N_0 \cong G \ \text{ in } \ K_{S_R}(\overline{B_R}, C). \tag{12.2.24}$$

However since $G(x) = \{\delta_0 v\}$ and $\delta_0 v \in C \backslash \overline{B_R}$ (see (12.2.20)) we have from Theorem 12.2.6 (ii) that

$$G \ \text{ is inessential in } \ K_{S_R}(\overline{B_R}, C).$$

Now Theorem 12.1.1 implies that

$$N_0 = F \ \text{ is inessential in } \ K_{S_R}(\overline{B_R}, C).$$

Thus (12.2.17) holds. □

Theorem 12.2.8. *Let $E = (E, \|\,.\,\|)$ be a Banach space, $C \subseteq E$ a cone and let $\|\,.\,\|$ be increasing with respect to C. Also r, R are constants with $0 < r < R$. Suppose*

$$F : \overline{\Omega_R} \cap C \to CK(C)$$

is a u.s.c., k–set contractive (here $0 \leq k < 1$) map and assume the following conditions hold:

$$\|y\| > \|x\| \ \text{ for all } \ y \in F(x) \ \text{ and } \ x \in \partial_E \Omega_R \cap C \tag{12.2.25}$$

and

$$\|y\| \leq \|x\| \ \text{ for all } \ y \in F(x) \ \text{ and } \ x \in \partial_E \Omega_r \cap C. \tag{12.2.26}$$

Then F *has a fixed point in*

$$C \cap \{x \in E: \ r \leq \|x\| \leq R\}.$$

PROOF: Notice (12.2.25) and (12.2.26) imply (12.2.18) and (12.2.19) are true (see the ideas used in Theorem 12.2.4). □

We now combine some of the theorems in this section to establish the existence of multiple fixed points.

Theorem 12.2.9. *Let* $E = (E, \|\,.\,\|)$ *be a Banach space,* $C \subseteq E$ *a cone and let* $\|\,.\,\|$ *be increasing with respect to* C. *Also* r, R *are constants with* $0 < r < R$. *Suppose*

$$F : \overline{\Omega_R} \cap C \to CK(C)$$

is a u.s.c., k*–set contractive (here* $0 \leq k < 1$*) map and assume the following conditions hold:*

(12.2.27) $$x \notin F(x) \quad \text{for all} \quad x \in \partial_E \Omega_r \cap C$$

(12.2.28) $$\|y\| > \|x\| \quad \text{for all} \quad y \in F(x) \quad \text{and} \quad x \in \partial_E \Omega_R \cap C$$

and

(12.2.29) $$\|y\| \leq \|x\| \quad \text{for all} \quad y \in F(x) \quad \text{and} \quad x \in \partial_E \Omega_r \cap C.$$

Then F *has at least two fixed points* x_0 *and* x_1 *with*

$$x_0 \in \Omega_r \cap C \quad \text{and} \quad x_1 \in C \cap \left(\overline{\Omega_R} \backslash \overline{\Omega_r}\right).$$

PROOF: Now Theorem 12.1.2 (note (12.2.29) implies $x \notin \lambda F(x)$ for all $\lambda \in [0,1)$ and $x \in \partial_E \Omega_r \cap C$) implies F has a fixed point

$$x_0 \in C \cap \overline{\Omega_r}.$$

In fact (12.2.27) implies $x_0 \in \Omega_r \cap C$. Also Theorem 12.2.8 implies F has a fixed point

$$x_1 \in C \cap \left(\overline{\Omega_R} \backslash \Omega_r\right).$$

In fact $x_1 \in C \cap \left(\overline{\Omega_R} \backslash \overline{\Omega_r}\right)$ (To see this note if this is not true then $x_1 \in C$ and $\|x_1\| = r$, so $x_1 \in \partial_E \Omega_r \cap C$. This contradicts (12.2.27)). □

Theorem 12.2.10. *Let $E = (E, \|.\|)$ be a Banach space, $C \subseteq E$ a cone and let $\|.\|$ be increasing with respect to C. Also L, r, R are constants with $0 < L < r < R$. Suppose*

$$F : \overline{\Omega_R} \cap C \to CK(C)$$

is a u.s.c., k-set contractive (here $0 \le k < 1$) map and assume the following conditions hold:

$$x \notin F(x) \quad \text{for all} \quad x \in \partial_E \Omega_r \cap C \tag{12.2.30}$$

$$\|y\| > \|x\| \quad \text{for all} \quad y \in F(x) \quad \text{and} \quad x \in \partial_E \Omega_L \cap C \tag{12.2.31}$$

$$\|y\| \le \|x\| \quad \text{for all} \quad y \in F(x) \quad \text{and} \quad x \in \partial_E \Omega_r \cap C \tag{12.2.32}$$

and

$$\|y\| > \|x\| \quad \text{for all} \quad y \in F(x) \quad \text{and} \quad x \in \partial_E \Omega_R \cap C. \tag{12.2.33}$$

Then F has at least two fixed points x_0 and x_1 with

$$x_0 \in C \cap (\Omega_r \backslash \Omega_L) \quad \text{and} \quad x_1 \in C \cap \left(\overline{\Omega_R} \backslash \overline{\Omega_r}\right).$$

PROOF: Theorem 12.2.4 implies that F has a fixed point

$$x_0 \in C \cap \left(\overline{\Omega_r} \backslash \Omega_L\right).$$

In fact (12.2.30) guarantees that $x_0 \in C \cap (\Omega_r \backslash \Omega_L)$. Theorem 12.2.8 guarantees the existence of x_1. □

12.3. $L^p[0,T]$ solutions to Fredholm integral inclusions.

In this section we discuss the nonlinear Fredholm integral inclusion

$$y(t) \in \int_0^T k(t,s)\, f(s, y(s))\, ds \quad \text{a.e.} \quad t \in [0,T] \tag{12.3.1}$$

where

$$k : [0,T] \times [0,T] \to \mathbf{R} \quad \text{and} \quad f : [0,T] \times \mathbf{R} \to CK(\mathbf{R}).$$

We would like to know what conditions one requires on k and f in order that the inclusion (12.3.1) has a positive solution $y \in L^p[0,T]$, where $1 \le$

$p < \infty$. Here by a positive solution y we mean $y(t) > 0$ for a.e. $t \in [0,T]$. Throughout this section $\|\,.\,\|_q$ denotes the usual norm on L^q for $1 \le q \le \infty$.

Theorem 12.3.1. *Let $k : [0,T]\times[0,T] \to \mathbf{R}$ and $f : [0,T]\times\mathbf{R} \to CK(\mathbf{R})$ and suppose the following conditions hold:*

$$\text{(12.3.2)} \qquad \begin{cases} \textit{the map } u \mapsto f(t,u) \textit{ is upper semicontinuous} \\ \textit{for a.e. } t \in [0,T] \end{cases}$$

$$\text{(12.3.3)} \qquad \begin{cases} \textit{the graph of } f \textit{ belongs to the } \sigma\textit{-field } \mathcal{L}\otimes\mathcal{B}(\mathbf{R}\times\mathbf{R}) \\ \textit{(here } \mathcal{L} \textit{ denotes the Lebesgue } \sigma\textit{-field on } [0,T] \\ \textit{and } \mathcal{B}(\mathbf{R}\times\mathbf{R}) = \mathcal{B}(\mathbf{R})\otimes\mathcal{B}(\mathbf{R}) \textit{ is the Borel} \\ \sigma\textit{-field in } \mathbf{R}\times\mathbf{R}) \end{cases}$$

$$\text{(12.3.4)} \qquad \begin{cases} \exists\, p_2,\ 1 \le p_2 < \infty,\ a_1 \in L^{p_2}[0,T] \textit{ and } a_2 > 0 \textit{ a constant,} \\ \textit{with } |f(t,y)| = \sup\{|z| :\ z \in f(t,y)\} \le a_1(t) + a_2\,|y|^{\frac{p}{p_2}} \\ \textit{for a.e. } t \in [0,T] \textit{ and all } y \in \mathbf{R} \end{cases}$$

$$\text{(12.3.5)} \qquad (t,s) \mapsto k(t,s) \textit{ is measurable}$$

$$\text{(12.3.6)} \qquad \begin{cases} \exists\, 0 < M \le 1,\ k_1 \in L^p[0,T],\ k_2 \in L^{p_1}[0,T], \textit{ here} \\ \frac{1}{p_1} + \frac{1}{p_2} = 1, \textit{ such that } 0 < k_1(t),\ k_2(t) \textit{ a.e. } t \in [0,T] \\ \textit{and } M\,k_1(t)\,k_2(s) \le k(t,s) \le k_1(t)\,k_2(s) \\ \textit{a.e. } t \in [0,T], \textit{ a.e. } s \in [0,T] \end{cases}$$

$$\text{(12.3.7)} \qquad \begin{cases} \textit{for a.e. } t \in [0,T] \textit{ and all } y \in (0,\infty), \\ u > 0 \textit{ for all } u \in f(t,y) \end{cases}$$

$$\text{(12.3.8)} \qquad \begin{cases} \exists\, q \in L^{p_2}[0,T] \textit{ and } \psi : [0,\infty) \to [0,\infty),\ \psi(u) > 0 \\ \textit{for } u > 0, \textit{ continuous and nondecreasing with} \\ \textit{for a.e. } t \in [0,T] \textit{ and } y > 0,\ u \ge q(t)\,\psi(y) \\ \textit{for all } u \in f(t,y) \end{cases}$$

$$\text{(12.3.9)} \qquad \exists\, \alpha > 0 \textit{ with } 1 < \frac{\alpha}{2^{\frac{p_2-1}{p_2}}\,\|k_1\|_p\,\|k_2\|_{p_1}\,(\|a_1\|_{p_2}^{p_2} + [a_2]^{p_2}\,\alpha^p)^{\frac{1}{p_2}}}$$

and

$$(12.3.10)\quad \exists\, \beta > 0,\ \beta \neq \alpha \ \ \textit{with}\ \ 1 > \frac{\beta}{M\,\|k_1\|_p \int_0^T k_2(s)\,q(s)\,\psi(a(s)\,\beta)\,ds}$$

where

$$(12.3.11)\qquad a(t) = M\,\frac{k_1(t)}{\|k_1\|_p}.$$

Then (12.3.1) *has at least one positive solution* $y \in L^p[0,T]$ *and either*
(A). $0 < \alpha < \|y\|_p < \beta$ *and* $y(t) \geq a(t)\,\alpha$ *a.e.* $t \in [0,T]$ *if* $\alpha < \beta$
or
(B). $0 < \beta < \|y\|_p < \alpha$ *and* $y(t) \geq a(t)\,\beta$ *a.e.* $t \in [0,T]$ *if* $\beta < \alpha$
holds.

PROOF: Let $E = (L^p[0,T], \|\,.\,\|_p)$ and

$$C = \{y \in L^p[0,T] :\ y(t) \geq a(t)\,\|y\|_p \ \text{ a.e. } \ t \in [0,T]\}\,.$$

It is easy to see that $C \subseteq E$ is a cone. Next let $A = K \circ N_f : C \to 2^E$, where the linear integral (single valued) operator K is given by

$$K\,y(t) = \int_0^T k(t,s)\,y(s)\,ds,$$

and the multivalued Nemytskij operator N_f is given by

$$N_f\,u = \{y \in L^{p_2}[0,T] :\ y(t) \in f(t,u(t)) \ \text{ a.e. } \ t \in [0,T]\}\,.$$

Remark. Note A is well defined since if $x \in C$ then (12.3.2)–(12.3.4) and [41] guarantee that $N_f\,x \neq \emptyset$.

We first show $A : C \to 2^C$. To see this let $x \in C$ and $y \in A\,x$. Then there exists a $v \in N_f\,x$ with

$$y(t) = \int_0^T k(t,s)\,v(s)\,ds \quad \text{for a.e. } \ t \in [0,T].$$

Now

$$|y(t)|^p \leq [k_1(t)]^p \left(\int_0^T k_2(s)\,v(s)\,ds\right)^p \quad \text{for a.e. } \ t \in [0,T]$$

so

$$\|y\|_p \leq \|k_1\|_p \int_0^T k_2(s)\, v(s)\, ds. \tag{12.3.12}$$

Combining this with (12.3.6) gives

$$y(t) \geq M \int_0^T k_1(t)\, k_2(s)\, v(s)\, ds \geq M\, \frac{k_1(t)}{\|k_1\|_p}\, \|y\|_p \;=\; a(t)\, \|y\|_p$$

for a.e. $t \in [0,T]$. Thus $y \in C$ so $A : C \to 2^C$. Also notice [41] guarantees that

$$A : C \to 2^C \quad \text{is upper semicontinuous.} \tag{12.3.13}$$

In addition note [127, 128, 137 pp. 47–49] implies

$$K : L^{p_2}[0,T] \to L^p[0,T] \quad \text{is completely continuous,}$$

and

$$N_f : L^p[0,T] \to 2^{L^{p_2}[0,T]} \quad \text{maps bounded sets into bounded sets.}$$

Consequently

$$A : C \to CK(C) \quad \text{is completely continuous.} \tag{12.3.14}$$

Let

$$\Omega_\alpha = \{y \in L^p[0,T] : \; \|y\|_p < \alpha\}$$

and

$$\Omega_\beta = \{y \in L^p[0,T] : \; \|y\|_p < \beta\}\,.$$

Assume that $\beta < \alpha$ (a similar argument holds if $\alpha < \beta$). It is immediate from (12.3.13) and (12.3.14) that

$$A : C \cap \overline{\Omega_\alpha} \to CK(C) \quad \text{is upper semicontinuous and compact.}$$

If we show

$$\|y\|_p \;<\; \|x\|_p \quad \text{for all} \quad y \in A\,x \quad \text{and} \quad x \in C \cap \partial\Omega_\alpha \tag{12.3.15}$$

and

$$\|y\|_p \;>\; \|x\|_p \quad \text{for all} \quad y \in A\,x \quad \text{and} \quad x \in C \cap \partial\Omega_\beta \tag{12.3.16}$$

are true, then Theorem 12.2.4 guarantees that the operator A has a fixed point in $C \cap \left(\overline{\Omega_\alpha} \setminus \Omega_\beta\right)$. This in turn implies that (12.3.1) has at least one solution $y \in L^p[0,T]$ with $\beta \le \|y\|_p \le \alpha$ and $y(t) \ge a(t)\,\beta$ for a.e. $t \in [0,T]$.

Suppose $x \in C \cap \partial\Omega_\alpha$, so $\|x\|_p = \alpha$, and $y \in A\,x$. Then there exists a $v \in N_f\,x$ with

$$y(t) = \int_0^T k(t,s)\,v(s)\,ds \quad \text{for a.e. } t \in [0,T].$$

Now (12.3.4) and (12.3.6) guarantee that

$$|y(t)| \le k_1(t) \int_0^T k_2(s)\,\left[|a_1(s)| + a_2\,|x(s)|^{\frac{p}{p_2}}\right] ds \quad \text{for a.e. } t \in [0,T].$$

This together with (12.3.9) yields

$$\begin{aligned}
\|y\|_p &\le \|k_1\|_p\,\|k_2\|_{p_1} \left(\int_0^T \left[|a_1(s)| + a_2\,|x(s)|^{\frac{p}{p_2}}\right]^{p_2} ds\right)^{\frac{1}{p_2}} \\
&\le \|k_1\|_p\,\|k_2\|_{p_1} \left(2^{p_2-1} \int_0^T \left[|a_1(s)|^{p_2} + [a_2]^{p_2}\,|x(s)|^p\right] ds\right)^{\frac{1}{p_2}} \\
&= 2^{\frac{p_2-1}{p_2}}\,\|k_1\|_p\,\|k_2\|_{p_1} \left(\|a_1\|_{p_2}^{p_2} + [a_2]^{p_2}\,\|x\|_p^p\right)^{\frac{1}{p_2}} \\
&= 2^{\frac{p_2-1}{p_2}}\,\|k_1\|_p\,\|k_2\|_{p_1} \left(\|a_1\|_{p_2}^{p_2} + [a_2]^{p_2}\,\alpha^p\right)^{\frac{1}{p_2}} \\
&< \alpha = \|x\|_p,
\end{aligned}$$

and so (12.3.15) is satisfied.

Now suppose $x \in C \cap \partial\Omega_\beta$, so $\|x\|_p = \beta$ and $x(t) \ge a(t)\,\beta$ for a.e. $t \in [0,T]$, and $y \in A\,x$. Then there exists a $v \in N_f\,x$ with

$$y(t) = \int_0^T k(t,s)\,v(s)\,ds \quad \text{for a.e. } t \in [0,T].$$

Notice (12.3.8) guarantees that $v(s) \ge q(s)\,\psi(x(s))$ for a.e. $s \in [0,T]$ and this together with (12.3.6) yields

$$y(t) \ge M\,k_1(t) \int_0^T k_2(s)\,q(s)\,\psi(x(s))\,ds \quad \text{for a.e. } t \in [0,T].$$

Combining with (12.3.10) gives

$$\begin{aligned}\|y\|_p &\geq M\,\|k_1\|_p \int_0^T k_2(s)\,q(s)\,\psi(x(s))\,ds \\ &\geq M\,\|k_1\|_p \int_0^T k_2(s)\,q(s)\,\psi(a(s)\,\beta)\,ds \\ &> \beta = \|x\|_p,\end{aligned}$$

and thus (12.3.16) is satisfied. Now apply Theorem 12.2.4. □

12.4. $C[0,T]$ solutions to Fredholm integral inclusions.

In this section we discuss the Fredholm integral inclusion

$$y(t) \in \int_0^T k(t,s)\,f(s,y(s))\,ds \quad \text{for } t \in [0,T] \tag{12.4.1}$$

where

$$k : [0,T] \times [0,T] \to \mathbf{R} \ \text{ and } \ f : [0,T] \times \mathbf{R} \to CK(\mathbf{R}).$$

We will use Theorem 12.2.4 or Theorem 12.2.8 to establish the existence of a nonnegative solution $y \in C[0,T]$ to (12.4.1). We will let $|\,.\,|_0$ denote the usual norm on $C[0,T]$ i.e. $|u|_0 = \sup_{[0,T]} |u(t)|$ for $u \in C[0,T]$.

Theorem 12.4.1. *Let $1 \le p < \infty$ and q, $1 < q \le \infty$, the conjugate to p, $k : [0,T] \times [0,T] \to \mathbf{R}$, $f : [0,T] \times \mathbf{R} \to CK(\mathbf{R})$ and assume the following conditions are satisfied:*

$$\text{for each } t \in [0,T], \text{ the map } s \mapsto k(t,s) \text{ is measurable} \tag{12.4.2}$$

$$\sup_{t\in[0,T]} \left(\int_0^T |k(t,s)|^q\,ds \right)^{\frac{1}{q}} < \infty \tag{12.4.3}$$

$$\int_0^T |k(t',s) - k(t,s)|^q\,ds \to 0 \ \text{ as } \ t \to t', \ \text{ for each } \ t' \in [0,T] \tag{12.4.4}$$

$$\text{for each } t \in [0,T],\ k(t,s) \ge 0 \ \text{ for a.e. } \ s \in [0,T] \tag{12.4.5}$$

$$\begin{cases} \text{for each measurable } u : [0,T] \to \mathbf{R}, \ \text{ the map} \\ t \mapsto f(t,u(t)) \ \text{ has measurable single valued selections} \end{cases} \tag{12.4.6}$$

(12.4.7) $$\begin{cases} \text{for a.e. } t \in [0,T], \text{ the map } u \mapsto f(t,u) \\ \text{is upper semicontinuous} \end{cases}$$

(12.4.8) $$\begin{cases} \text{for each } r > 0, \ \exists\, h_r \in L^p[0,T] \text{ with } |f(t,y)| \le h_r(t) \\ \text{for a.e. } t \in [0,T] \text{ and every } y \in \mathbf{R} \text{ with } |y| \le r \end{cases}$$

(12.4.9) $$\begin{cases} \text{for a.e. } t \in [0,T] \text{ and all } y \in (0,\infty), \\ u > 0 \text{ for all } u \in f(t,y) \end{cases}$$

(12.4.10) $$\begin{cases} \exists\, g \in L^q[0,T] \text{ with } g : [0,T] \to (0,\infty) \\ \text{and with } k(t,s) \le g(s) \text{ for } t \in [0,T] \end{cases}$$

(12.4.11) $$\begin{cases} \exists\, \delta,\ \epsilon,\ 0 \le \delta < \epsilon \le T \text{ and } M,\ 0 < M < 1, \\ \text{with } k(t,s) \ge M\, g(s) \text{ for } t \in [\delta,\epsilon] \end{cases}$$

(12.4.12) $$\begin{cases} \exists\, h \in L^p[0,T] \text{ with } h : [0,T] \to (0,\infty), \text{ and } w \ge 0 \\ \text{continuous and nondecreasing on } (0,\infty) \text{ with} \\ |f(t,y)| \le h(t)\, w(y) \text{ for a.e. } t \in [0,T] \text{ and} \\ \text{all } y \in (0,\infty) \end{cases}$$

(12.4.13) $$\begin{cases} \exists\, \tau \in L^p[\delta,\epsilon] \text{ with } \tau > 0 \text{ a.e. on } [\delta,\epsilon] \text{ and with} \\ \text{for a.e. } t \in [\delta,\epsilon] \text{ and } y \in (0,\infty),\ u \ge \tau(t)\, w(y) \\ \text{for all } u \in f(t,y) \end{cases}$$

(12.4.14) $$\exists\, \alpha > 0 \text{ with } 1 < \frac{\alpha}{w(\alpha)\ \sup_{t\in[0,T]} \int_0^T k(t,s)\, h(s)\, ds}$$

and

(12.4.15) $$\exists\, \beta > 0,\ \beta \ne \alpha \text{ with } 1 > \frac{\beta}{w(M\beta) \int_\delta^\epsilon \tau(s)\, k(\sigma,s)\, ds};$$

here $\sigma \in [0,T]$ *is such that*

(12.4.16) $$\int_\delta^\epsilon \tau(s)\, k(\sigma,s)\, ds = \sup_{t\in[0,T]} \int_\delta^\epsilon \tau(s)\, k(t,s)\, ds.$$

Then (12.4.1) *has at least one nonnegative solution* $y \in C[0,T]$ *and either*
(A). $0 < \alpha < |y|_0 < \beta$ *and* $y(t) \geq M\,\alpha$ *for* $t \in [\delta, \epsilon]$ *if* $\alpha < \beta$
or
(B). $0 < \beta < |y|_0 < \alpha$ *and* $y(t) \geq M\,\beta$ *for* $t \in [\delta, \epsilon]$ *if* $\beta < \alpha$
holds.

PROOF: Let $E = (C[0,T], |\,.\,|_0)$ and

$$C = \left\{ y \in C[0,T] : \; y(t) \geq 0 \;\text{ for }\; t \in [0,T] \;\text{ and }\; \min_{t \in [\delta,\epsilon]} y(t) \geq M\,|y|_0 \right\}.$$

Also let

$$A = K \circ N_f : C \to 2^E,$$

where $K : L^p[0,T] \to C[0,T]$ and $N_f : C[0,T] \to 2^{L^p[0,T]}$ are given by

$$K\,y(t) = \int_0^T k(t,s)\,y(s)\,ds$$

and

$$N_f\,u = \{y \in L^p[0,T] : \; y(t) \in f(t, u(t)) \;\text{ a.e. }\; t \in [0,T]\}\,.$$

Remark. Note A is well defined since if $x \in C$ then [54, 67] guarantee that $N_f\,x \neq \emptyset$.

We first show $A : C \to 2^C$. To see this let $x \in C$ and $y \in A\,x$. Then there exists a $v \in N_f\,x$ with

$$y(t) = \int_0^T k(t,s)\,v(s)\,ds \quad \text{for } \; t \in [0,T].$$

This together with (12.4.10) yields

$$|y(t)| \leq \int_0^T g(s)\,v(s)\,ds \quad \text{for } \; t \in [0,T]$$

and so

$$|y|_0 \leq \int_0^T g(s)\,v(s)\,ds. \tag{12.4.17}$$

On the other hand (12.4.11) and (12.4.17) yields

$$\min_{t \in [\delta,\epsilon]} y(t) = \min_{t \in [\delta,\epsilon]} \int_0^T k(t,s)\,v(s)\,ds \geq M \int_0^T g(s)\,v(s)\,ds \geq M\,|y|_0,$$

so $y \in C$. Thus $A : C \to 2^C$. It is immediate from Chapter 10 (Lemma 10.4.2) that $A : C \to 2^C$ has closed graph. Also the Arzela–Ascoli theorem guarantees that $A : C \to 2^C$ is completely continuous. As a result

$A : C \to CK(C)$ is upper semicontinuous and completely continuous.

To see the upper semicontinuity, let B be a closed subset of C and

$$W = \{u \in C : \; A(u) \cap B \neq \emptyset\}.$$

We must show W is a closed subset of C. Let (y_n) be a sequence in W converging to y in $C[0,1]$. Let $v_n \in A(y_n) \cap B$. Now $A : C \to 2^C$ has closed graph and is completely continuous, so there exists a subsequence of (v_n) converging to $v \in A(y) \cap B$. This $y \in W$, so W is a closed subset of C.

Let

$$\Omega_\alpha = \{u \in C[0,T] : \; |u|_0 < \alpha\}$$

and

$$\Omega_\beta = \{u \in C[0,T] : \; |u|_0 < \beta\}.$$

Without loss of generality assume $\beta < \alpha$. If we show

$$|y|_0 < |x|_0 \quad \text{for all} \quad y \in A\,x \quad \text{and} \quad x \in C \cap \partial\Omega_\alpha \tag{12.4.18}$$

and

$$|y|_0 > |x|_0 \quad \text{for all} \quad y \in A\,x \quad \text{and} \quad x \in C \cap \partial\Omega_\beta \tag{12.4.19}$$

are true, then Theorem 12.2.4 guarantees the result.

Suppose $x \in C \cap \partial\Omega_\alpha$, so $|x|_0 = \alpha$, and $y \in A\,x$. Then there exists $v \in N_f\,x$ with

$$y(t) = \int_0^T k(t,s)\,v(s)\,ds \quad \text{for} \quad t \in [0,T].$$

Now (12.4.12) implies that for $t \in [0,T]$ we have

$$\begin{aligned} |y(t)| &\leq \int_0^T k(t,s)\,h(s)\,w(x(s))\,ds \\ &\leq w(|x|_0) \int_0^T k(t,s)\,h(s)\,ds \\ &\leq w(\alpha) \sup_{t\in[0,T]} \int_0^T k(t,s)\,h(s)\,ds. \end{aligned}$$

This together with (12.4.14) yields

$$|y|_0 \le w(\alpha) \sup_{t\in[0,T]} \int_0^T k(t,s)\,h(s)\,ds \;<\; \alpha \;=\; |x|_0,$$

so (12.4.18) holds.

Next suppose $x \in C \cap \partial\Omega_\beta$, so $|x|_0 = \beta$ and $M\,\beta \le x(t) \le \beta$ for $t \in [\delta, \epsilon]$, and $y \in A\,x$. Then there exists $v \in N_f\,x$ with

$$y(t) = \int_0^T k(t,s)\,v(s)\,ds \quad \text{for} \quad t \in [0,T].$$

Notice (12.4.13) and (12.4.15) imply

$$\begin{aligned}
y(\sigma) &= \int_0^T k(\sigma,s)\,v(s)\,ds \\
&\ge \int_\delta^\epsilon k(\sigma,s)\,v(s)\,ds \\
&\ge \int_\delta^\epsilon k(\sigma,s)\,\tau(s)\,w(x(s))\,ds \\
&\ge w(M\,\beta) \int_\delta^\epsilon k(\sigma,s)\,\tau(s)\,ds \\
&> \beta = |x|_0.
\end{aligned}$$

Thus $|y|_0 > |x|_0$, so (12.4.19) holds. Now apply Theorem 12.2.4. □

CHAPTER 13

THE PETRYSHYN LEGGETT WILLIAMS FIXED POINT THEOREM WITH APPLICATIONS TO INTEGRAL INCLUSIONS

13.1. Introduction.

In this chapter we present a generalization of the Petryshyn Leggett Williams fixed point theorem using an axiomatic index theory. In [96] Leggett and Williams presented criteria which guaranteed the existence of at least three fixed points for single valued continuous, compact maps. Eight years later in 1987 Petryshyn [143], using a modification of the arguments in [96], extended the results in [96] to convex, compact valued upper semicontinuous, condensing maps. In Section 13.2 we discuss maps which obey an axiomatic index theory, so in particular the results apply to all multivalued maps in the literature which have a well defined fixed point index (see [27] and the reference therein). In Section 13.3 we will use the results in Section 13.2 to establish results which guarantee the existence of three nonnegative solutions to integral inclusions.

13.2. Fixed point theory.

Let X be a closed, convex set of some Banach space $E = (E, \|\,.\,\|)$. Suppose for every open subset U of X and every upper semicontinuous map $A : \overline{U^X} \to 2^X$ (here 2^X denotes the family of nonempty subsets of X) which satisfies property (B) (to be specified later) with $x \notin Ax$ for $x \in \partial_X U$ (here $\overline{U^X}$ and $\partial_X U$ denote the closure and boundary of U in X respectively) there exists an integer, denoted by $i_X(A, U)$, satisfying the following properties:

(P1). if $x_0 \in U$ then $i_X(\hat{x}_0, U) = 1$ (here $\hat{x}_0$ denotes the map whose constant value is x_0);

(P2). for every pair of disjoint open subsets U_1, U_2 of U such that A has no fixed points on $\overline{U^X} \setminus (U_1 \cup U_2)$,

$$i_X(A, U) = i_X(A, U_1) + i_X(A, U_2);$$

(P3). for every upper semicontinuous map $H : [0,1] \times \overline{U^X} \to 2^X$ which satisfies property (B) and $x \notin H(t,x)$ for $(t,x) \in [0,1] \times \partial_X U$,

$$i_X(H(1,\,.\,),U) = i_X(H(0,\,.\,),U);$$

(P4). if Y is a closed, convex set of X and $A(\overline{U^X}) \subseteq Y$, then

$$i_X(A,U) = i_Y(A, U \cap Y).$$

Also assume the family

$$\{\, i_X(A,U) : \quad \begin{array}{l} X \text{ a closed, convex set of a Banach space } E, \\ U \text{ open in } X,\ A : \overline{U^X} \to 2^X \text{ is upper semicontinuous,} \\ \text{satisfies property (B) and } x \notin Ax \text{ on } \partial_X U \,\} \end{array}$$

is uniquely determined by the properties (P1)–(P4).

We note that property (B) is any property on the map so that the fixed point index is well defined. In the applications in this chapter, property (B) will mean that the map is compact with convex compact values. Other examples of map with a well defined fixed point index (for example, property (B) could mean that the map is countably condensing with convex compact values) can be found in the literature.

If the above hold notice also that

(P5). for every open subset V of U such that A has no fixed points on $\overline{U^X} \setminus V$,

$$i_X(A,U) = i_X(A,V);$$

and

(P6). if $i_X(A,U) \neq 0$ then A has at least one fixed point in U.

To see (P5) let $U_1 = U$ and $U_2 = \emptyset$ in (P2) to get

$$i_X(A,\emptyset) = 0.$$

Next let $U_1 = V$ and $U_2 = \emptyset$ in (P2) to get

$$i_X(A,U) = i_X(A,V).$$

To see (P6) note if A has no fixed point in U, then letting $V = \emptyset$ in (P5) gives

$$i_X(A,U) = i_X(A,\emptyset) = 0.$$

In this sections $C \subset E$ will be a cone. By a concave nonnegative continuous functional ψ on C we mean a continuous mapping $\psi : C \to [0,\infty)$ with

$$\psi(\lambda x+(1-\lambda) y) \geq \lambda \psi(x)+(1-\lambda) \psi(y) \quad \text{for all } x,\, y \in C \text{ and } \lambda \in [0,1].$$

Let K, L, $r > 0$ be constants with C and ψ as defined above. We let

$$C_K = \{y \in C : \ \|y\| < K\},$$

$$S_K = \{y \in C : \ \|y\| = K\},$$

$$\overline{C_K} = \{y \in C : \ \|y\| \leq K\},$$

and

$$C(\psi, r, L) = \{y \in C : \ \psi(y) \geq r \ \text{ and } \ \|y\| \leq L\}.$$

Theorem 13.2.1. *Let $E = (E, \|\,.\,\|)$ be a Banach space and X a closed, convex set of E. Suppose for every open subset U of X and every upper semicontinuous map $A : \overline{U^X} \to 2^X$ which satisfies property (B) with $x \notin A\,x$ for $x \in \partial_X U$ there exists an integer $i_X(A,U)$ satisfying (P1)–(P4). In addition assume the family*

$$\{ i_X(A,U) : \quad \begin{array}{l} X \ \text{a closed, convex set of a Banach space } E, \\ U \ \text{open in } X,\ A : \overline{U^X} \to 2^X \ \text{is upper semicontinuous,} \\ \text{satisfies property (B) and } x \notin A\,x \ \text{on } \partial_X U \} \end{array}$$

is uniquely determined by the properties (P1)–(P4). Let $C \subset E$ be a cone in E and $R > 0$ a constant. Suppose there exists a concave nonnegative continuous functional ψ on C with

$$\psi(y) \leq \|y\| \ \ \text{for} \ \ y \in \overline{C_R}$$

and let $F : \overline{C_R} \to 2^{\overline{C_R}}$ be an upper semicontinuous map which satisfies property (B). Assume there are numbers r, L and K with $0 < r < L < K \leq R$ such that

(H1). $\{x \in C(\psi, L, K) : \ \psi(x) > L\} \neq \emptyset$ and if $x \in C(\psi, L, K)$ then $\psi(y) > L$ for all $y \in F\,x$;

(H2). if $x \in \overline{C_r}$ then $\|y\| < r$ for all $y \in F\,x$;

(H3). if $x \in C(\psi, L, R)$ with $\|y\| > K$ for some $y \in F\,x$ then $\psi(y) > L$;

(H4). the mapping $H_1 : [0,1] \times Y \to 2^Y$, given by $H_1(t,x) = t\,F\,x$, satisfies property (B) (here Y is $\overline{C_r}$ or $\overline{C_R}$);

(H5). there exists $x_0 \in \{x \in C(\psi, L, K) : \psi(x) > L\}$ such that the mapping $H_2 : [0,1] \times C(\psi, L, R) \to 2^{\overline{C_R}}$, given by $H_2(t,x) = (1-t)\,Fx + t\,x_0$, satisfies property (B).

Then F has at least three fixed points y_1, y_2 and y_3 in $\overline{C_R}$. Furthermore we have

$$y_1 \in C_r, \quad y_2 \in \{y \in C(\psi, L, R) : \psi(y) > L\}$$

and

$$y_3 \in \overline{C_R} \setminus \left(C(\psi, L, R) \cup \overline{C_r}\right).$$

PROOF: Let

$$U_1 = \{x \in \overline{C_R} : \|x\| < r\}$$

and

$$U_2 = \{x \in C(\psi, L, R) : \psi(x) > L\}.$$

Notice U_1 and U_2 are nonempty, disjoint open subsets of $\overline{C_R}$ (to see that $U_1 \cap U_2 = \emptyset$ notice if $x \in U_2$ then $\psi(x) > L$ so $\|x\| \geq \psi(x) > L > r$). Also

$$\partial_{\overline{C_R}} U_2 = \{x \in C(\psi, L, R) : \psi(x) = L\}$$

and so $\overline{U_2}$ (the closure of U_2 in $\overline{C_R}$) $= C(\psi, L, R)$. We begin by showing

$$i_{\overline{C_R}}(F, U_1) = 1. \tag{13.2.1}$$

Remark 13.2.1. Notice $(H2)$ implies F has no fixed points on S_r.

Notice (P4) implies (note $F : \overline{C_r} \to 2^{\overline{C_r}}$),

$$i_{\overline{C_R}}(F, U_1) = i_{\overline{C_r}}(F, U_1). \tag{13.2.2}$$

Also from (P5) (note $(H2)$ implies F has no fixed point in $\overline{C_r} \setminus U_1$) we have

$$i_{\overline{C_r}}(F, \overline{C_r}) = i_{\overline{C_r}}(F, U_1); \tag{13.2.3}$$

note $\overline{C_r}$ is an open set in $\overline{C_r}$. Let $H_1 : [0,1] \times \overline{C_r} \to 2^{\overline{C_r}}$ be given by

$$H_1(t,x) = t\,F\,x.$$

Notice $H_1 : [0,1] \times \overline{C_r} \to 2^{\overline{C_r}}$ is upper semicontinuous, satisfies property (B) (from $(H4)$) and trivially (note $\partial_{\overline{C_r}} \overline{C_r} = \emptyset$)

$$x \notin H(t,x) \text{ for } (t,x) \in [0,1] \times \partial_{\overline{C_r}} \overline{C_r}.$$

Thus (P1) and (P3) yields

$$i_{\overline{C_r}}(F, \overline{C_r}) = i_{\overline{C_r}}(\hat{0}, \overline{C_r}) = 1,$$

and this together with (13.2.2) and (13.2.3) gives (13.2.1).

Also we have (recall $H_1 : [0,1] \times \overline{C_R} \to 2^{\overline{C_R}}$, given by $H_1(t,x) = t\,F\,x$, is upper semicontinuous, and satisfies property (B)),

$$i_{\overline{C_R}}(F, \overline{C_R}) = 1. \tag{13.2.4}$$

Now choose

$$x_0 \in \{x \in C(\psi, L, K) :\ \psi(x) > L\}$$

such that $(H5)$ is satisfied. Notice $\psi(x_0) > L$. Let $H_2 : [0,1] \times \overline{U_2} \to 2^{\overline{C_R}}$ be given by

$$H_2(t,x) = (1-t)\,F\,x\, +\, t\,x_0.$$

Now $H_2 : [0,1] \times \overline{U_2} \to 2^{\overline{C_R}}$ is upper semicontinuous and satisfies property (B) (from $(H5)$). We now claim

$$x \notin H_2(t,x) \quad \text{for} \quad (t,x) \in [0,1] \times \partial_{\overline{C_R}} U_2. \tag{13.2.5}$$

Remark 13.2.2. If (13.2.5) is true, note $x \notin F\,x$ for $x \in \partial_{\overline{C_R}} U_2$.

If (13.2.5) was false then there exists $(t,x) \in [0,1] \times \partial_{\overline{C_R}} U_2$ with

$$x \in (1-t)\,F\,x\, +\, t\,x_0.$$

Now since $x \in \partial_{\overline{C_R}} U_2$ we have $\psi(x) = L$. Also $x = (1-t)\,y + t\,x_0$ for some $y \in F\,x$. There are two cases to consider.

Case (a). Suppose $\|y\| > K$.

Then $(H3)$ implies $\psi(y) > L$ and so we have

$$L = \psi(x) = \psi((1-t)\,y + t\,x_0) \geq (1-t)\,\psi(y) + t\,\psi(x_0) > L,$$

a contradiction.

Case (b). Suppose $\|y\| \leq K$.

Now

$$\|x\| = \|(1-t)\,y + t\,x_0\| \leq (1-t)\,K + t\,K = K.$$

Thus $\|x\| \leq K$ and $\psi(x) = L$ so $x \in C(\psi, L, K)$. Now $(H1)$ guarantees that $\psi(z) > L$ for all $z \in F\,x$, so

$$L = \psi(x) = \psi((1-t)\,y + t\,x_0) \geq (1-t)\,\psi(y) + t\,\psi(x_0) > L,$$

a contradiction.

Thus (13.2.5) is true. Now (P3) and (P1) imply

$$i_{\overline{C_R}}(F, U_2) = i_{\overline{C_R}}(\hat{x}_0, U_2) = 1. \tag{13.2.6}$$

From Remarks 13.2.1 and 13.2.2 we note that F has no fixed points on

$$\partial_{\overline{C_R}}(U_1 \cup U_2) = \partial_{\overline{C_R}} U_1 \cup \partial_{\overline{C_R}} U_2.$$

Also notice $\overline{U_1} \cap \overline{U_2} = \emptyset$ since if $x \in \overline{U_2}$ then $\|x\| \geq \psi(x) \geq L > r$. Now (13.2.1), (13.2.6) and (P2) imply

$$i_{\overline{C_R}}(F, U_1 \cup U_2) = i_{\overline{C_R}}(F, U_1) + i_{\overline{C_R}}(F, U_2) = 2. \tag{13.2.7}$$

Now since

$$\overline{C_R} = \left(\overline{U_1 \cup U_2}\right) \cup \left(\overline{C_R} \setminus \left(\overline{U_1 \cup U_2}\right)\right)$$

we have that

$$U_1 \cup U_2 \quad \text{and} \quad \overline{C_R} \setminus \left(\overline{U_1 \cup U_2}\right)$$

are disjoint open sets in $\overline{C_R}$ and F has no fixed points in

$$\overline{C_R} \setminus \left((U_1 \cup U_2) \cup \left(\overline{C_R} \setminus \left(\overline{U_1 \cup U_2}\right)\right)\right).$$

Now (P2) implies

$$i_{\overline{C_R}}(F, \overline{C_R}) = i_{\overline{C_R}}(F, U_1 \cup U_2) + i_{\overline{C_R}}\left(F, \overline{C_R} \setminus \left(\overline{U_1 \cup U_2}\right)\right),$$

and this together with (13.2.4) and (13.2.7) yield

$$i_{\overline{C_R}}\left(F, \overline{C_R} \setminus \left(\overline{U_1 \cup U_2}\right)\right) = -1. \tag{13.2.8}$$

Now (13.2.1), (13.2.6), (13.2.8) and (P6) guarantee that F has fixed points in U_1, U_2 and $\overline{C_R} \setminus \left(\overline{U_1 \cup U_2}\right) = \overline{C_R} \setminus \left(\overline{U_1} \cup \overline{U_2}\right)$. □

Remark 13.2.3. Notice from the proof that $(H1)$ and $(H3)$ could be replaced by the less restrictive conditions:

$(H1)^\star$. $\{x \in C(\psi, L, K): \psi(x) > L\} \neq \emptyset$ and if $x \in \{z \in C(\psi, L, K): \psi(z) = L\}$ and $y \in Fx$ with $\|y\| \le K$ then $\psi(y) > L$;

$(H3)^\star$. if $x \in \{z \in C(\psi, L, K): \psi(z) = L\}$ with $\|y\| > K$ for some $y \in Fx$ then $\psi(y) > L$;

From Theorem 13.2.1 we can immediately deduce the Petryshyn Leggett Williams Theorem

Theorem 13.2.2. *Let $E = (E, \|\,.\,\|)$ be a Banach space, $C \subset E$ a cone in E and $R > 0$ a constant. Suppose there exists a concave nonnegative continuous functional ψ on C with*

$$\psi(y) \le \|y\| \quad \textit{for} \quad y \in \overline{C_R}$$

and let $F: \overline{C_R} \to K(\overline{C_R})$ be an upper semicontinuous, compact (or more generally condensing) map (here $K(\overline{C_R})$ denotes the family of nonempty, convex, compact subsets of $\overline{C_R}$). Also assume there are numbers r, L and K with $0 < r < L < K \le R$ such that $(H1)$, $(H2)$ and $(H3)$ hold. Then F has at least three fixed points y_1, y_2 and y_3 in $\overline{C_R}$. Furthermore we have

$$y_1 \in C_r, \quad y_2 \in \{y \in C(\psi, L, R): \psi(y) > L\}$$

and

$$y_3 \in \overline{C_R} \setminus \left(C(\psi, L, R) \cup \overline{C_r}\right).$$

PROOF: In this theorem property (B) means the map is compact (or more generally condensing) with nonempty, convex, compact values. If X is a closed, convex set of E, U an open subset of X and $A: \overline{U^X} \to K(X)$ is an upper semicontinuous, compact (or more generally condensing) map then there exists [143] an integer $i_X(A, U)$ satisfying (P1)–(P4) and moreover

$$\begin{aligned} \{\, i_X(A,U): \quad & X \text{ a closed, convex set of a Banach space } E, \\ & U \text{ open in } X,\ A: \overline{U^X} \to K(X) \text{ is upper semicontinuous,} \\ & \text{compact (or more generally condensing) map and} \\ & x \notin Ax \text{ on } \partial_X U \,\} \end{aligned}$$

is uniquely determined by the properties (P1)–(P4). Trivially $(H4)$ and $(H5)$ are true here so the result follows from Theorem 13.2.1. □

Remark 13.2.4. Note $(H1)$ and $(H3)$ could be replaced by $(H1)^\star$ and $(H3)^\star$ in Theorem 13.2.2.

Many other standard fixed point theorems in a cone could be extended in a similar manner for upper semicontinuous maps which satisfy property (B). We illustrate the ideas involved by extending Krasnoselskii's compression and expansion in a cone theorem.

Theorem 13.2.3. *Let $E = (E, \|\,.\,\|)$ be a Banach space and X a closed, convex set of E. Suppose for every open subset U of X and every upper semicontinuous map $A : \overline{U^X} \to 2^X$ which satisfies property (B) with $x \notin A\,x$ for $x \in \partial_X U$ there exists an integer $i_X(A, U)$ satisfying (P1)–(P4). In addition assume the family*

$$\{\, i_X(A, U) : \quad X \text{ a closed, convex set of a Banach space } E, \; U \text{ open in } X,\; A : \overline{U^X} \to 2^X \text{ is upper semicontinuous, satisfies property (B) and } x \notin A\,x \text{ on } \partial_X U \,\}$$

is uniquely determined by the properties (P1)–(P4). Let $C \subset E$ be a cone in E, $r_1 > 0$, $r_2 > 0$, $r_1 \neq r_2$ with

$$R = \max\{r_1, r_2\} \quad \text{and} \quad r = \min\{r_1, r_2\}.$$

Let $F : \overline{C_R} \to 2^C$ be an upper semicontinuous map which satisfies property (B) and assume

$$\mu = \sup\{\|F\,x\| : x \in \overline{C_{r_2}}\} < \infty.$$

Moreover suppose F satisfies

(H6). $x \notin \lambda\,F\,x$ for $\lambda \in [0, 1]$ and $x \in S_{r_1}$;

(H7). $\exists v \in C \setminus \{0\}$ with $x \notin F\,x + \delta\,v$ for $\delta \geq 0$ and $x \in S_{r_2}$;

(H8). the mapping $H_3 : [0, 1] \times \overline{C_{r_1}} \to 2^C$, given by $H_3(t, x) = t\,F\,x$, satisfies property (B);

(H9). $\exists \lambda_1 > \frac{(r_2+\mu)}{\|v\|}$ such that the mapping $H_4 : [0, 1] \times \overline{C_{r_2}} \to 2^C$, given by $H_4(t, x) = F\,x + t\,\lambda_1\,v$, satisfies property (B).

Then F has at least one fixed point y with $r < \|y\| < R$.

PROOF: First we claim

$$i_C(F, C_{r_1}) = 1. \tag{13.2.9}$$

Let $H_3 : [0, 1] \times \overline{C_{r_1}} \to 2^C$ be given by $H_3(t, x) = t\,F\,x$. Notice $H_3 : [0, 1] \times \overline{C_{r_1}} \to 2^C$ is upper semicontinuous and satisfies property (B) (from ($H8$)). Also ($H6$) guarantees that $x \notin H_3(t, x)$ for $(t, x) \in [0, 1] \times S_{r_1}$. Thus (P1) and (P3) imply

$$i_C(F, C_{r_1}) = i_C(\hat{0}, C_{r_1}) = 1$$

so (13.2.9) is true.

Next we show

$$i_C(F, C_{r_2}) = 0. \tag{13.2.10}$$

Let

$$\lambda_1 > \frac{(r_2 + \mu)}{\|v\|}$$

be chosen as in $(H9)$ and let $H_4 : [0,1] \times \overline{C_{r_2}} \to 2^C$ be given by $H_4(t,x) = F\,x + t\,\lambda_1\,v$. Notice $H_4 : [0,1] \times \overline{C_{r_2}} \to 2^C$ is upper semicontinuous and satisfies property (B) (from $(H9)$). Also $(H7)$ guarantees that $x \notin H_4(t,x)$ for $(t,x) \in [0,1] \times S_{r_2}$. Thus (P3) implies

$$i_C(F, C_{r_2}) = i_C(F + \lambda_1\,v, C_{r_2}). \tag{13.2.12}$$

If we now show

$$i_C(F + \lambda_1\,v, C_{r_2}) = 0 \tag{13.2.13}$$

then (13.2.10) holds. Suppose

$$i_C(F + \lambda_1\,v, C_{r_2}) \neq 0.$$

Then (P6) implies that there exists $x \in C_{r_2}$ with $x \in F\,x + \lambda_1\,v$. Now since $x = y + \lambda_1\,v$ for some $y \in F\,x$ then

$$\|x\| \geq \lambda_1\,\|v\| - \|y\| \geq \lambda_1\,\|v\| - \mu > r_2,$$

a contradiction.

Now since F has no fixed points on $\overline{C_R} \setminus C_R$ (this follows from $(H6)$ if $R = r_1$ and $(H7)$ if $R = r_2$) we have from (P2) that

$$i_C(F, C_R) = i_C(F, C_r) + i_C(F, C_R \setminus \overline{C_r}). \tag{13.2.14}$$

Now if $R = r_1$ then $r = r_2$ so (13.2.9), (13.2.10) and (13.2.14) imply

$$i_C(F, C_R \setminus \overline{C_r}) = 1. \tag{13.2.15a}$$

On the other hand if $R = r_2$ then $r = r_1$ so (13.2.9), (13.2.10) and (13.2.14) imply

$$i_C(F, C_R \setminus \overline{C_r}) = -1. \tag{13.2.15b}$$

In either case $i_C(F, C_R \setminus \overline{C_r}) \neq 0$ so (P6) guarantees the result. □

13.3. Integral inclusions.

To illustrate how the results in Section 13.2 can be applied in practice we consider the integral inclusion

$$y(t) \in h(t) + \int_0^1 k(t,s)\, F(s, y(s))\, ds \quad \text{for } t \in [0,1]; \tag{13.3.1}$$

here $k : [0,1] \times [0,1] \to \mathbf{R}$, $h : [0,1] \to \mathbf{R}$ and $F : [0,1] \times \mathbf{R} \to K([0,\infty))$. The following conditions will be assumed:

$$F : [0,1] \times \mathbf{R} \to K([0,\infty)) \tag{13.3.2}$$

$$t \mapsto F(t,x) \text{ is measurable for every } x \in \mathbf{R} \tag{13.3.3}$$

$$x \mapsto F(t,x) \text{ is upper semicontinuous for a.e. } t \in [0,1] \tag{13.3.4}$$

$$\begin{cases} \text{for each } b > 0 \text{ there exists } h_b \in L^1[0,1] \\ \text{with } |F(t,x)| \leq h_b(t) \text{ for a.e. } t \in [0,1] \\ \text{and every } x \in \mathbf{R} \text{ with } |x| \leq b \end{cases} \tag{13.3.5}$$

$$\begin{cases} \text{there exists } f : [0,\infty) \to [0,\infty) \text{ continuous and} \\ \text{nondecreasing, a function } q : [0,1] \to [0,\infty), \\ q \in L^1[0,1], \text{ and a constant } A_0,\ 0 < A_0 \leq 1 \text{ with} \\ A_0\, q(s)\, f(y) \leq u \leq q(s)\, f(y) \text{ for any } u \in F(s,y) \\ \text{with } (s,y) \in [0,1] \times [0,\infty) \end{cases} \tag{13.3.6}$$

$$\begin{cases} \text{for each } t \in [0,1],\ k(t,s) \text{ is measurable on } [0,1] \text{ and} \\ k(t) = ess\ sup\, |k(t,s)|,\ 0 \leq s \leq 1, \text{ is bounded on } [0,1] \end{cases} \tag{13.3.7}$$

$$\begin{cases} \text{the map } t \mapsto k_t \text{ is continuous from } [0,1] \text{ to} \\ L^\infty[0,1] \text{ (here } k_t(s) = k(t,s)) \end{cases} \tag{13.3.8}$$

$$\text{for each } t \in [0,1],\ k(t,s) \geq 0 \text{ for a.e. } s \in [0,1] \tag{13.3.9}$$

$$\text{(13.3.10)} \quad \begin{cases} \exists\, r > 0 \ \text{ with } \ |h|_0 + f(r)\ \sup_{t\in[0,1]} \int_0^1 q(s)\,k(t,s)\,ds \ < r \\ (\text{here } \ |h|_0 = \sup_{t\in[0,1]} |h(t)|) \end{cases}$$

$$\text{(13.3.11)} \quad \begin{cases} \exists\, M,\ 0 < M < 1,\ \kappa \in L^1[0,1] \ \text{ and an} \\ \text{interval } \ [a,b] \subseteq [0,1],\ a < b, \ \text{ such that} \\ k(t,s) \geq \ M\,\kappa(s) \geq 0 \ \text{ for } \ t \in [a,b] \\ \text{and a.e. } \ s \in [0,1] \end{cases}$$

$$\text{(13.3.12)} \qquad k(t,s) \leq \ \kappa(s),\ t \in [0,1], \ \text{ a.e. } \ s \in [0,1]$$

$$\text{(13.3.13)} \quad \begin{cases} h \in C[0,1] \ \text{ with } \ h(t) \geq 0 \ \text{ for } \ t \in [0,1] \\ \text{and } \ \min_{t\in[a,b]} h(t) \geq M\,A_0\,|h|_0 \end{cases}$$

$$\text{(13.3.14)} \quad \begin{cases} \exists\, L \ > \ r \ \text{ with} \\ \min_{t\in[a,b]} \left[h(t) \ + \ A_0\, f(L) \ \int_a^b q(s)\,k(t,s)\,ds \right] \ > \ L \end{cases}$$

and

$$\text{(13.3.15)} \quad \begin{cases} \exists\, R \geq \ L\,(M\,A_0)^{-1} \ \text{ with} \\ |h|_0 + f(R) \ \sup_{t\in[0,1]} \int_0^1 q(s)\,k(t,s)\,ds \ \leq \ R. \end{cases}$$

Theorem 13.3.1. *Suppose* (13.3.2) – –(13.3.15) *hold. Then* (13.3.1) *has three nonnegative solution* y_1, y_2 *and* y_3 *in* $C[0,1]$ *with*

$$|y_1|_0 < r, \quad y_2(t) > L \ \ \textit{for} \ \ t \in [a,b]$$

and

$$|y_3|_0 > r \ \ \textit{with} \ \ \min_{t\in[a,b]} y_3(t) < L.$$

PROOF: Let

$$E = (C[0,1], |\,.\,|_0) \quad \text{and} \quad C = \{u \in C[0,1] : \ u(t) \geq 0 \ \text{ for } \ t \in [0,1]\}.$$

Now let $A : C \to 2^C$ be defined by

$$\text{(13.3.16)} \qquad A\,y(t) = h(t) + \int_0^1 k(t,s)\,F(s,y(s))\,ds \quad \text{for } \ t \in [0,1],$$

here $y \in C$. It is immediate from Chapter 10 (Lemma 10.4.2) that $A : C \to 2^C$ has closed graph. Also the Arzela–Ascoli theorem guarantees that $A : C \to 2^C$ is completely continuous. As a result

$$A : C \to K(C) \quad \text{is upper semicontinuous and completely continuous.}$$

To see the upper semicontinuity, let B be a closed subset of C and

$$W = \{u \in C : \; A(u) \cap B \neq \emptyset\}.$$

We must show W is a closed subset of C. Let (y_n) be a sequence in W converging to y in $C[0,1]$. Let $v_n \in A(y_n) \cap B$. Now $A : C \to 2^C$ has closed graph and is completely continuous, so there exists a subsequence of (v_n) converging to $v \in A(y) \cap B$. This $y \in W$, so W is a closed subset of C.

For $y \in C$ let

$$\psi(y) = min_{t \in [a,b]} \, y(t).$$

Next choose and <u>fix</u> K so that

$$L\,(M\,A_0)^{-1} \le K \le R; \tag{13.3.17}$$

this is possible since $R \ge L\,(M\,A_0)^{-1}$. Let

$$C_r = \{y \in C : \; |y|_0 < r\},$$

$$C_R = \{y \in C : \; |y|_0 < R\}$$

$$C(\psi, L, K) = \{y \in C : \psi(y) \ge L \text{ and } |y|_0 \le K\},$$

and

$$C(\psi, L, R) = \{y \in C : \psi(y) \ge L \text{ and } |y|_0 \le R\}.$$

We first show condition $(H2)$ of Theorem 13.2.1 holds. Let $x \in \overline{C_r}$ with $y \in A\,x$. Then there exists $v \in F(s,x)$ with

$$y(t) = h(t) + \int_0^1 k(t,s)\, v(s)\, ds \quad \text{for } t \in [0,1].$$

In addition (13.3.6) guarantees that $v(s) \le q(s)\, f(x(s))$ for $s \in [0,1]$. Now (13.3.6) and (13.3.10) imply for $t \in [0,1]$ that

$$|y(t)| \;\le\; |h|_0 + \int_0^1 k(t,s)\, v(s)\, ds$$

$$\begin{aligned}
&\le |h|_0 + \int_0^1 k(t,s)\, q(s)\, f(x(s))\, ds \\
&\le |h|_0 + f(r) \int_0^1 k(t,s)\, q(s)\, ds \\
&\le |h|_0 + f(r) \sup_{t\in[0,1]} \int_0^1 q(s)\, k(t,s)\, ds \\
&< r.
\end{aligned}$$

Consequently $|y|_0 < r$ so $(H2)$ in Theorem 13.2.1 is satisfied. Next we show $A : \overline{C_R} \to 2^{\overline{C_R}}$. Let $x \in \overline{C_R}$ and $y \in A\,x$. Then there exists $v \in F(s,x)$ with

$$y(t) = h(t) + \int_0^1 k(t,s)\, v(s)\, ds \quad \text{for} \quad t \in [0,1].$$

Moreover $v(s) \le q(s)\, f(x(s))$ for $s \in [0,1]$ and so (13.3.14) guarantees that

$$|y|_0 \le |h|_0 + f(R) \sup_{t\in[0,1]} \int_0^1 q(s)\, k(t,s)\, ds \;\le\; R,$$

so $A : \overline{C_R} \to 2^{\overline{C_R}}$. To show that $(H1)$ of Theorem 13.2.1 holds first notice that if

$$u(t) \;=\; \frac{L+K}{2} \quad \text{for} \quad t \in [0,1]$$

then $u \in \{y \in C(\psi, L, K) : \; \psi(y) > L\}$. Also if $x \in C(\psi, L, K)$ then

$$\psi(x) = \min_{t\in[a,b]} x(t) \ge L \quad \text{and} \quad |x|_0 \le K,$$

so

$$x(t) \in [L, K] \quad \text{for} \quad t \in [a,b].$$

Let $y \in A\,x$. Then there exists $v \in F(s,x)$ with

$$y(t) = h(t) + \int_0^1 k(t,s)\, v(s)\, ds \quad \text{for} \quad t \in [0,1].$$

In addition (13.3.6) guarantees that

$$v(s) \ge A_0\, q(s)\, f(x(s)) \;\ge\; A_0\, q(s)\, f(L) \quad \text{for} \quad s \in [a,b].$$

This together with (13.3.14) yields

$$\psi(y) \;=\; \min_{t\in[a,b]} \left(h(t) + \int_0^1 k(t,s)\, v(s)\, ds \right)$$

$$
\begin{aligned}
&\geq \min_{t\in[a,b]} \left(h(t) + \int_a^b k(t,s)\, v(s)\, ds \right) \\
&\geq \min_{t\in[a,b]} \left(h(t) \,+\, A_0\, f(L) \int_a^b q(s)\, k(t,s)\, ds \right) \\
&> L,
\end{aligned}
$$

so condition $(H1)$ of Theorem 13.2.1 is satisfied. Finally to see that $(H3)$ of Theorem 13.2.1 holds let $x \in C(\psi, L, R)$ and $y \in A\, x$ with $|y|_0 > K$. Then there exists $v \in F(s,x)$ with

$$
y(t) = h(t) + \int_0^1 k(t,s)\, v(s)\, ds \quad \text{for} \quad t \in [0,1].
$$

Also $A_0\, q(s)\, f(x(s)) \leq v(s) \leq q(s)\, f(x(s))$ for $s \in [0,1]$. Notice (13.3.12) implies for $t \in [0,1]$ that

$$
|y(t)| \,\leq\, |h|_0 + \int_0^1 \kappa(s)\, v(s)\, ds \,\leq\, |h|_0 + \int_0^1 \kappa(s)\, q(s)\, f(x(s))\, ds.
$$

Thus

$$
|y|_0 \,\leq\, |h|_0 + \int_0^1 \kappa(s)\, q(s)\, f(x(s))\, ds
$$

and this together with (13.3.11), (13.3.13) and (13.3.17) yields

$$
\begin{aligned}
\psi(y) &= \min_{t\in[a,b]} \left(h(t) + \int_0^1 k(t,s)\, v(s)\, ds \right) \\
&\geq M\, A_0\, |h|_0 \,+\, \int_0^1 M\, \kappa(s)\, A_0\, q(s)\, f(x(s))\, ds \\
&\geq M\, A_0\, |y|_0 \,>\, M\, A_0\, K \,\geq\, L.
\end{aligned}
$$

Thus condition $(H3)$ of Theorem 13.2.1 holds. Apply Theorem 13.2.2 to complete the proof. □

CHAPTER 14

MULTIVALUED NONLINEAR EQUATIONS ON THE HALF LINE: A FIXED POINT APPROACH

14.1. Introduction.

This chapter is concerned with the existence of one (or more) solutions to the nonlinear integral inclusion

$$y(t) \in \int_0^\infty K(t,s)\, F(s,y(s))\, ds \quad \text{for} \quad t \in [0,\infty). \tag{14.1.1}$$

Here

$$K : [0,\infty) \times [0,\infty) \to \mathbf{R} \;\text{ and }\; F : [0,\infty) \times \mathbf{R} \to CK(\mathbf{R})$$

with $CK(\mathbf{R})$ denoting the family of nonempty, convex, compact subsets of $\mathbf{R}$. It was observed quite some time ago that the classical fixed point theory in Fréchet spaces (see [65]) limit considerably the type of equation we can discuss on the half line. In Section 14.2 and Section 14.3 we present applicable fixed point theorems which will be used to examine general equations on the noncompact interval. Also we will show how the fixed point theory derived in Section 14.2 and Section 14.3 can be used to discuss the existence of one or more $C[0,\infty)$ solutions to (14.1.1).

For the remainder of this section we gather together some definitions and known results. Let (X,d) be a metric space and Ω_X the bounded subsets of X. The Kuratowski measure of noncompactness is the map $\alpha : \Omega_X \to [0,\infty]$ defined by (here $B \in \Omega_X$),

$$\alpha(B) = \inf\{r > 0 : \; B \subseteq \cup_{i=1}^n B_i \;\text{ and }\; diam\,(B_i) \le r\}.$$

Let S be a nonempty subset of X. For each $x \in X$, define $d(x,S) = \inf_{y \in S} d(x,y)$. Now suppose $G : S \to 2^X$; here 2^X denotes the family of nonempty subsets of X. Then (i). $G : S \to 2^X$ is k–set contractive (here $k \ge 0$) if $\alpha(G(W)) \le k\,\alpha(W)$ for all nonempty, bounded sets W of S; (ii). $G : S \to 2^X$ is condensing if G is 1–set contractive and

$\alpha(G(W)) < \alpha(W)$ for all bounded sets W of S with $\alpha(W) \neq 0$; (iii). $G : S \to 2^X$ is hemicompact if each sequence $(x_n)_1^\infty$ in S has a convergent subsequence whenever $d(x_n, G\,x_n) \to 0$ as $n \to \infty$.

We now state three results from the literature which will be used in Section 14.2 and Section 14.3. The proof of the first result can be found in [158], the second in Chapter 10, and the third in Chapter 12.

Theorem 14.1.1. *Let (X, d) be a metric space, D a nonempty, complete subset of X, and $G : D \to 2^X$ a condensing map with $G(D)$ bounded. Then G is hemicompact.*

Theorem 14.1.2. *Let C be a closed, convex subset of a Banach space with U an open subset of C and $0 \in U$. Suppose $F : \overline{U} \to CK(C)$ is an upper semicontinuous (u.s.c.), k-set contractive (here $0 \leq k < 1$) map with $F(\overline{U})$ bounded; here $CK(C)$ denotes the family of nonempty, compact, convex subsets of C. Then either*

$(A1)$. there exists $x \in \overline{U}$ with $x \in F\,(x)$; or

$(A2)$. there exists $u \in \partial_C U$ and $\lambda \in (0, 1)$ with $u \in \lambda\, F\,(u)$.

Let $E = (E, \|\,.\,\|)$ and for $\rho > 0$,

$$\Omega_\rho = \{x \in E : \ \|x\| < \rho\}\,.$$

Theorem 14.1.3. *Let $E = (E, \|\,.\,\|)$ be a Banach space, $C \subseteq E$ a cone and let $\|\,.\,\|$ be increasing with respect to C. Also r, R are constants with $0 < r < R$. Suppose $F : \overline{\Omega_R} \cap C \to CK(C)$ is a u.s.c., k-set contractive (here $0 \leq k < 1$) map and assume the following conditions hold:*

$$\|y\| \geq \|x\| \ \ \textit{for all} \ \ y \in F\,(x) \ \ \textit{and} \ \ x \in \partial_E \Omega_R \cap C \tag{14.1.1}$$

and

$$\|y\| \leq \|x\| \ \ \textit{for all} \ \ y \in F\,(x) \ \ \textit{and} \ \ x \in \partial_E \Omega_r \cap C. \tag{14.1.2}$$

Then F has a fixed point in $C \cap \{x \in E : \ r \leq \|x\| \leq R\}$.

14.2. Existence of one solution.

Before we discuss (14.1.1) we present the fixed point result that we will use in this section. Let $N = \{1, 2, ...\}$. We assume E is a Fréchet space endowed with a family of seminorms $\{|\,.\,|_n : \ n \in N\}$ with

$$|x|_1 \leq |x|_2 \leq .... \quad \text{for all } \ x \in E.$$

Also assume for each $n \in N$ that $(E_n, |\,.\,|_n)$ is a Banach space and suppose

$$E_1 \supseteq E_2 \supseteq \ldots\ldots\ldots\ldots$$

with $E = \cap_{n=1}^{\infty} E_n$ and $|x|_n \leq |x|_{n+1}$ for all $x \in E_{n+1}$ (here $n \in N$). We now present a general result which guarantees that the inclusion

$$y \in F\,y \tag{14.2.1}$$

has a solution in E.

The main points needed to establish the existence of solutions to (14.2.1) are the following:

(1). the existence of upper semicontinuous maps $F_n : \overline{U_n} \to CK(E_n)$;

(2). the sequence of maps $\{F_n\}$ has the property that a convergent sequence of fixed points $\{y_n\}$ of $\{F_n\}$ converges to a fixed point of F;

(3). the assumptions on F_n are such that Theorem 14.1.2 can be applied.

Definition 14.2.1.. Fix $k \in N$. If $x,\, y \in E_k$ then we say $x = y$ in E_k if $|x - y|_k = 0$ (i.e. if $x - y = 0$; here 0 is the zero in E_k).

Definition 14.2.2.. If $x,\, y \in E$ then we say $x = y$ in E if $x = y$ in E_k for each $k \in N$.

Definition 14.2.3.. Fix $k \in N$. We say $x \in F\,y$ in E_k if there exists $w \in F\,y$ with $x = w$ in E_k.

Theorem 14.2.1. *For each $n \in N$, let U_n be an open, bounded subset of E_n with $0 \in U_n$ and*

$$\overline{U_1} \supseteq \overline{U_2} \supseteq \ldots\ldots\ldots\,;$$

here $\overline{U_n}$ denotes the closure of U_n in E_n. Suppose the following conditions are satisfied:

$$\begin{cases} \textit{for each } n \in N,\ F_n : \overline{U_n} \to CK(E_n) \\ \textit{is an upper semicontinuous map} \end{cases} \tag{14.2.2}$$

$$\begin{cases} \textit{for each } n \in N,\ y \notin \lambda\, F_n\, y \ \textit{ in } E_n \textit{ for all } \lambda \in (0,1) \textit{ and} \\ y \in \partial\, U_n \ (\textit{ here } \partial\, U_n \textit{ denotes the boundary of } U_n \textit{ in } E_n) \end{cases} \tag{14.2.3}$$

$$\begin{cases} \textit{for each } n \in N, \textit{ the map } \mathcal{K}_n : \overline{U_n} \to 2^{E_n}, \textit{ given by} \\ \mathcal{K}_n\, y = \cup_{m=n}^{\infty} F_m\, y \ \textit{(see Remark 14.2.1), is } k\textit{-set} \\ \textit{contractive } (0 \leq k < 1) \end{cases} \tag{14.2.4}$$

and

$$(14.2.5)\quad \begin{cases} \textit{if there exists a } w \in E \textit{ and a sequence } \{y_n\}_{n\in N} \\ \textit{with } y_n \in \overline{U_n} \textit{ and } y_n \in F_n\, y_n \textit{ in } E_n \textit{ such that} \\ \textit{for every } k \in N \textit{ there exists a subsequence} \\ S \subseteq \{k+1, k+2, ...\} \textit{ of } N \textit{ with } y_n \to w \textit{ in } E_k \\ \textit{as } n \to \infty \textit{ in } S, \textit{ then } w \in F\,w \textit{ in } E. \end{cases}$$

Then (14.2.1) *has a solution in* E *(in fact in* $\cap_{n=1}^{\infty} \overline{U_n}$*).*

Remark 14.2.1. The definition of $\mathcal{K}_n$ is as follows. If $y \in \overline{U_n}$ and $y \notin \overline{U_{n+1}}$ then $\mathcal{K}_n\, y = F_n\, y$, whereas if $y \in \overline{U_{n+1}}$ and $y \notin \overline{U_{n+2}}$ then $\mathcal{K}_n\, y = F_n\, y \cup F_{n+1}\, y$, and so on.

Remark 14.2.2. We note that although (14.2.4) and (14.2.5) seem technical, they are in fact easily checked in applications (see Theorem 14.2.2).

PROOF: Fix $n \in N$. Theorem 14.1.2 guarantees that

$$y \in F_n\, y \text{ has a solution } y_n \text{ in } \overline{U_n}.$$

Lets look at $\{y_n\}_{n\in N}$. Now Theorem 14.1.1 (with $X = E_1$, $G = \mathcal{K}_1$, $D = \overline{U_1}$ and note $d(y_n, \mathcal{K}_1\, y_n) = 0$ for each $n \in N$ since $|x|_1 \le |x|_n$ for all $x \in E_n$ and $y_n \in F_n\, y_n$ in E_n; here $d(x, S) = \inf_{y\in S} |x - y|_1$ if S is a nonempty subset of X) guarantees that there exists a subsequence $N_1^\star$ of N and $z_1 \in E_1$ with $y_n \to z_1$ in E_1 as $n \to \infty$ in $N_1^\star$. Let $N_1 = N_1^\star/\{1\}$. Look at $\{y_n\}_{n\in N_1}$. Now Theorem 14.1.1 (with $X = E_2$, $G = \mathcal{K}_2$, $D = \overline{U_2}$ and note $d(y_n, \mathcal{K}_2\, y_n) = 0$ for each $n \in N_1$; here $d(x, S) = \inf_{y\in S} |x - y|_2$ if S is a nonempty subset of X) guarantees that there exists a subsequence $N_2^\star$ of N_1 and $z_2 \in E_2$ with $y_n \to z_2$ in E_2 as $n \to \infty$ in $N_2^\star$. Note $|z_2 - z_1|_1 = 0$ since $N_2^\star \subseteq N_1^\star$ and $E_1 \supseteq E_2$. Thus $z_2 = z_1$ in E_1. Let $N_2 = N_2^\star/\{2\}$. Proceed inductively to obtain subsequences of integers

$$N_1^\star \supseteq N_2^\star \supseteq ......, \quad N_k^\star \subseteq \{k, k+1, ...\}$$

and $z_k \in E_k$ with $y_n \to z_k$ in E_k as $n \to \infty$ in $N_k^\star$. Note $z_{k+1} = z_k$ in E_k for $k = 1, 2, ...$ Also let $N_k = N_k^\star/\{k\}$.

Fix $k \in N$. Let $y = z_k$ in E_k. Notice y is well defined and $y \in E_k$ for each $k = 1, 2....$ Fix $k \in N$. Now $y_n \in F_n\, y_n$ in E_n for $n \in N_k$ and $y_n \to y$ in E_k as $n \to \infty$ in N_k (since $y = z_k$ in E_k). This together with (14.2.5) implies $y \in F\, y$ in E. □

Remark 14.2.3. Suppose F is defined on E_1 with $F_n = F|_{E_n}$ for each $n \in N$. Then (14.2.5) is automatically satisfied. To see this suppose there exists $w \in E$ and for every $k \in N$ there exists a subsequence $S \subseteq \{k+1, k+2, ...\}$ of N and a sequence $\{y_n\}_{n \in S}$ with $y_n \in \overline{U_n}$, $y_n \in F\, y_n$ in E_n for $n \in S$ and with $y_n \to w$ in E_k as $n \to \infty$ in S. Now from (14.2.2), $F(= F_k) : \overline{U_k} \to CK(E_k)$ is an upper semicontinuous map so we have $w \in F\, w$ in E_k for each $k = 1, 2, ...$ i.e. $w \in F\, w$ in E.

Remark 14.2.4. Notice (14.2.3) can be replaced by any assumption that guarantees that F_n has a fixed point in $\overline{U_n}$ for each $n \in N$.

We now use Theorem 14.2.1 to obtain a general existence theorem for (14.1.1).

Theorem 14.2.2. *Let $1 \le p < \infty$ be a constant and $1 < q \le \infty$ the conjugate to p. Assume the following conditions are satisfied:*

(14.2.6) $\quad$ *for each $t \in [0, \infty)$, the map $s \mapsto K(t,s)$ is measurable*

$$\sup_{t \in [0,\infty)} \left(\int_0^\infty |K(t,s)|^q \, ds \right)^{\frac{1}{q}} < \infty \tag{14.2.7}$$

$$\begin{cases} \int_0^\infty |K(t',s) - K(t,s)|^q \, ds \to 0 \quad as \quad t \to t', \\ for\ each \quad t' \in [0,\infty) \end{cases} \tag{14.2.8}$$

$$\begin{cases} F : [0,\infty) \times \mathbf{R} \to CK(\mathbf{R}) \quad is\ a \quad L^p\text{-}Carath\acute{e}odory\ function: \\ by\ this\ we\ mean \\ (a).\ for\ each\ measurable \quad u : [0,\infty) \to \mathbf{R}, \quad the\ map \\ x \mapsto F(x, u(x)) \quad has\ measurable\ single\ valued\ selections \\ (b).\ for\ a.e. \quad x \in [0,\infty), \quad the\ map \quad u \mapsto F(x,u) \quad is \\ upper\ semicontinuous \\ (c).\ for\ each \quad r > 0 \quad there\ exists \quad h_r \in L^p[0,\infty) \quad with \\ |F(x,y)| \le h_r(x) \quad for\ a.e. \quad x \in [0,\infty) \quad and\ all \quad y \in \mathbf{R} \\ with \quad |y| \le r; \quad here \quad |F(x,y)| = \sup\{|v| : \ v \in F(x,y)\} \end{cases} \tag{14.2.9}$$

$$\begin{cases} there\ exists\ a\ function \quad \psi : [0,\infty) \to [0,\infty) \quad continuous \\ and\ nonincreasing\ and\ a \quad \phi \in L^p[0,\infty) \quad with \\ |F(s,y)| \le \phi(s)\, \psi(|y|) \quad for\ all \quad y \in \mathbf{R} \quad and \\ a.e. \quad s \in [0,\infty) \end{cases} \tag{14.2.10}$$

and

$$\begin{cases} \exists\, r > 0 \quad with \quad r > K_1\, \psi(r) \quad where \\ K_1 = \sup_{t \in [0,\infty)} \int_0^\infty |K(t,s)|\, \phi(s)\, ds. \end{cases} \tag{14.2.11}$$

Then (14.1.1) *has a solution* $y \in C[0,\infty)$ *with* $|y|_n \le r$ *for each* $n \in N$ *(here we have* $|y|_n = \sup_{t\in[0,n]} |y(t)|$*).*

Remark 14.2.5. Note (14.2.9) (a) could be replaced by: the map $x \mapsto F(x,u)$ is measurable for all $u \in \mathbf{R}$.

PROOF: Let $n \in N$ and

$$F_n\, y(t) = \int_0^n K(t,s)\, F(s,y(s))\, ds \quad \text{for} \quad t \in [0,n],$$

here $y \in C[0,n]$. We wish to apply Theorem 14.2.1 with

$$E_n = C[0,n] \quad \text{and} \quad U_n = \{x \in C[0,n]: \ |x|_n < r\}.$$

Let $K_n : L^p[0,n] \to C[0,n]$ and $\mathcal{F}_n : C[0,n] \to L^p[0,n]$ be given by

$$K_n\, y(t) = \int_0^n K(t,s)\, y(s)\, ds$$

and

$$\mathcal{F}_n\,(y) = \{u \in L^p[0,n]: \ u(t) \in F(t,y(t)) \ \text{ for a.e. } \ t \in [0,n]\}.$$

Notice (14.2.6), (14.2.7) together with a standard result from the literature [67, 93 or 129] (see also Theorem 12.4.1) immediately guarantees that

$$F_n = K_n \circ \mathcal{F}_n : \overline{U_n} \to CK(E_n) \ \text{ is upper semicontinuous for each } \ n \in N,$$

so (14.2.2) holds. To see (14.2.3) fix $n \in N$ and suppose there exists $y \in \partial U_n$ (i.e. $|y|_n = r$) and $\lambda \in (0,1)$ with $y \in \lambda\, F_n\, y$. Since $\mathcal{F}_n\,(y) \neq \emptyset$ [54, 67, 93 or 146] there exists $v \in \mathcal{F}_n\,(y)$ with

$$y(t) = \lambda \int_0^n K(t,s)\, v(s)\, ds \quad \text{for} \quad t \in [0,n].$$

In addition (14.2.10) guarantees that $|v(s)| \le \phi(s)\, \psi(|y(s)|)$ for a.e. $s \in [0,n]$, and so for $t \in [0,n]$ we have

$$|y(t)| \le \psi\,(|y|_n) \int_0^n |K(t,s)|\, \phi(s)\, ds \le \psi\,(|y|_n)\ K_1.$$

Consequently

$$|y|_n \le \psi\,(|y|_n)\ K_1. \tag{14.2.12}$$

Now (14.2.11) and (14.2.12) imply $|y|_n \neq r$, a contradiction since $y \in \partial U_n$. Thus (14.2.3) holds. To show (14.2.4) fix $n \in N$. We will show $\mathcal{K}_n : \overline{U_n} \to 2^{E_n}$, given by $\mathcal{K}_n\, y = \cup_{m=n}^{\infty} F_m\, y$ (see Remark 14.2.1), is compact. Let $y \in \overline{U_n}$. Without loss of generality assume there exists $l \in \{0, 1, 2, ...\}$ with $y \in \overline{U_{n+l}}$ and $y \notin \overline{U_{n+l+1}}$. Then by definition (see Remark 14.2.1), $\mathcal{K}_n\, y = \cup_{m=n}^{n+l} F_m\, y$. Since $y \in \overline{U_{n+l}}$ then (14.2.9) guarantees that there exists a $h_r \in L^p[0,\infty)$ with $|F(s, y(s))| \leq h_r(s)$ for a.e. $s \in [0, n+l]$. Now for $j \in \{0, 1, ..., l\}$ we have for $t \in [0, n]$ that

$$
\begin{aligned}
|F_{n+j}\, y(t)| &\leq \int_0^{n+j} h_r(s)\, |K(t,s)|\, ds \\
&\leq \left(\int_0^{n+j} [h_r(s)]^p\, ds \right)^{\frac{1}{p}} \left(\int_0^{n+j} |K(t,s)|^q\, ds \right)^{\frac{1}{q}} \\
&\leq \left(\int_0^{\infty} [h_r(s)]^p\, ds \right)^{\frac{1}{p}} \sup_{t \in [0,\infty)} \left(\int_0^{\infty} |K(t,s)|^q\, ds \right)^{\frac{1}{q}}.
\end{aligned}
$$

Thus for $j \in \{0, 1, ..., l\}$ we have

$$
|F_{n+j}\, y|_n \leq \left(\int_0^{\infty} [h_r(s)]^p\, ds \right)^{\frac{1}{p}} \sup_{t \in [0,\infty)} \left(\int_0^{\infty} |K(t,s)|^q\, ds \right)^{\frac{1}{q}}
$$

and so

$$
|\mathcal{K}_n\, y|_n \leq \left(\int_0^{\infty} [h_r(s)]^p\, ds \right)^{\frac{1}{p}} \sup_{t \in [0,\infty)} \left(\int_0^{\infty} |K(t,s)|^q\, ds \right)^{\frac{1}{q}}. \tag{14.2.13}
$$

In addition for $t_1,\ t_2 \in [0, n]$ and $j \in \{0, 1, ..., l\}$ we have

$$
\begin{aligned}
|F_{n+j}\, y(t_1) - F_{n+j}\, y(t_2)| &\leq \int_0^{n+j} h_r(s)\, |K(t_1,s) - K(t_2,s)|\, ds \\
&\leq \left(\int_0^{\infty} [h_r(s)]^p\, ds \right)^{\frac{1}{p}} \\
&\quad \times \left(\int_0^{\infty} |K(t_1,s) - K(t_2,s)|^q\, ds \right)^{\frac{1}{q}}
\end{aligned}
$$

and so

$$
\begin{aligned}
|\mathcal{K}_n\, y(t_1) - \mathcal{K}_n\, y(t_1)| &\leq (\textstyle\int_0^{\infty} [h_r(s)]^p\, ds)^{\frac{1}{p}} \\
&\quad \times (\textstyle\int_0^{\infty} |K(t_1,s) - K(t_2,s)|^q\, ds)^{\frac{1}{q}}.
\end{aligned} \tag{14.2.14}
$$

Now (14.2.8), (14.2.13) and (14.2.14) guarantee that $\{\mathcal{K}_n y : y \in \overline{U_n}\}$ is uniformly bounded and equicontinuous on $[0, n]$. The Arzela–Ascoli theorem implies that $\mathcal{K}_n : \overline{U_n} \to 2^{E_n}$ is compact, so (14.2.4) holds. It remains to show (14.2.5) is satisfied. Suppose there exists $w \in C[0, \infty)$ and a sequence $\{y_n\}_{n \in N}$ with $y_n \in \overline{U_n}$ and $y_n(t) \in F_n y_n(t)$, $t \in [0, n]$, such that for every $k \in N$ there exists a subsequence $S \subseteq \{k+1, k+2, ...\}$ of N with $y_n \to w$ in $C[0, k]$ as $n \to \infty$ in S. If we show

$$w(t) \in \int_0^\infty K(t, s)\, F(s, w(s))\, ds \quad \text{for} \quad t \in [0, \infty)$$

then (14.2.5) will hold. Fix $t \in [0, \infty)$. Consider $k \geq t$ and $n \in S$ (as described above). Then $y_n(x) \in F_n y_n(x)$, $x \in [0, n]$, for $n \in S$. Now there exists $v_n \in \mathcal{F}_n(y_n)$ with

$$y_n(x) = \int_0^n K(x, s)\, v_n(s)\, ds \quad \text{for} \quad x \in [0, n],$$

and so

$$y_n(t) - \int_0^k K(t, s)\, v_n(s)\, ds = \int_k^n K(t, s)\, v_n(s)\, ds. \tag{14.2.15}$$

Now (14.2.9) guarantees that there exists a $h_r \in L^p[0, \infty)$ with $|v_n(s)| \leq h_r(s)$ for a.e. $s \in [0, n]$. Then

$$\left| y_n(t) - \int_0^k K(t, s)\, v_n(s)\, ds \right| \leq \int_k^n h_r(s)\, |K(t, s)|\, ds$$

and so

$$\left| y_n(t) - \int_0^k K(t, s)\, v_n(s)\, ds \right| \leq \int_k^\infty h_r(s)\, |K(t, s)|\, ds. \tag{14.2.16}$$

Consider $\{v_n\}_{n \in S}$. A standard result from the literature [146 Proposition 1.4] guarantees that $\mathcal{F}_k : C[0, k] \to L^p[0, k]$ is upper semicontinuous with respect to the weak topology (w–u.s.c.) and also weakly completely continuous. Now since $v_n \in \mathcal{F}_k(y_n)$ for $n \in S$, there exists a $u_k \in L^p[0, k]$ and a subsequence of S (without loss of generality assume its S) with v_n converging weakly to u_k (i.e. $v_n \rightharpoonup u_k$ in $L^p[0, k]$) as $n \to \infty$ in S. Now $y_n \to w$ in $C[0, k]$ and $v_n \rightharpoonup u_k$ in $L^p[0, k]$ as $n \to \infty$ in S together with $v_n \in \mathcal{F}_k(y_n)$ for $n \in S$ and the fact that $\mathcal{F}_k : C[0, k] \to L^p[0, k]$ is w–u.s.c. guarantees

$$u_k \in \mathcal{F}_k(w). \tag{14.2.17}$$

Note as well that $|w|_k \le r$ since $|y_n|_k \le r$ for $n \in S$, and also we have $|u_k(x)| \le h_r(x)$ for a.e. $x \in [0,k]$. Let $n \to \infty$ through S in (14.2.16) to obtain

$$\left| w(t) - \int_0^k K(t,s)\, u_k(s)\, ds \right| \le \int_k^\infty h_r(s)\, |K(t,s)|\, ds. \tag{14.2.18}$$

Similarly we can show that there exists $u_{k+1} \in L^p[0,k+1]$ and a subsequence of S, say S_1, with $v_n \rightharpoonup u_{k+1}$ in $L^p[0,k+1]$ as $n \to \infty$ in S_1 and with $u_{k+1} \in \mathcal{F}_{k+1}(w)$. Of course this implies $v_n \rightharpoonup u_{k+1}$ in $L^p[0,k]$ as $n \to \infty$ in S_1 so $u_{k+1}(x) = u_k(x)$ for a.e. $x \in [0,k]$. In addition note $|u_{k+1}(x)| \le h_r(x)$ for a.e. $x \in [0,k+1]$. Continue and construct u_{k+2}, $u_{k+3}, \ldots\ldots$. For $l \in \{k, k+1, \ldots\} \equiv P$ let $u_l^\star(x)$ be any extension to $[0,\infty)$ of u_l with $|u_l^\star(x)| \le h_r(x)$ for a.e. $x \in (l,\infty)$. Also let

$$\begin{aligned} \mathcal{F}_l^\star(w) &= \{v \in L^p[0,\infty) : \ v(x) \in F(x,w(x)) \text{ for a.e. } x \in [0,l], \\ &\qquad |v(x)| \le h_r(x) \text{ for a.e. } x \in [0,\infty)\}. \end{aligned}$$

Now $\{u_l^\star\}_{l \in P}$ is a weakly compact sequence in $L^p[0,\infty)$ (see [26 or 57]) so there exists a subsequence which converges weakly to a function $u \in L^p[0,\infty)$. Note $u(x) = u_k(x)$ for a.e. $x \in [0,k]$ since $u_{k+m}(x) = u_k(x)$ for a.e. $x \in [0,k]$, here $m \in N$. This together with (14.2.18) yields

$$\left| w(t) - \int_0^k K(t,s)\, u(s)\, ds \right| \le \int_k^\infty h_r(s)\, |K(t,s)|\, ds. \tag{14.2.19}$$

Let

$$\mathcal{F}(w) = \{v \in L^p[0,\infty) : \ v(x) \in F(x,w(x)) \ \text{ for a.e. } \ x \in [0,\infty)\}$$

(note $|w|_k \le r$ for each $k \in N$ so $w \in BC[0,\infty)$). We next <u>claim</u> that

$$\mathcal{F}(w) = \cap_{l \in N}\, \mathcal{F}_l^\star(w)$$

(and $\mathcal{F}(w)$ is nonempty, closed and convex). Note first $\sup_{t \in [0,\infty)} |w(t)| \le r$ so $|F(x,w(x))| \le h_r(x)$ for a.e. $x \in [0,\infty)$. Let w_k be the restriction to the interval $[0,k]$, $k \in N$, of w. Note [67, 93 or 129] that

$$\mathcal{F}_k(w_k) = \{v \in L^p[0,k] : \ v(x) \in F(x,w_k(x)) \ \text{ for a.e. } \ x \in [0,k]\}$$

is closed in $L^p[0,k]$ for all $k \in N$. Let

$$\begin{aligned} \mathcal{F}_k^\star(w_k) &= \{v \in L^p[0,\infty) : \ v \in \mathcal{F}_k(w_k) \ \text{ for } \ x \in [0,k] \\ &\qquad \text{and } \ v(x) = 0 \ \text{ for } \ x > k\}. \end{aligned}$$

It is immediate that $\mathcal{F}_k^\star(w_k)$ is a closed set in $L^p[0,\infty)$ for each $k \in N$. Let

$$\begin{aligned} R_k &= \{v \in L^p[0,\infty) : \; v(x) = 0 \;\; \text{for} \;\; x \in [0,k], \\ &\quad |v(x)| \le h_r(x) \;\; \text{for a.e.} \;\; x \in (k,\infty)\} \end{aligned}$$

and notice it is clear that

$$\mathcal{F}_k^\star(w) = \mathcal{F}_k^\star(w_k) \oplus R_k.$$

It is immediate that $\mathcal{F}_k^\star(w)$ is a closed set in $L^p[0,\infty)$. Also for each $k \in N$ we have $\mathcal{F}(w) \subseteq \mathcal{F}_k^\star(w)$ and so

$$\mathcal{F}(w) \subseteq \cap_{l \in N} \mathcal{F}_l^\star(w).$$

On the other hand if $v \in \mathcal{F}_l^\star(w)$ for each $l \in N$ then $v(x) \in F(x, w(x))$ for a.e. $x \in [0,\infty)$ and so

$$\cap_{l \in N} \mathcal{F}_l^\star(w) \subseteq \mathcal{F}(w).$$

Thus $\mathcal{F}(w) = \cap_{l \in N} \mathcal{F}_l^\star(w)$ and also $\mathcal{F}(w)$ is a closed subset of $L^p[0,\infty)$. Thus our claim is established. Now since u belongs to $\cap_{l \in N} \mathcal{F}_l^\star(w)$ (note for each $l \in N$, $u \in \mathcal{F}_l^\star(w)$) we have $u \in \mathcal{F}(w)$. Let $k \to \infty$ in (14.2.19) to obtain

$$w(t) = \int_0^\infty K(t,s)\, u(s)\, ds$$

and so

$$w(t) \in \int_0^\infty K(t,s)\, F(s, w(s))\, ds.$$

Thus (14.2.5) holds. Now apply Theorem 14.2.1. $\square$

Remark 14.2.6. Notice that it is possible to prove Theorem 14.2.2 also using Ky Fan's fixed point theorem.

Remark 14.2.7. Notice (14.2.11) can be replaced by any condition which guarantees the existence of a constant $r > 0$ with (14.2.3) holding (here F_n, U_n are as in Theorem 14.2.2).

For completeness we state the single valued version of Theorem 14.2.2. We will consider the integral equation

$$y(t) = \int_0^\infty K(t,s)\, f(s, y(s))\, ds \quad \text{for} \;\; t \in [0,\infty). \tag{14.2.20}$$

Theorem 14.2.3. *Let $1 \le p \le \infty$ be a constant and $1 \le q \le \infty$ the conjugate to p. Suppose (14.2.6)–(14.2.8) hold. In addition assume the following conditions are satisfied:*

$$(14.2.21) \quad \begin{cases} f : [0,\infty) \times \mathbf{R} \to \mathbf{R} \ \text{ is a } L^p\text{-Carathéodory function:} \\ \text{by this we mean} \\ \text{(a). the map } x \mapsto f(x,u) \text{ is measurable for all } y \in \mathbf{R} \\ \text{(b). for a.e. } x \in [0,\infty), \text{ the map } u \mapsto f(x,u) \text{ is} \\ \text{continuous} \\ \text{(c). for each } r > 0 \text{ there exists } h_r \in L^p[0,\infty) \text{ with} \\ |f(x,y)| \le h_r(x) \text{ for a.e. } x \in [0,\infty) \text{ and all} \\ y \in \mathbf{R} \text{ with } |y| \le r \end{cases}$$

$$(14.2.22) \quad \begin{cases} \text{there exists a function } \psi : [0,\infty) \to [0,\infty) \text{ continuous} \\ \text{and nonincreasing and a } \phi \in L^p[0,\infty) \text{ with} \\ |f(s,y)| \le \phi(s)\,\psi(|y|) \text{ for all } y \in \mathbf{R} \text{ and} \\ \text{a.e. } s \in [0,\infty) \end{cases}$$

and

$$(14.2.23) \quad \begin{cases} \exists\, r > 0 \text{ with } r > K_1\,\psi(r) \text{ where} \\ K_1 = \sup_{t\in[0,\infty)} \int_0^\infty |K(t,s)|\,\phi(s)\,ds. \end{cases}$$

Then (14.2.20) *has a solution $y \in C[0,\infty)$ with $|y|_n \le r$ for each $n \in N$.*

Example 14.2.1. Consider the nonlinear integral equation

$$(14.2.24) \qquad y(t) = \int_0^\infty K(t,s)\,e^{-s}\left[|y(s)|^\beta + A\right] ds \quad \text{for } t \in [0,\infty)$$

with $A > 0$ and $\beta > 0$. Assume (14.2.6), (14.2.7) with $q = 1$, (14.2.8) with $q = 1$, and

$$(14.2.25) \quad \begin{cases} \sup_{x\in[0,\infty)} \left(\frac{x}{x^\beta + A}\right) > K_1 \quad \text{where} \\ \\ K_1 = \sup_{t\in[0,\infty)} \int_0^\infty |K(t,s)|\,e^{-s}\,ds \end{cases}$$

hold. Then (14.2.24) has a solution $y \in C[0,\infty)$.

Remark 14.2.8. Note if $0 < \beta < 1$ then the supremum in (14.2.25) is infinity, and so (14.2.25) is automatically satisfied.

We will apply Theorem 14.2.3 with

$$f(s,y) = e^{-s}\,(y^\beta + A),\ \ \psi(y) = y^\beta + A \ \text{ and } \ \phi(s) = e^{-s}.$$

Notice (14.2.21) and (14.2.22) are clearly satisfied. In addition notice that (14.2.25) guarantees that there exists $r > 0$ with

$$\frac{r}{r^\beta + A} > K_1,$$

so (14.2.23) is satisfied with this r.

14.3. Existence of two solutions.

Let $N = \{1, 2, ...\}$. In this section we assume E is a Fréchet space endowed with a family of seminorms $\{|\,.\,|_n : n \in N\}$ with

$$|x|_1 \leq |x|_2 \leq \text{..........} \quad \text{for all } x \in E.$$

Also assume for each $n \in N$ that $(E_n, |\,.\,|_n)$ is a Banach space and suppose

$$E_1 \supseteq E_2 \supseteq \text{...........}$$

with $E = \cap_{n=1}^{\infty} E_n$ and $|x|_n \leq |x|_{n+1}$ for all $x \in E_{n+1}$. For each $n \in N$ let C_n be a cone in E_n and assume $|\,.\,|_n$ is increasing with respect to C_n. In addition assume

$$C_1 \supseteq C_2 \supseteq \text{...........} \,.$$

For $\rho > 0$ and $n \in N$ let

$$U_{n,\rho} = \{x \in E_n : |x|_n < \rho\} \quad \text{and} \quad \Omega_{n,\rho} = U_{n,\rho} \cap C_n.$$

Notice

$$\partial_{C_n} \Omega_{n,\rho} = \partial_{E_n} U_{n,\rho} \cap C_n \quad \text{and} \quad \overline{\Omega_{n,\rho}} = \overline{U_{n,\rho}} \cap C_n$$

(the first closure is with respect to C_n whereas the second is with respect to E_n). In addition notice since $|x|_n \leq |x|_{n+1}$ for all $x \in E_{n+1}$ that

$$\Omega_{1,\rho} \supseteq \Omega_{2,\rho} \supseteq \text{.......}$$

and

$$\overline{\Omega_{1,\rho}} \supseteq \overline{\Omega_{2,\rho}} \supseteq \text{.......} \,.$$

We now establish a general result which guarantees that the inclusion

$$y \in F\, y \tag{14.3.1}$$

has two solutions in E.

The main points needed to establish the existence of solutions to (14.3.1) are the following:

(1). the existence of upper semicontinuous maps $F_n : \overline{U_{n,R}} \cap C_n \to CK(C_n)$;
(2). the sequence of maps $\{F_n\}$ has the property that a convergent sequence of fixed points $\{y_n\}$ of $\{F_n\}$ converges to a fixed point of F;
(3). the assumptions on F_n are such that the Krasnoselskii–Petryshyn theorem [26, 67] in a Banach space can be applied.

We note here that F_n need not be the restriction of F to E_n.

Definition 14.3.1.. Fix $k \in N$. If $x, y \in E_k$ then we say $x = y$ in E_k if $|x - y|_k = 0$ (i.e. if $x - y = 0$; here 0 is the zero in E_k).

Definition 14.3.2.. If $x, y \in E$ then we say $x = y$ in E if $x = y$ in E_k for each $k \in N$.

Definition 14.3.3.. Fix $k \in N$. We say $x \in F y$ in E_k if there exists $w \in F y$ with $x = w$ in E_k.

Theorem 14.3.1. *Let L, γ, r, R be constants with $0 < L < \gamma < r < R$. Assume the following conditions are satisfied:*

$$\left\{ \begin{array}{l} \text{for each } n \in N,\ F_n : \overline{U_{n,R}} \cap C_n \to CK(C_n) \\ \text{is a upper semicontinuous map} \end{array} \right. \tag{14.3.2}$$

$$\left\{ \begin{array}{l} \text{for each } n \in N,\ |y|_n \le |x|_n \text{ for all} \\ y \in F_n(x) \text{ and } x \in \partial_{E_n} U_{n,L} \cap C_n \end{array} \right. \tag{14.3.3}$$

$$\left\{ \begin{array}{l} \text{for each } n \in N,\ |y|_n \le |x|_n \text{ for all} \\ y \in F_n(x) \text{ and } x \in \partial_{E_n} U_{n,r} \cap C_n \end{array} \right. \tag{14.3.4}$$

$$\left\{ \begin{array}{l} \text{for each } n \in N,\ |y|_n \ge |x|_n \text{ for all} \\ y \in F_n(x) \text{ and } x \in \partial_{E_n} U_{n,R} \cap C_n \end{array} \right. \tag{14.3.5}$$

$$\left\{ \begin{array}{l} \text{for each } n \in N, \text{ the map } \mathcal{K}_n : \overline{U_{n,R}} \cap C_n \to 2^{C_n} \\ \text{given by } \mathcal{K}_n y = \cup_{m=n}^{\infty} F_m y \text{ is } k\text{-set} \\ \text{contractive (here } 0 \le k < 1) \end{array} \right. \tag{14.3.6}$$

$$\left\{ \begin{array}{l} \text{for every } k \in N \text{ and any subsequence } A \subseteq \{k, k+1, ...\} \\ \text{if } x \in C_n,\ n \in A, \text{ is such that } R \ge |x|_n \ge r \\ \text{then } |x|_k \ge \gamma \end{array} \right. \tag{14.3.7}$$

$$(14.3.8) \qquad \begin{cases} \textit{if there exists a } v \in E \textit{ and a sequence } \{u_n\}_{n\in N} \\ \textit{with } u_n \in \overline{U_{n,L}} \cap C_n \textit{ and } u_n \in F_n u_n \textit{ in } E_n \textit{ such} \\ \textit{that for every } k \in N \textit{ there exists a subsequence} \\ S \subseteq \{k+1, k+2, \dots\} \textit{ of } N \textit{ with } u_n \to v \textit{ in } E_k \\ \textit{as } n \to \infty \textit{ in } S, \textit{ then } v \in F v \textit{ in } E \end{cases}$$

and

$$(14.3.9) \qquad \begin{cases} \textit{if there exists a } z \in E \textit{ and a sequence } \{w_n\}_{n\in N} \\ \textit{with } w_n \in \left(\overline{U_{n,R}} \setminus U_{n,r}\right) \cap C_n \textit{ and } w_n \in F_n w_n \textit{ in } E_n \\ \textit{such that for every } k \in N \textit{ there exists a subsequence} \\ P \subseteq \{k+1, k+2, \dots\} \textit{ of } N \textit{ with } w_n \to z \textit{ in } E_k \\ \textit{as } n \to \infty \textit{ in } P, \textit{ then } z \in F z \textit{ in } E. \end{cases}$$

Then (14.3.1) *has at least two solutions* x_0 *and* x_1 *with*

$$x_0 \in \cap_{n=1}^{\infty} \left(\overline{U_{n,L}} \cap C_n\right) \quad \textit{and} \quad x_1 \in \cap_{n=1}^{\infty} \left(\left(\overline{U_{n,R}} \setminus U_{n,\gamma}\right) \cap C_n\right).$$

Remark 14.3.1. The definition of $\mathcal{K}_n$ in (14.3.6) is as follows. If $y \in \overline{U_{n,R}} \cap C_n$ and $y \notin \overline{U_{n+1,R}} \cap C_{n+1}$ then $\mathcal{K}_n y = F_n y$, whereas if $y \in \overline{U_{n+1,R}} \cap C_{n+1}$ and $y \notin \overline{U_{n+2,R}} \cap C_{n+2}$ then $\mathcal{K}_n y = F_n y \cup F_{n+1} y$, and so on.

Remark 14.3.2. If F is defined on E_1 and $F_n = F|_{E_n}$ for each $n \in N$ then (14.3.8) and (14.3.9) are automatically satisfied.

PROOF: Fix $n \in N$. Note (14.3.3) implies

$$x \notin \lambda F_n(x) \text{ for all } \lambda \in (0,1) \text{ and } x \in \partial_{E_n} U_{n,L} \cap C_n.$$

To see this suppose there exists $x \in \partial_{E_n} U_{n,L} \cap C_n$ and $\lambda \in (0,1)$ with $x \in \lambda F_n(x)$. Then there exists $y \in F_n(x)$ with $x = \lambda y$ and so

$$L = |x|_n = |\lambda|\,|y|_n < |y|_n \le |x|_n = L,$$

a contradiction. Theorem 14.1.2 guarantees that

$$y \in F_n y \text{ has a fixed point } u_n \in \overline{U_{n,L}} \cap C_n.$$

In particular

$$|u_n|_n \le L.$$

Theorem 14.1.3 (note (14.3.4) and (14.3.5)) guarantees that

$$y \in F_n y \text{ has a fixed point } w_n \in \left(\overline{U_{n,R}} \setminus U_{n,r}\right) \cap C_n.$$

In particular

$$r \le |w_n|_n \le R.$$

Lets look at $\{u_n\}_{n\in N}$. Note $u_n \in \overline{U_{1,L}}$ for each $n \in N$. To see this notice $|u_n|_n \le L$ and $|x|_1 \le |x|_n$ for all $x \in E_n$ implies $|u_n|_1 \le L$. Now Theorem 14.1.1 (with $X = E_1$, $G = \mathcal{K}_1$, $D = \overline{U_{1,L}} \cap C_1$ and note $d_1(u_n, \mathcal{K}_1\, u_n) = 0$ for each $n \in N$ since $|x|_1 \le |x|_n$ for all $x \in E_n$ and $u_n \in F_n\, u_n$ in E_n; here $d_1(x,S) = \inf_{y\in S} |x-y|_1$ if S is a nonempty subset of X) guarantees that there exists a subsequence $N_1^\star$ of N and $v_1 \in \overline{U_{1,L}} \cap C_1$ with $u_n \to v_1$ in E_1 as $n \to \infty$ in $N_1^\star$. Notice in particular that $|v_1|_1 \le L$. Lets now look at $\{w_n\}_{n\in N}$. Note $w_n \in \overline{U_{1,R}} \backslash U_{1,\gamma}$ for each $n \in N$. To see this notice $|w_n|_n \le R$ and $|x|_1 \le |x|_n$ for all $x \in E_n$ implies $|w_n|_1 \le R$. Thus $w_n \in \overline{U_{1,R}}$ for each $n_0 \in N$. On the other hand $|w_n|_n \ge r$, $w_n \in C_n$ together with (14.3.7) implies $|w_n|_1 \ge \gamma$. Now Theorem 14.1.1 (with $X = E_1$, $G = \mathcal{K}_1$, $D = \left(\overline{U_{1,R}} \backslash U_{1,\gamma}\right) \cap C_1$ and note $d_1(u_n, \mathcal{K}_1\, u_n) = 0$ for each $n \in N$) guarantees that there exists a subsequence $P_1^\star$ of N and $z_1 \in \left(\overline{U_{1,R}} \backslash U_{1,\gamma}\right) \cap C_1$ with $w_n \to z_1$ in E_1 as $n \to \infty$ in $P_1^\star$. Notice in particular that $\gamma \le |z_1|_1 \le R$.

Let $N_1 = N_1^\star \setminus \{1\}$ and $P_1 = P_1^\star \setminus \{1\}$. Look at $\{u_n\}_{n\in N_1}$. Notice $u_n \in \overline{U_{2,L}}$ for each $n \in N_1$. Now Theorem 14.1.1 with $X = E_2$, $G = \mathcal{K}_2$, $D = \overline{U_{2,L}} \cap C_2$ and note $d_2(u_n, \mathcal{K}_2\, u_n) = 0$ for each $n \in N_1$; here $d_2(x,S) = \inf_{y\in S} |x-y|_2$ if S is a nonempty subset of X) guarantees that there exists a subsequence $N_2^\star$ of N_1 and $v_2 \in \overline{U_{2,L}} \cap C_2$ with $u_n \to v_2$ in E_2 as $n \to \infty$ in $N_2^\star$. Notice in particular that $|v_2|_2 \le L$. Note $|v_2 - v_1|_0 = 0$ since $N_2^\star \subseteq N_1$ and $E_1 \supseteq E_2$. Thus $v_2 = v_1$ in E_1. Lets now look at $\{w_n\}_{n\in P_1}$. Notice it is easy to see, using $r \le |w_n|_n \le R$ and (14.3.7), that $w_n \in \overline{U_{2,R}} \backslash U_{2,\gamma}$ for each $n \in P_1$. Now Theorem 14.1.1 (with $X = E_2$, $G = \mathcal{K}_2$, $D = \left(\overline{U_{2,R}} \backslash U_{2,\gamma}\right) \cap C_2$ and note $d_2(u_n, \mathcal{K}_2\, u_n) = 0$ for each $n \in P_1$) guarantees that there exists a subsequence $P_2^\star$ of P_1 and $z_2 \in \left(\overline{U_{2,R}} \backslash U_{2,\gamma}\right) \cap C_2$ with $w_n \to z_2$ in E_2 as $n \to \infty$ in $P_2^\star$. Notice in particular that $\gamma \le |z_2|_2 \le R$ and $z_2 = z_1$ in E_1.

Let $N_2 = N_2^\star \setminus \{2\}$ and $P_2 = P_2^\star \setminus \{2\}$. Proceed inductively to obtain subsequences of integers

$$N_1^\star \supseteq N_2^\star \supseteq \ldots\ldots, \quad N_k^\star \subseteq \{k, k+1, \ldots.\}$$

and

$$P_1^\star \supseteq P_2^\star \supseteq \ldots\ldots, \quad P_k^\star \subseteq \{k, k+1, \ldots.\}$$

and $v_k \in \overline{U_{k,L}} \cap C_k$, $z_k \in \left(\overline{U_{k,R}} \backslash U_{k,\gamma}\right) \cap C_k$ with $u_n \to v_k$ in E_k as $n \to \infty$ in $N_k^\star$ and $w_n \to z_k$ in E_k as $n \to \infty$ in $P_k^\star$. Note $v_{k+1} = v_k$

in E_k and $z_{k+1} = z_k$ in E_k for $k = 1, 2, \ldots.$ Also let $N_k = N_k^\star \setminus \{k\}$ and $P_k = P_k^\star \setminus \{k\}$.

Let $y_1 = v_k$ in E_k and $y_2 = z_k$ in E_k. Notice y_1 and y_2 are well defined and $y_1, y_2 \in E_k$ for each $k = 1, 2 \ldots.$ Fix $k \in N$. Now $u_n \in F_n u_n$ in E_n for $n \in N_k$ and $u_n \to y_1$ in E_k as $n \to \infty$ in N_k (since $y_1 = v_k$ in E_k). This together with (14.3.8) implies $y_1 \in F y_1$. Similarly $w_n \in F_n w_n$ in E_n for $n \in N_k$ and $w_n \to y_2$ in E_k as $n \to \infty$ in P_k together with assumption (14.3.9) guarantees that $y_2 \in F y_2$. $\square$

We now use Theorem 14.3.1 to obtain the existence of two solutions to (14.1.1). Two results will be presented.

Theorem 14.3.2. *Let $1 \le p < \infty$ be a constant and $1 < q \le \infty$ the conjugate to p. Suppose (14.2.6)–(14.2.9) hold. In addition assume the following conditions are satisfied:*

$$\text{for each } t \in [0,\infty),\ k(t,s) \ge 0 \text{ for a.e. } s \in [0,\infty) \tag{14.3.10}$$

$$F : [0,\infty) \times [0,\infty) \to CK([0,\infty)) \tag{14.3.11}$$

$$\begin{cases} \text{there exists a function } \psi : [0,\infty) \to [0,\infty) \text{ continuous} \\ \text{and nondecreasing, a constant } A_0,\ 0 < A_0 \le 1, \text{ and a} \\ \phi \in L^p[0,\infty) \text{ with } A_0\,\phi(s)\,\psi(y) \le u \le \phi(s)\,\psi(y), \text{ for any} \\ u \in F(s,y), \text{ for all } y \in [0,\infty) \text{ and a.e. } s \in [0,\infty) \end{cases} \tag{14.3.12}$$

$$K_2 \equiv \inf_{t \in [0,\infty)} \int_0^1 \phi(s)\,K(t,s)\,ds > 0 \tag{14.3.13}$$

$$\begin{cases} \text{there exists a continuous function } \tau : (0,1) \to (0,\infty) \\ \text{such that for any } 0 < m < 1 \text{ and } u \ge 0 \text{ we have} \\ \psi(m\,u) \ge \tau(m)\,\psi(u) \end{cases} \tag{14.3.14}$$

$$\begin{cases} \exists\, 0 < M < 1 \text{ with } \frac{M}{\tau(M)} \le A_0\,\frac{K_2}{K_1} \text{ where} \\ K_1 \equiv \sup_{t \in [0,\infty)} \int_0^\infty \phi(s)\,K(t,s)\,ds \end{cases} \tag{14.3.15}$$

$$\exists\, r > 0 \text{ with } r > K_1\,\psi(r) \tag{14.3.16}$$

$$\exists\, L > 0,\ 0 < L < r \text{ with } L > K_1\,\psi(L) \tag{14.3.17}$$

(14.3.18) $$M\,r > L$$

and

(14.3.19) $$\exists\, R > r \;\; with \;\; R < A_0\, K_2\, \psi(M\,R).$$

Then (14.1.1) *has two solutions* $y_1,\, y_2 \in C[0,\infty)$ *with* $y_1 \geq 0,\; y_2 \geq 0$ *on* $[0,\infty)$ *and with*

$$|y_1|_n \leq L < |y_2|_n \leq R$$

for each $n \in N$.

PROOF: Let $n \in N$ and

$$F_n\, y(t) = \int_0^n K(t,s)\, F(s, y(s))\, ds \quad \text{for} \quad t \in [0,n],$$

here $y \in C[0,n]$. Let $E_n = (C[0,n], |\,.\,|_n)$ and

$$C_n = \{y \in C[0,n]: \; y(t) \geq M\,|y|_n \;\; \text{for} \;\; t \in [0,n]\}.$$

Also from (14.3.18) let $\gamma,\; L < \gamma < r$, be chosen so that

(14.3.20) $$M\,r \geq \gamma.$$

For $\rho = L,\, r$ or R let

$$U_{n,\rho} = \{x \in C[0,n]: \; |x|_n < \rho\}.$$

We wish to apply Theorem 14.3.1. Let $K_n : L^p[0,n] \to C[0,n]$ and $\mathcal{F}_n : C[0,n] \to L^p[0,n]$ be given by

$$K_n\, y(t) = \int_0^n K(t,s)\, y(s)\, ds$$

and

$$\mathcal{F}_n\,(y) = \{u \in L^p[0,n]: \; u(t) \in F(t, y(t)) \;\; \text{for a.e.} \;\; t \in [0,n]\}.$$

Now [67, 93 or 129] guarantees that $F_n = K_n \circ \mathcal{F}_n : \overline{U_{n,R}} \cap C_n \to CK(E_n)$ is upper semicontinuous for each $n \in N$. We claim that

(14.3.21) $$F_n : \overline{U_{n,R}} \cap C_n \to CK(C_n) \quad \text{for each} \quad n \in N.$$

If (14.3.21) holds then of course (14.3.2) is true. T o see (14.3.21) fix $n \in N$ and take $x \in \overline{U_{n,R}} \cap C_n$ with $y \in F_n\, x$. Then $x(t) \geq M\, |x|_n$ for $t \in [0, n]$ and there exists [54, 67, 93 or 146] $v \in \mathcal{F}_n\, x$ with

$$y(t) = \int_0^n K(t, s)\, v(s)\, ds \quad \text{for} \quad t \in [0, n].$$

Now (14.3.12) implies for $t \in [0, n]$ that

$$\begin{aligned} |y(t)| &\leq \int_0^n K(t, s)\, \phi(s)\, \psi(x(s))\, ds \\ &\leq \psi\left(|x|_n\right) \int_0^\infty K(t, s)\, \phi(s)\, ds \\ &\leq K_1\, \psi\left(|x|_n\right) \end{aligned}$$

and so

$$|y|_n \leq K_1\, \psi\left(|x|_n\right). \tag{14.3.22}$$

On the other hand (14.3.12) guarantees that $v(s) \geq A_0\, \phi(s)\, \psi(x(s))$ for a.e. $s \in [0, n]$. Now $x \in C_n$ (so $x(t) \geq M\, |x|_n$ for $t \in [0, n]$) together with (14.3.14), (14.3.22) and (14.3.15) gives for $t \in [0, n]$,

$$\begin{aligned} y(t) &\geq \int_0^n K(t, s)\, A_0\, \phi(s)\, \psi(x(s))\, ds \\ &\geq A_0\, \psi\left(M\, |x|_n\right) \int_0^n K(t, s)\, \phi(s)\, ds \\ &\geq A_0\, \tau(M)\, \psi\left(|x|_n\right) \int_0^n K(t, s)\, \phi(s)\, ds \\ &\geq A_0\, \tau(M)\, K_2\, \psi\left(|x|_n\right) \\ &\geq \frac{K_2}{K_1}\, A_0\, \tau(M)\, |y|_n \geq M\, |y|_n, \end{aligned}$$

so $y \in C_n$. Thus (14.3.21) (and consequently (14.3.2)) holds. To see (14.3.3) fix $n \in N$ and let $x \in \partial_{E_n} U_{n,L} \cap C_n$ with $y \in F_n\, x$. Then for $t \in [0, n]$,

$$\begin{aligned} |y(t)| &\leq \int_0^n K(t, s)\, \phi(s)\, \psi(x(s))\, ds \\ &\leq \psi\left(|x|_n\right) \int_0^n K(t, s)\, \phi(s)\, ds \\ &\leq K_1\, \psi(L), \end{aligned}$$

and this together with (14.3.17) yields

$$|y|_n \le \psi(L)\, K_1 \;<\; L \;=\; |x|_n.$$

Thus (14.3.3) holds. A similar argument shows (14.3.4) is true. To see (14.3.5) fix $n \in N$ and let $x \in \partial_{E_n} U_{n,R} \cap C_n$ with $y \in F_n\, x$. Note $|x|_n = R$ and $M\, R \le x(t) \le R$ for $t \in [0,n]$. Now for $t \in [0,n]$ we have

$$\begin{aligned} y(t) \;&\ge\; A_0 \int_0^n K(t,s)\, \phi(s)\, \psi(x(s))\, ds \\ &\ge\; A_0\, \psi(M\, R) \int_0^n K(t,s)\, \phi(s)\, ds \\ &\ge\; A_0\, K_2\, \psi(M\, R). \end{aligned}$$

This inequality together with (14.3.19) gives

$$y(t) \ge A_0\, K_2\, \psi(M\, R) \;>\; R \;=\; |x|_n \quad \text{for} \;\; t \in [0,n],$$

and so $|y|_n > |x|_n$ i.e. (14.3.5) is true. Essentially the same reasoning as in Theorem 14.2.2 guarantees that (14.3.6) holds. To see (14.3.7) fix $k \in N$ and take any subsequence $A \subseteq \{k, k+1, ...\}$. Now if $x \in C_n$, $n \in A$, is such that $R \ge |x|_n \ge r$ then $x(t) \ge M\, |x|_n \ge M\, r$ for $t \in [0,n]$ and this together with (14.3.20) gives $x(t) \ge \gamma$ for $t \in [0,n]$. Thus $|x|_k = \sup_{t\in[0,k]} |x(t)| \ge \gamma$, so (14.3.7) holds. Essentially the same reasoning as in Theorem 14.2.2 guarantees that (14.3.8) and (14.3.9) hold. Now apply Theorem 14.3.1. □

Remark 14.3.3. There is an obvious analogue of Theorem 14.3.2 for the single valued case (i.e. (14.2.20)). We leave the detail to the reader.

We now present a result when (14.3.13) may not hold.

Theorem 14.3.3. *Let* $1 \le p < \infty$ *be a constant and* $1 < q \le \infty$ *the conjugate to* p. *Suppose* (14.2.6)–(14.2.9), (14.3.10) *and* (14.3.11) *hold. In addition assume the following conditions are satisfied:*

$$(14.3.23) \qquad \begin{cases} \textit{there exists } 0 < M < 1,\ \kappa \in L^p[0,\infty) \textit{ and an} \\ \textit{interval } [a,b] \subseteq [0,\infty),\ a < b \textit{ with } K(t,s) \ge M\,\kappa(s) \\ \textit{for } t \in [a,b] \textit{ and a.e. } s \in [0,\infty) \end{cases}$$

$$(14.3.24) \qquad K(t,s) \le \kappa(s) \;\; \textit{for} \;\; t \in [0,\infty) \;\; \textit{and a.e.} \;\; s \in [0,\infty)$$

$$(14.3.25)\qquad \begin{cases} \text{there exists a function } \psi : [0,\infty) \to [0,\infty) \text{ continuous} \\ \text{and nondecreasing, and a } \phi \in L^p[0,\infty) \text{ with} \\ u \le \phi(s)\,\psi(y), \text{ for any } u \in F(s,y), \text{ for all} \\ y \in [0,\infty) \text{ and a.e. } s \in [0,\infty) \end{cases}$$

$$(14.3.26)\qquad \begin{cases} \text{there exists a } \tau \in L^p[a,b] \text{ with } u \ge \tau(s)\,\psi(y), \text{ for} \\ \text{any } u \in F(s,y), \text{ for all } y \in [0,\infty) \text{ and a.e. } s \in [a,b] \end{cases}$$

$$(14.3.27)\qquad \begin{cases} \exists\, r > 0 \text{ with } r > K_1\,\psi(r) \text{ where} \\ K_1 = \sup_{t\in[0,\infty)} \int_0^\infty K(t,s)\,\phi(s)\,ds \end{cases}$$

$$(14.3.28)\qquad \exists\, L,\ 0 < L < r \text{ with } L > K_1\,\psi(L)$$

$$(14.3.29)\qquad M\,r > L$$

and

$$(14.3.30)\qquad \begin{cases} \exists\, R > r, \text{ with } R < K_3\,\psi(M\,R) \text{ where} \\ K_3 = \sup_{t\in[0,b]} \int_a^b K(t,s)\,\tau(s)\,ds. \end{cases}$$

Choose $n_1 \in N$ *with* $n_1 \ge b$ *and let* $N_1 = \{n_1, n_1+1, ...\}$. *Then* (14.1.1) *has two solutions* $y_1, y_2 \in C[0,\infty)$ *with* $y_1 \ge 0$, $y_2 \ge 0$ *on* $[0,\infty)$ *and with*

$$|y_1|_n \le L < |y_2|_n \le R$$

for each $n \in N_1$.

PROOF: Let $n \in N_1$ with F_n, E_n as in Theorem 14.3.2. Let

$$C_n = \left\{ y \in C[0,n] : \ y(t) \ge 0 \ \text{ for } \ t \in [0,n] \ \text{ and } \ \min_{t\in[a,b]} y(t) \ge M\,|y|_n \right\}.$$

Also choose γ, $L < \gamma < r$, with

$$(14.3.31)\qquad M\,r \ge \gamma.$$

Let K_n, $\mathcal{F}_n$, $U_{n,\rho}$ ($\rho = r$, L or R) be as in Theorem 14.3.2. Now [67, 93 or 129] guarantees that $F_n = K_n \circ \mathcal{F}_n : \overline{U_{n,R}} \cap C_n \to CK(E_n)$ is upper semicontinuous for each $n \in N_1$. We claim that

$$(14.3.32)\qquad F_n : \overline{U_{n,R}} \cap C_n \to CK(C_n) \ \text{ for each } \ n \in N_1.$$

If (14.3.32) holds then of course (14.3.2) is true for $n \in N_1$. To see (14.3.32) fix $n \in N_1$ and take $x \in \overline{U_{n,R}} \cap C_n$ with $y \in F_n\, x$. Then $x(t) \geq M\, |x|_n$ for $t \in [a, b]$. Also there exists a $v \in \mathcal{F}_n\, x$ with

$$y(t) = \int_0^n K(t,s)\, v(s)\, ds \quad \text{for} \quad t \in [0, n].$$

This together with (14.3.24) yields

$$|y(t)| \leq \int_0^n \kappa(s)\, v(s)\, ds \quad \text{for} \quad t \in [0, n]$$

and so

$$|y|_n \leq \int_0^n \kappa(s)\, v(s)\, ds. \tag{14.3.33}$$

On the other hand (14.3.23) and (14.3.33) imply

$$\min_{t\in[a,b]} y(t) = \min_{t\in[a,b]} \int_0^n K(t,s)\, v(s)\, ds \geq M \int_0^n \kappa(s)\, v(s)\, ds \geq M\, |y|_n,$$

so $y \in C_n$. Thus (14.3.32) (and consequently (14.3.2) for $n \in N_1$) is true. Essentially the same reasoning as in Theorem 14.3.2 guarantees that (14.3.3) and (14.3.4) hold for $n \in N_1$. To see (14.3.5) with $n \in N_1$, fix $n \in N_1$ and let $x \in \partial_{E_n} U_{n,R} \cap C_n$ with $y \in F_n\, x$. Note $|x|_n = R$ and $M\, R \leq x(s)$ for $s \in [a, b]$. Now there exists a $v \in \mathcal{F}_n\, x$ with

$$y(t) = \int_0^n K(t,s)\, v(s)\, ds \quad \text{for} \quad t \in [0, n].$$

Notice (14.3.26) and (14.3.30) yield

$$\begin{aligned} |y|_n = \sup_{t\in[0,n]} |y(t)| \quad &= \quad \sup_{t\in[0,n]} \int_0^n K(t,s)\, v(s)\, ds \\ &\geq \quad \sup_{t\in[0,n]} \int_a^b K(t,s)\, v(s)\, ds \\ &\geq \quad \sup_{t\in[0,n]} \int_a^b K(t,s)\, \tau(s)\, \psi(x(s))\, ds \\ &\geq \quad \psi(M\, R) \sup_{t\in[0,n]} \int_a^b K(t,s)\, \tau(s)\, ds \\ &\geq \quad K_3\, \psi(M\, R) \; > \; R \; = \; |x|_n, \end{aligned}$$

so (14.3.5) is true for $n \in N_1$. Essentially the same reasoning as in Theorem 14.3.2 guarantees that (14.3.6) is true $n \in N_1$. To see (14.3.7) for $n \in N_1$, fix $k \in N_1$ and take any subsequence $A \subseteq \{k, k+1, ...\}$. Now if $x \in C_n$, $n \in A$, is such that $R \geq |x|_n \geq r$ then $x(t) \geq M\,|x|_n \geq M\,r$ for $t \in [a,b]$ and this together with (14.3.31) gives $\min_{t\in[a,b]} x(t) \geq \gamma$. Thus $|x|_k = \sup_{t\in[0,k]} |x(t)| \geq \gamma$, so (14.3.7) holds for $n \in N_1$. Essentially the same reasoning as in Theorem 14.2.2 guarantees that (14.3.8) and (14.3.9) hold (with N replaced by N_1). Now apply Theorem 14.3.1 (with N replaced by N_1). □

Remark 14.3.4. Again there is an obvious analogue (the detail is left to the reader) of Theorem 14.3.3 for the single valued case (i.e. (14.2.20)).

CHAPTER 15

THE SOLUTION SET OF INTEGRAL INCLUSIONS ON THE HALF LINE

15.1. Introduction.

This chapter discusses the topological structure of the solution set of the Volterra integral inclusion

$$y(t) \in \int_0^t k(t,s)\, F(s, y(s))\, ds \quad \text{for} \quad t \in [0, \infty). \tag{15.1.1}$$

Throughout this chapter

$$k : [0, \infty) \times [0, t] \to \mathbf{R} \quad \text{and} \quad F : [0, \infty) \times \mathbf{R}^n \to CK(\mathbf{R}^n);$$

here $CK(\mathbf{R}^n)$ denotes the family of nonempty, compact, convex subsets of $\mathbf{R}^n$. Our chapter was motivated by ideas presented by Andres, Gabor and Gorniewicz [23], and Agarwal and O'Regan [4]. For the remainder of this introduction we gather together some definitions and facts from the literature. Let X and Y be two metric spaces and $F : [0, \infty) \times X \to 2^Y$ (here 2^Y denotes the family of nonempty subsets of Y). F is **almost uppersemicontinuous** (almost u.s.c.) if for every $\epsilon > 0$ there exists a measurable set $A_\epsilon \subset [0, \infty)$ such that $m([0, \infty) \backslash A_\epsilon) < \epsilon$ and $F|_{A_\epsilon \times X}$ is u.s.c. (here m stands for the Lebesgue measure).

Theorem 15.1.1. [23] *Let X be a separable metric space and suppose $F : [0, \infty) \times X \to CK(\mathbf{R}^n)$ is a Carathéodory map (By this we mean: (i). $t \mapsto F(t, y)$ is measurable for all $y \in \mathbf{R}^n$ and (ii). $y \mapsto F(t, y)$ is u.s.c. for all $t \in [0, \infty)$). Then there exists an almost u.s.c. map $\Phi : [0, \infty) \times X \to CK(\mathbf{R}^n)$ such that*

(a). $\Phi(t, x) \subset F(t, x)$ for every $(t, x) \in [0, \infty) \times X$ and

(b). if $\triangle \subset [0, \infty)$ is measurable, $u : \triangle \to \mathbf{R}^n$ and $v : \triangle \to X$ are measurable maps and $u(t) \in F(t, v(t))$ for almost all $t \in \triangle$, then $u(t) \in \Phi(t, v(t))$ for almost all $t \in \triangle$.

A single valued map $f : [0,\infty) \times X \to Y$ is said to be **measurable–locally Lipschitz** if for every $x \in X$ there exists a neighborhood U_x of x in X and a $L_x \in L^1[0,\infty)$ (here $L_x : [0,\infty) \to [0,\infty)$) such that

$$|f(t,x_1) - f(t,x_2)| \le L_x(t)\,|x_1 - x_2|$$

for every $t \in [0,\infty)$ and $x_1,\ x_2 \in U_x$, and also $f(\,.\,,x)$ is measurable for every $x \in X$. A map $G : [0,\infty) \times \mathbf{R}^n \to 2^{\mathbf{R}^n}$ is said to be **integrably bounded** if there exists a $\mu \in L^1[0,\infty)$ (here $\mu : [0,\infty) \to [0,\infty)$) such that $|y| \le \mu(t)$ for every $x \in \mathbf{R}^n$, $t \in [0,\infty)$ and $y \in G(t,x)$ [and we write $|G(t,x)| \le \mu(t)$ for every $x \in \mathbf{R}^n$, $t \in [0,\infty)$ for convenience].

Let X and Y be metric spaces with $F : X \to 2^Y$ and $f : X \to Y$. We say f is a **selection** of F (and we write $f \subset F$) if $f(x) \in F(x)$ for every $x \in X$.

Definition 15.1.1. $F : [0,\infty) \times X \to 2^Y$ is Ca–selectionable if there exists a Carathéodory selection of F.

Definition 15.1.2. $F : [0,\infty) \times X \to 2^Y$ is mLL–selectionable if there exists a measurable–locally Lipschitz selection of F.

Theorem 15.1.2. [23] *Let E_1 and E_2 be two separable Banach spaces and $\Phi : [0,\infty) \times E_1 \to CK(E_2)$ be an almost u.s.c. map. Then Φ is σ-Ca–selectionable i.e.*

$$\Phi(t,x) = \cap_{k=1}^{\infty} \Phi_k(t,x) \quad \textit{and} \quad \Phi_{k+1}(t,x) \subset \Phi_k(t,x)$$

for almost all $t \in [0,\infty)$ and all $x \in \mathbf{R}^n$ and Φ_k is Ca-selectionable for every $k = \{1,2,....\}$. The maps $\Phi_k : [0,\infty) \times E_1 \to 2^{E_2}$ are almost u.s.c. and

$$\Phi_k(t,x) \subset \overline{conv}\ (\cup_{x \in E_1} \Phi(t,x)) \quad \textit{for all} \ \ (t,y) \in [0,\infty) \times E_1.$$

In addition, if Φ is integrably bounded then Φ is σ–mLL–selectionable i.e. it is an intersection of a decreasing sequence of mLL–selectionable mappings.

Definition 15.1.3. A nonempty set A is **contractible** provided there exists $x_0 \in A$ and a homotopy $H : A \times [0,1] \to A$ such that $H(x,1) = x$ and $H(x,0) = x_0$ for every $x \in A$.

Definition 15.1.4. A set A is called an R_δ set provided there exists a decreasing sequence $\{A_n\}_1^{\infty}$ of nonempty compact, contractible sets such that $A = \cap \{A_n :\ n = 1,2,...\}$.

Finally we state the following well known existence–uniqueness result from the literature (see [137 Theorem 5.2.5] or [94] for existence and [2] for uniqueness) for the integral equation

$$y(t) = \int_0^t k(t,s)\, f(s, y(s))\, ds, \quad t \in [0,\infty). \tag{15.1.2}$$

Theorem 15.1.3. *Let $k : [0,\infty) \times [0,t] \to \mathbf{R}$ and $f : [0,\infty) \times \mathbf{R}^n \to \mathbf{R}^n$. Suppose the following conditions hold:*

$$t \mapsto f(t,y) \ \text{ is measurable for every } \ y \in \mathbf{R}^n \tag{15.1.3}$$

$$\begin{cases} \text{there exists } \ \mu \in L^1[0,\infty) \ \text{ with } \ |f(t,y)| \le \mu(t) \\ \text{for a.e. } \ t \in [0,\infty) \ \text{ and all } \ y \in \mathbf{R}^n \end{cases} \tag{15.1.4}$$

$$\begin{cases} k : [0,\infty) \times [0,t] \to \mathbf{R} \ \text{ is continuous and} \\ k(t) = \sup_{s\in[0,t]} |k(t,s)| \ \text{ is bounded on } \ [0,\infty) \end{cases} \tag{15.1.5}$$

and

$$\begin{cases} \text{for every } \ x \in \mathbf{R}^n \ \text{ there exists a neighborhood} \\ U_x \ \text{ of } \ x \ \text{ in } \ \mathbf{R}^n \ \text{ and a } \ L_x \in L^1[0,\infty) \\ \text{(here } \ L_x : [0,\infty) \to [0,\infty)\text{) with} \\ |f(t,x_1) - f(t,x_2)| \le L_x(t)\, |x_1 - x_2| \\ \text{for every } \ t \in [0,\infty) \ \text{ and } \ x_1,\, x_2 \in U_x. \end{cases} \tag{15.1.6}$$

Then there exists a unique solution $y \in C[0,\infty)$ (in fact $y \in BC[0,\infty)$ because of (15.1.4) *and* (15.1.5)*) to* (15.1.2).

Remark. We could replace (15.1.5) in Theorem 15.1.3 (and throughout Section 15.2) by a less restrictive assumption (see [137 Theorem 5.2.5]).

15.2. Solution set.

This section is concerned with the integral inclusion

$$y(t) \in \int_0^t k(t,s)\, F(s, y(s))\, ds \quad \text{for } \ t \in [0,\infty), \tag{15.2.1}$$

where $k : [0,\infty) \times [0,t] \to \mathbf{R}$ and $F : [0,\infty) \times \mathbf{R}^n \to 2^{\mathbf{R}^n}$. Let $S(F)$ denote the solution set of (15.2.1).

Remark. In this section by a solution to (15.2.1) we mean a $y \in BC[0,\infty)$ which satisfies (15.2.1).

We begin by establishing a result for the case when F is integrably bounded. This result together with a trick involving the Urysohn function will enable us to discuss the case when F satisfies a more general growth condition.

Theorem 15.2.1. *Let* $k : [0,\infty) \times [0,t] \to \mathbf{R}$ *and* $F : [0,\infty) \times \mathbf{R}^n \to 2^{\mathbf{R}^n}$ *and suppose* (15.1.5) *holds. In addition assume the following conditions are satisfied:*

$$\text{(15.2.2)} \quad \begin{cases} F \text{ is integrably bounded i.e. there exists a } \mu \in L^1[0,\infty) \\ \text{such that } |y| \le \mu(t) \text{ for every } x \in \mathbf{R}^n,\ t \in [0,\infty) \\ \text{and } y \in F(t,x) \end{cases}$$

and

$$\text{(15.2.3)} \quad \begin{cases} F : [0,\infty) \times \mathbf{R}^n \to 2^{\mathbf{R}^n} \text{ is } mLL - \text{selectionable i.e.} \\ \text{there exists } f : [0,\infty) \times \mathbf{R}^n \to \mathbf{R}^n \text{ with } f \subset F \text{ such} \\ \text{that for every } x \in \mathbf{R}^n \text{ there exists a neighborhood} \\ U_x \text{ of } x \text{ in } \mathbf{R}^n \text{ and a } L_x \in L^1[0,\infty) \\ \text{(here } L_x : [0,\infty) \to [0,\infty)\text{) with} \\ |f(t,x_1) - f(t,x_2)| \le L_x(t)\,|x_1 - x_2| \\ \text{for every } t \in [0,\infty) \text{ and } x_1,\ x_2 \in U_x, \text{ and} \\ \text{also } f(\,.\,,x) \text{ is measurable for every } x \in \mathbf{R}^n. \end{cases}$$

Then $S(F)$ *is contractible.*

PROOF: Let f be a measurable–locally Lipschitz selection of F. Now Theorem 15.1.3 implies

$$y(t) = u_0 + \int_{t_0}^{t} k(t,s)\, f(s,y(s))\, ds, \quad t \in [0,\infty)$$

has exactly one solution $y[t_0;u_0]$ for every $u_0 \in \mathbf{R}^n$ and $t_0 \in [0,\infty)$. Let

$$A = S(F).$$

If we define a homotopy $H : A \times [0,1] \to A$ such that

$$H(x,s) = \begin{cases} x \text{ when } s = 1 \text{ and } x \in A \\ y[0;0] \text{ when } s = 0 \text{ and } x \in A \end{cases}$$

then we are finished. Put $\alpha(s) = \tan\left(\frac{\pi}{2}\, s\right)$ and

$$H(x,s)(t) = \begin{cases} x(t), \quad 0 \le t \le \alpha(s), \ s < 1 \\ y[\alpha(s); x(\alpha(s))]\,(t), \quad \alpha(s) \le t < \infty, \ s < 1 \\ x(t), \quad 0 \le t < \infty, \ s = 1. \end{cases}$$

Note $y[\alpha(s); x(\alpha(s))]$, for fixed $s < 1$ and fixed $x \in A$, is the unique solution of

$$y(t) = x(\alpha(s)) + \int_{\alpha(s)}^{t} k(t,z)\, f(z, y(z))\, dz, \quad t \in [0, \infty).$$

We have immediately that $H(x,1) = x$ for $x \in A$ and $H(x,0) = y[0;0]$ (note $H(x,0)(t) = y[0; x(0)]\,(t) = y[0;0]\,(t)$, $0 \le t < \infty$ since $x \in A$). Notice also that

$$H : A \times [0,1] \to A.$$

To see this fix $x \in A$ and let $s < 1$ be fixed. Then with

$$\eta(t) = \begin{cases} x(t), & 0 \le t \le \alpha(s) \\ y[\alpha(s); x(\alpha(s))]\,(t), & \alpha(s) \le t < \infty \end{cases}$$

it is easy to check that $\eta(t) \in \int_0^t k(t,z)\, F(z, \eta(z))\, dz$ (consider $0 \le t \le \alpha(s)$ and $\alpha(s) \le t < \infty$ seperately) since $x \in A$ and $\eta(\alpha(s)) = x(\alpha(s))$. This H is a continuous homotopy contracting $S(F)$ to the point $y[0;0]$. □

Suppose $\Phi : [0,\infty) \times \mathbf{R}^n \to 2^{\mathbf{R}^n}$ is the intersection of the decreasing sequence $\Phi_k : [0,\infty) \times \mathbf{R}^n \to 2^{\mathbf{R}^n}$ i.e.

$$\Phi(t,x) = \cap_{k=1}^{\infty} \Phi_k(t,x) \quad \text{and} \quad \Phi_{k+1}(t,x) \subset \Phi_k(t,x)$$

for almost all $t \in [0,\infty)$ and all $x \in \mathbf{R}^n$. If $x \in \cap_{k=1}^{\infty} S(\Phi_k)$ then for **each** $k \in \{1, 2, ...\}$ and each $t \in [0,\infty)$ we have

$$x(t) \in \int_0^t k(t,s)\, \Phi_k(s, x(s))\, ds.$$

Consequently $x \in S(\Phi)$ and so

$$S(\Phi) = \cap_{k=1}^{\infty} S(\Phi_k). \tag{15.2.4}$$

Theorem 15.2.2. *Let* $k : [0,\infty) \times [0,t] \to \mathbf{R}$ *and* $F : [0,\infty) \times \mathbf{R}^n \to CK(\mathbf{R}^n)$ *and assume*

$$y \mapsto F(t,y) \text{ is u.s.c. for a.e. } t \in [0,\infty) \tag{15.2.5}$$

and

$$t \mapsto F(t,y) \text{ is measurable for every } y \in \mathbf{R}^n \tag{15.2.6}$$

hold. In addition suppose (15.1.5) *and* (15.2.2) *are satisfied. Then* $S(F)$ *is an* R_δ *set.*

PROOF: From Theorem 15.1.1 there exists an almost u.s.c. map $\Phi : [0,\infty) \times \mathbf{R}^n \to CK(\mathbf{R}^n)$ with

$$S(\Phi) = S(F).$$

Also Theorem 15.1.2 implies Φ is σ–mLL–selectionable so we can find a sequence of maps Φ_k, $k \in \{1,2,...\}$, as described in Theorem 15.1.2. Notice Theorem 15.2.1 and (15.2.4) implies that $S(\Phi)$ is the interection of a decreasing sequence of contractible spaces (Of course Theorem 15.1.3 guarantees that $S(\Phi_k)$ is nonempty for each $k \in \{1,2,...\}$). Finally it is easy to see from the Arzela–Ascoli Theorem that $S(\Phi_k)$ is compact for every $k \in \{1,2,...\}$. Consequently $S(\Phi)$ is an R_δ set. □

We now discuss the more general situation. Assume for the remainder of this section that (15.1.5), (15.2.5) and (15.2.6) hold. In addition assume the following are satisfied:

$$\text{(15.2.7)} \qquad \begin{cases} \text{for each } r > 0 \text{ there exists a function } h_r \in L^1[0,\infty) \\ \text{with } |F(t,x)| \le h_r(t) \text{ for } t \in [0,\infty) \text{ and every} \\ x \in \mathbf{R}^n \text{ with } |x| \le r \end{cases}$$

and

$$\text{(15.2.8)} \qquad \begin{cases} \text{there exists } M > 0 \text{ with } |y|_\infty < M \text{ for any} \\ \text{possible solution } y \in BC[0,\infty) \text{ to (15.2.1);} \\ \text{here } |y|_\infty = \sup_{t\in[0,\infty)} |y(t)|. \end{cases}$$

Let $\epsilon > 0$ be given and let $\tau_\epsilon : \mathbf{R}^n \to [0,1]$ be the Urysohn function for

$$\left(\overline{B}(0,M)\,,\ \mathbf{R}^n \backslash B(0,M+\epsilon)\right)$$

such that

$$\tau_\epsilon(x) = 1 \text{ if } |x| \le M \text{ and } \tau_\epsilon(x) = 0 \text{ if } |x| \ge M + \epsilon.$$

Let $F^\star(t,x) = \tau_\epsilon(x)\,F(t,x)$ and consider the inclusion

$$\text{(15.2.9)} \qquad y(t) \in \int_0^t k(t,s)\,F^\star(s,y(s))\,ds \quad \text{for } t \in [0,\infty).$$

Let $S_\epsilon(F^\star)$ denote the solution set of (15.2.9).

Theorem 15.2.3. *Let* $k : [0,\infty) \times [0,t] \to \mathbf{R}$ *and* $F : [0,\infty) \times \mathbf{R}^n \to CK(\mathbf{R}^n)$ *and suppose* (15.1.5), (15.2.5), (15.2.6), (15.2.7) *and* (15.2.8) *hold. Let* $\epsilon > 0$ *be given and assume*

$$(15.2.10) \qquad \begin{cases} |w|_\infty < M \ \text{ for any possible solution} \\ w \in BC[0,\infty) \ \text{ to (15.2.9).} \end{cases}$$

Then $S(F)$ *is an* R_δ *set.*

PROOF: Notice (15.2.8) and (15.2.10) imply

$$S(F) = S_\epsilon(F^\star).$$

Its easy to see that $F^\star : [0,\infty) \times \mathbf{R}^n \to CK(\mathbf{R}^n)$ satisfies (15.2.5) and (15.2.6) (with F replaced by $F^\star$). Also (15.2.7) and the definition of τ_ϵ implies $F^\star$ is integrably bounded. Now Theorem 15.2.2 implies that $S_\epsilon(F^\star)$ is an R_δ set. □

We now illustrate the generality of Theorem 2.3 with the following very applicable result.

Theorem 15.2.4. *Let* $k : [0,\infty) \times [0,t] \to \mathbf{R}$ *and* $F : [0,\infty) \times \mathbf{R}^n \to CK(\mathbf{R}^n)$ *and assume* (15.1.5), (15.2.5), (15.2.6) *and* (15.2.7) *hold. In addition suppose the following conditions are satisfied:*

$$(15.2.11) \qquad \begin{cases} \text{there exists } \alpha \in L^1[0,\infty) \text{ and } g : [0,\infty) \to (0,\infty) \\ \text{a nondecreasing continuous function such that} \\ |k(t,s)\, F(s,u)| \le \alpha(s)\, g(|u|) \ \text{ for a.e } s \in [0,t], \\ \text{a.e. } t \in [0,\infty), \ \text{ and all } u \in \mathbf{R}^n \end{cases}$$

and

$$(15.2.12) \qquad \int_0^\infty \alpha(s)\, ds < \int_0^\infty \frac{dx}{g(x)}.$$

Then $S(F)$ *is an* R_δ *set.*

PROOF: Let $\epsilon > 0$ be given,

$$M_0 = I^{-1}\left(\int_0^\infty \alpha(s)\, ds\right), \quad \text{where } I(z) = \int_0^z \frac{dx}{g(x)}, \quad \text{and } M = M_0 + 1.$$

We will show any possible solution u of (15.2.1) satisfies $|u|_\infty \le M_0$ and any possible solution y of (15.2.9) satisfies $|y|_\infty \le M_0$. If this is true then Theorem 15.2.3 guarantees the result.

Suppose u is any possible solution of (15.2.1). Then

$$|u(t)| \leq \int_0^t \alpha(s)\, g(|u(s)|)\, ds \equiv w(t) \quad \text{for } t \in [0,\infty).$$

Now $w'(t) = \alpha(t)\, g(|u(t)|) \leq \alpha(t)\, g(w(t))$ a.e. on $[0,\infty)$ and so

$$\int_0^{w(x)} \frac{ds}{g(s)} = \int_0^x \frac{w'(z)}{g(w(z))}\, dz \leq \int_0^x \alpha(z)\, dz \leq \int_0^\infty \alpha(z)\, dz$$

for $x \in [0,\infty)$. Thus $w(x) \leq M_0$ for $x \in [0,\infty)$ and so $|u(x)| \leq M_0$ for $x \in [0,\infty)$.

Next let y be any possible solution of (15.2.9). For $t \in [0,\infty)$, since $\tau_\epsilon : \mathbf{R}^n \to [0,1]$, we have

$$|y(t)| \leq \int_0^t \alpha(s)\, g(|y(s)|)\, ds$$

and once again we have $|y(x)| \leq M_0$ for $x \in [0,\infty)$. □

Remark. Of course the ideas and results in this chapter apply to the Volterra integral inclusion

$$y(t) \in \int_0^t k(t,s)\, F(s, y(s))\, ds \quad \text{for } t \in [0,T];$$

here $T > 0$ is fixed.

REFERENCES

[1]. R.A. Adams, Sobolev Spaces, *Academic Press*, 1975.

[2]. R.P. Agarwal and V. Lakshmikantham, Uniqueness and Nonuniqueness Criteria for Ordinary Differential Equations, *World Scientific*, Singapore, 1993.

[3]. R.P. Agarwal, M. Meehan and D. O'Regan, Positive L^p and continuous solutions for Fredholm integral inclusions, *Set Valued Mappings and Applications in Nonlinear Analysis* (edited by R.P. Agarwal and D. O'Regan), *Gordon and Breach Publishers, to appear.*

[4]. R.P. Agarwal and D. O'Regan, The solution set of integral inclusions on the half line, *Integral and Integrodifferential Equations: Theory, Methods and Applications* (edited by R.P. Agarwal and D. O'Regan), *Gordon and Breach Publishers*, 2000, 1–7.

[5]. R.P. Agarwal and D. O'Regan, Periodic solutions of integrodifferential inclusions, *Integral and Integrodifferential Equations: Theory, Methods and Applications* (edited by R.P. Agarwal and D. O'Regan), *Gordon and Breach Publishers*, 2000, 9–19.

[6]. R.P. Agarwal and D. O'Regan, Existence criteria for operator inclusions in abstract spaces, *J. Comput. Appl. Math.*, **113**(2000), no.1–2, 183–193.

[7]. R.P. Agarwal and D. O'Regan, A note on the existence of multiple fixed points for multivalued maps with applications, *J. Differential Equations*, **160**(2000), no.2, 389–403.

[8]. R.P. Agarwal and D. O'Regan, Lidstone continuous and discrete boundary value problems, *Mem. Differential Equations Math. Phys.*, **19**(2000), 107–125.

[9]. R.P. Agarwal and D. O'Regan, Multiple solutions for higher order difference equations, *Comput. Math. Appl.*, **37**(1999), no.9, 39–48.

[10]. R.P. Agarwal and D. O'Regan, Nonlinear operator inclusions on the half line, *Math. Comput. Modelling*, **32**(2000), 1305–1310.

[11]. R.P. Agarwal and D. O'Regan, Existence of three solutions to integral and discrete equations via the Leggett Williams fixed point theorem, *Rocky Mountain J. Math., to appear.*

[12]. R.P. Agarwal and D. O'Regan, Fixed points in Fréchet spaces and variational inequalities, *Nonlinear Anal., to appear.*

[13]. R.P. Agarwal and D. O'Regan, A fixed point theorem of Leggett Williams type with applications to single and multivalued maps, *to appear.*

[14]. R.P. Agarwal and D. O'Regan, Periodic solutions to nonlinear integral equations on the infinite interval modelling infectious disease, *Nonlinear Analysis,* **40**(2000), 21–35.

[15]. R.P. Agarwal and D. O'Regan, Existence theory for single and multiple periodic and almost periodic solutions of nonlinear integral equations, *Ukranian Math. J., to appear.*

[16]. R.P. Agarwal and D. O'Regan, A generalization of the Petryshyn Leggett Williams fixed point theorem with applications to integral inclusions, *Appl. Math. Lett., to appear.*

[17]. R.P. Agarwal and D. O'Regan, Multivalued nonlinear equations on the half line: a fixed point approach, *to appear.*

[18]. R.P. Agarwal, D. O'Regan and P.J.Y. Wong, Positive solutions of differential, difference and integral equations, *Kluwer Acad. Publ.*, Dordrecht, 1998.

[19]. R.R. Akhmerov, M.I. Kamenskii, A.S. Potapov, A.E. Rodkina and B.N. Sadovskii, Measures of Noncompactness and Condensing Operators, *Birkhäuser*, Basel, 1992.

[20]. C. Aliprantis and K.C. Border, Infinite Dimensional Analysis, *Springer Verlag*, Berlin, 1985.

[21]. H. Amann, On the number of solutions of nonlinear equations in ordered Banach spaces, *J. Functional Anal.,* **14**(1973), 346–384.

[22]. D. Anderson, R.I. Avery and A.C. Peterson, Three positive solutions to a discrete focal boundary value problem, *J. Comput. Appl. Math,* **88**(1998), 103–118.

[23]. J. Andres, G. Gabor and L. Gorniewicz, Boundary value problems on infinite intervals, *Trans. Amer. Math. Soc., to appear.*

[24]. J. Appel, E. De Pascale, H.T. Nguyen and P.P. Zabreiko, Nonlinear integral inclusions of Hammerstein type, *Topol. Methods Nonlinear Anal.,* **5**(1995), 111–124.

[25]. N. Aronszajn, Le correspondant topologique de l'unicité dans la

théorie des équations différentielles, *Annals of Math.*, **43**(1942), 730–738.

[26]. J.P. Aubin and A. Cellina, Differential Inclusions, *Springer-Verlag*, Berlin-Heidelberg-New York-Tokyo, 1984.

[27]. R. Bader and W. Kryszewski, Fixed point index for compositions of set valued maps with proximally ∞–connected values on arbitrary ANR's, *Set Valued Anal.*, **2**(1994), 459–480.

[28]. J. Banas, Integrable solutions of Hammerstein and Urysohn integral equations, *J. Austral. Math. Soc.*, **46**(1989), 61–68.

[29]. J. Banas and W.G. El–Sayed, Measures of noncompactness and solvability of an integral equation in the class of functions of locally bounded variation, *J. Math. Anal. Appl.*, **167**(1992), 133–151.

[30]. J. Banas and K. Goebel, Measures of Noncompactness in Banach Spaces, *Marcel Dekker*, New York, 1980.

[31]. J. Banas and J. Rivero, On the measures of weak noncompactness, *Ann. Math. Pura Appl.*, **151**(1988), 213–224.

[32]. D.M. Bedivan and D. O'Regan, The set of solutions for abstract Volterra operators in $L^p([0,a],\mathbf{R}^m)$, *Appl. Math. Lett.*, **12**(1999), no.8, 7–11.

[33]. D.M. Bedivan and D. O'Regan, Fixed point sets for abstract Volterra operators on Fréchet spaces, *Appl. Anal.*, **76**(2000), 131–152.

[34]. H. Brezis, Analyse Fonctionnelle, Theorie et Applications, *Masson*, Paris, 1993.

[35]. F. Browder and C. Gupta, Topological degree and nonlinear mappings of analytic type in Banach spaces, *J. Math. Anal. Appl.*, **26**(1969), 390–402.

[36]. P.J. Bushell, On a class of Volterra and Fredholm nonlinear integral equations, *Math. Proc. Camb. Phil. Soc.*, **79**(1976), 329–335.

[37]. P.J. Bushell and W. Okrasinski, Nonlinear Volterra integral equations with convolution kernel, *J. London Math. Soc.*, **41**(1990), no.2, 503–510.

[38]. P.J. Bushell and W. Okrasinski, Uniqueness of solutions for a class of nonlinear Volterra integral equations with convolution kernel, *Math. Proc. Camb. Phil. Soc.*, **106**(1989), 547–552.

[39]. T. Cardinali and N.S. Papageorgiou, Hammerstein integral inclusions

in reflexive Banach spaces, *to appear.*

[40]. M. Cecchi, M. Marini and P. Zecca, Existence of bounded solutions for multivalued differential systems, *Nonlinear Anal.*, **9**(1985), 775–786.

[41]. A. Cellina, A. Fryszkowski and T. Rzezuchowski, Upper semicontinuity of Nemytskij operators, *Ann. Mat. Pura Appl. (4)*, **160**(1991), 321–330.

[42]. M. Cichon, On bounded weak solutions of a nonlinear differential equation in a Banach space, *Funct. Approx. Comment. Math.*, **21**(1992), 27–35.

[43]. C. Coddington, An Introduction to Ordinary Differential Equations, *Prentice-Hall, Inc.*, New Jersey, 1961.

[44]. J.B. Conway, A Course in Functional Analysis, *Springer Verlag*, Berlin, 1990.

[45]. K.L. Cooke and J.L. Kaplan, A periodicity threshold theorem for epidemics and population growth, *Mat. Biosc.*, **31**(1976), 87–104.

[46]. C. Corduneanu, Functional differential equations and their applications, In volume: Ordinary Differential Equations and their Applications, Bologna, (1995), 61–81.

[47]. C. Corduneanu, Integral Equations and Applications, *Cambridge Univ. Press*, New York, 1990.

[48]. C. Corduneanu, Integral Equations and Stability of Feedback Systems, *Academic Press,* 1973.

[49]. C. Corduneanu, Admissibility with respect to an integral operator and applications, *SIAM Studies Appl. Math.,* **5**(1969), 55–63.

[50]. C. Corduneanu, Almost Periodic Functions, *Interscience,* New York, 1968.

[51]. C. Corduneanu, Abstract Volterra equations: a survey, *Math. Comput. Modelling, to appear.*

[52]. J.F. Couchouron and M. Kamenski, A unified topological point of view for integro-differential inclusions, in *Differential Inclusions and Optimal Control*, Lecture Notes in Nonlinear Analysis, Vol. 2, 1998, 123–137.

[53]. F.S. De Blasi, On a property of the unit sphere in Banach spaces, *Bull. Math. Soc. Sci. Math. Roum.*, **21**(1977), 259–262.

[54]. K. Deimling, Multivalued Differential Equations, *Walter de Gruyter*, Berlin-New York, 1992.

[55]. K. Deimling, Nonlinear Functional Analysis, *Springer Verlag*, Berlin, 1985.

[56]. J. Dugundji and A. Granas, Fixed Point Theory, *Monografie Mathematyczne, PWN*, Warsaw, 1982.

[57]. N. Dunford and J.T. Schwartz, Linear Operators, *Interscience/Wiley*, New York, 1958.

[58]. R.E. Edwards, Functional Analysis, Theory and Applications, *Holt, Rinehart and Winston*, New York, 1965.

[59]. G. Emmanuele, Integrable solutions of Hammerstein integral equations, *Appl. Anal.,* **50**(1993), no.3–4, 277–284.

[60]. G. Emmanuele, Integrable solutions of a functional–integral equation, *J. Integral Equations Appl.,* **4**(1992), no.1, 89–94.

[61]. G. Emmanuele, Measures of weak noncompactness and fixed point theorems, *Bull. Math. Soc. Sci. Math. R. S. Roumaine,* **25**(1981), no.4, 353–358.

[62]. R. Engelking, General Topology, *Heldermann Verlag*, Berlin, 1989.

[63]. L.H. Erbe, S. Hu and H. Wang, Multiple positive solutions of some boundary value problems, *J. Math. Anal. Appl.,* **184**(1994), 640–648.

[64]. L.H. Erbe and H. Wang, On the existence of positive solutions of ordinary differential equations, *Proc. Amer. Math. Soc.,* **120**(1994), 743–748.

[65]. P.M. Fitzpatrick and W.V. Petryshyn, Fixed point theorems for multivalued noncompact acyclic mappings, *Pacific J. Math.,* **54**(1974), no.2, 17–23.

[66]. K. Floret, Weakly Compact Sets, *Lecture notes in mathematics,* **801**, Springer Verlag, Berlin, 1980.

[67]. M. Frigon, Théorèmes d'existence de solutions d'inclusions differentielles, in *Topological Methods in Differential Equations and Inclusions* (A. Granas and M. Frigon eds.), NATO ASI Series C, Vol. 472, *Kluwer Academic Publishers*, Dordrecht-Boston-London, 1995, 51–87.

[68]. M. Furi and P. Pera, A continuation method on locally convex spaces

and applications to ordinary differential equations on noncompact intervals, *Ann. Polon. Math.*, **47**(1987), 331–346.

[69]. R.F. Geitz, Pettis integration, *Proc. Amer. Math. Soc.*, **82**(1981), 81–86.

[70]. L. Gorniewicz, Topological approach to differential inclusions, *Topological methods in differential inclusions*, (edited by A. Granas and M. Frigon), *Kluwer Acad. Publ.*: Math. and Physics Sciences **472**(1995), 129–190.

[71]. L. Gorniewicz and S. Slosarski, Topological essentiality and differential inclusions, *Bull. Austral. Math. Soc.*, **45**(1992), 177–193.

[72]. A. Granas, Sur la méthode de continuité de Poincare, *C.R. Acad. Sci. Paris*, **282**(1976), 983–985.

[73]. G. Gripenberg, On the uniqueness of solutions of Volterra equations, *J. Integral Equations Appl.*, **2**(1990), no.3, 421–430.

[74]. G. Gripenberg, Unique solutions of some Volterra integral equations, *Math. Scand.*, **48**(1981), 59–67.

[75]. G. Gripenberg, S.O. Londen and O. Staffans, Volterra Integral and Functional Equations, *Encyclopedia of Mathematics and its Applications 34, Camb. Univ. Press,* New York, 1990.

[76]. R.B. Guenther and J.W. Lee, Some existence results for nonlinear integral equations via topological transversality, *J. Integral Equations Appl.*, **5**(1993), 195–209.

[77]. D. Guo, V. Lakshmikantham and X. Liu, Nonlinear Integral Equations in Abstract Spaces, *Kluwer Academic Publishers*, Dordrecht, 1996.

[78]. D. Guo and V. Lakshmikantham, Nonlinear problems in abstract cones, *Academic Press*, San Diego, 1988.

[79]. H.P. Heinz, On the behaviour of measures of noncompactness with respect to differentiation and integration of vector-valued functions, *Nonlinear Anal.*, **7**(1983), 1351–1371.

[80]. C.J. Himmelberg, Fixed points of compact multifunctions, *J. Math. Anal. Appl.*, **38**(1972), 205–207.

[81]. S. Hu and N.S. Papageorgiou, Handbook of Multivalued Analysis, Vol. I: Theory, *Kluwer Academic Publishers*, Dordrecht-Boston-London, 1997.

[82]. G.J.O. Jameson, An elementary proof of the Arens and Borsuk extension theorems, *J. London Math. Soc.,* **14**(1976), 364–368.

[83]. R. Kannan and D. O'Regan, A note on the solution set of integral inclusions, *J. Integral Equations Appl.,* **12**(2000), no.1, 85–94.

[84]. R. Kannan and D. O'Regan, A note on singular boundary value problems with solutions in weighted spaces, *Nonlinear Anal.,* **37**(1999), no.6, Ser. A: Theory, Methods, 791–796.

[85]. L.V. Kantorovich and G.P. Akilov, Functional Analysis, *Pergamon*, Oxford, 1982.

[86]. I. Kiguradze, On a singular Cauchy problem for systems of linear ordinary differential equations, *Diff. Eqns.,* **32** (1996), no.2, 173–180.

[87]. I. Kiguradze and Z. Sokhadze, The Cauchy problem for evolution singular function- differential equations, *Diff. Eqns.,* **33** (1997), no.1, 47–58.

[88]. G. Kothe, Topological Vector Spaces I, *Springer Verlag*, New York, 1983.

[89]. M.A. Krasnoselskii, Topological Methods in the Theory of Nonlinear Integral Equations, *Macmillan Co.*, New York, 1964.

[90]. M.A. Krasnoselskii, Positive Solutions of Operator Equations, *Noordhoff Ltd.,* Groningen, 1964.

[91]. Z. Kubáček, On the structure of fixed point sets of some compact maps in the Fréchet space, *Math. Bohemica,* **118**(1993), 343–358.

[92]. I. Kubiaczyk and S. Szufla, Knaeser's theorem for weak solutions of ordinary differential equations in Banach spaces, *Publ. Inst. Math. (Beograd),* **32**(1982), 99–103.

[93]. A. Lasota and Z. Opial, An application of the Kututani-Ky Fan theorem in the theory of ordinary differential equations, *Bull. Acad. Polon. Sci. Ser. Sci. Math. Astron. Phys.,* **13**(1965), 781–786.

[94]. J.W. Lee and D. O'Regan, Existence principles for nonlinear equations on semi-infinite and half-open intervals, *Advances in Nonlinear Dynamics* (edited by S.V. Sivasundarem and A.A. Martynuk), *Gordon and Breach Science Publ. Ltd.,* Amsterdam, 1997, 355–364.

[95]. R.W. Leggett and L.R. Williams, A fixed point theorem with application to an infectious disease model, *J. Math. Anal. Appl.,* **76**(1980),

91–97.

[96]. R.W. Leggett and L.R. Williams, Multiple positive fixed points of nonlinear operators on ordered Banach spaces, *Indiana Univ. Math. J.*, **28**(1979), 673–688.

[97]. W. Lian, F. Wong and C. Yeh, On the existence of positive solutions of nonlinear second order differential equations, *Proc. Amer. Math. Soc.*, **124**(1996), 1117–1126.

[98]. J. Mawhin, Topological degree methods in nonlinear boundary value problems, *AMS Regional Conf. Series in Math*, **40**, Providence, 1979.

[99]. M. Meehan and D. O'Regan, Continuums of solutions in a Fréchet space of abstract Volterra equations, *Appl. Anal.*, **74**(2000), no.1–2, 95–112.

[100]. M. Meehan and D. O'Regan, Multiple nonnegative solutions of nonlinear integral equations on compact and semi-infinite intervals, *Appl. Anal.*, **74**(2000), no.3–4, 413–427.

[101]. M. Meehan and D. O'Regan, Positive solutions of singular and non-singular Fredholm integral equations, *J. Math. Anal. Appl.*, **240**(1999), 416–432.

[102]. M. Meehan and D. O'Regan, Existence theory for nonlinear Fredholm and Volterra integral equations on half open intervals, *Nonlinear Anal.*, **35**(1999), no.3, Ser. A: Theory, Methods, 355–387.

[103]. M. Meehan and D. O'Regan, Periodic and almost periodic solutions of integral equations, *Appl. Math. Comput.*, **105**(1999), no.2–3, 121–136.

[104]. M. Meehan and D. O'Regan, Existence theory for nonlinear Volterra integrodifferential and integral equations, *Nonlinear Anal., Theory, Methods and Appl.*, **31**(1998), no.3–4, 317–341.

[105]. M. Meehan and D. O'Regan, Positive solutions of singular integral equations, *J. Integral Equations Appl.*, **12**(2000), 271–280.

[106]. M. Meehan and D. O'Regan, Positive L^p solutions of Hammerstein integral equations, *Arch. Math., to appear.*

[107]. M. Meehan and D. O'Regan, A note on positive solutions of Fredholm integral equations and related boundary value problems, *Proc. Roy. Irish Acad.*, **100A**(2000), 73–84.

[108]. M. Meehan and D. O'Regan, Positive solutions of Volterra integral

equations using integral inequalities, *J. Inequalities Appl., to appear.*

[109]. M. Meehan and D. O'Regan, A note on singular Volterra functional-differential equations, *Proc. Roy. Irish Acad., to appear.*

[110]. M. Meehan and D. O'Regan, Positive L^p solutions of Fredholm integral equations, *Proc. Dynamic Systems ans Applications, Vol. III* (edited by G.S. Ladde and M. Sambandham), *Dynamic Publishers, to appear.*

[111]. A.R. Mitchell and C.K.L. Smith, An existence theorem for weak solutions of differential equations in Banach spaces, in "Nonlinear Equations in Abstract Spaces" (edited by V. Lakshmikantham), *Academic Press* (1978), 387–404.

[112]. H. Mönch, Boundary value problems for nonlinear ordinary differential equations of second order in Banach spaces, *Nonlinear Anal.*, **4**(1980), 985–999.

[113]. H. Mönch and G.F. Von Harten, On the Cauchy problem for ordinary differential equations in Banach spaces, *Arch. Math. (Basel)*, **39**(1982), 153–160.

[114]. D. O'Regan, Fixed point theory of Mönch type for weakly sequentially upper semicontinuous maps, *Bull. Austral. Math. Soc.*, **61**(2000), no.3, 439–449.

[115]. D. O'Regan, Fixed point theorems for weakly sequentially closed maps, *Arch. Math. (Brno)*, **36**(2000), no.1, 61–70.

[116]. D. O'Regan, Weak solutions of ordinary differential equations in Banach spaces, *Appl. Math. Lett.*, **12**(1999), no.1, 101–105.

[117]. D. O'Regan, Nonlinear alternatives for multivalued maps with applications to operator inclusions in abstract spaces, *Proc. Amer. Math. Soc.*, **127**(1999), no.12, 3557–3564.

[118]. D. O'Regan, A note on the topological structure of the solution set of abstract Volterra equations, *Proc. Roy. Irish Acad.*, **99**(1999), Sect. A, 67–74.

[119]. D. O'Regan, Fixed points for set valued mappings in locally convex linear topological spaces, *Math. Comput. Modelling*, **28**(1998), no.1, 45–55.

[120]. D. O'Regan, Coincidence principles and fixed point theory for mappings in locally convex spaces, *Rocky Mountain J. Math.*, **28**(1998), no.4,

1407–1445.

[121]. D. O'Regan, Continuation methods based on essential and 0-epi maps, *Acta Appl. Math.*, **54**(1998), no.3, 319–330.

[122]. D. O'Regan, Periodic solutions for operator equations in $\mathbf{R}^n$, *Dynam. Systems Appl.*, **7**(1998), no.4, 519–528.

[123]. D. O'Regan, Fixed point theory for weakly contractive maps with applications to operator equations in Banach spaces relative to the weak topology, *Z. Anal. Anwendungen,* **17**(1998), no.2, 281–296.

[124]. D. O'Regan, Volterra and Urysohn integral equations in Banach spaces, *J. Appl. Math. Stochastic Anal.*, **11**(1998), no.4, 449–464.

[125]. D. O'Regan, Multivalued integral equations in finite and infinite dimensions, *Commun. Appl. Anal.*, **2**(1998), no.4, 487–496.

[126]. D. O'Regan, A note on solutions in $L^1[0,1]$ to Hammerstein integral equations, *J. Integral Equations Appl.*, **9**(1997), no.2, 165–178.

[127]. D. O'Regan, A topological approach to integral inclusions, *Proc. Roy. Irish Acad. Sect. A,* **97**(1997), no.1, 101–111.

[128]. D. O'Regan, Some fixed point theorems for concentrative mappings between locally convex linear topological spaces, *Nonlinear Anal.*, **27**(1996), 1437–1446.

[129]. D. O'Regan, Integral inclusions of upper semicontinuous or lower semicontinuous type, *Proc. Amer. Math. Soc.*, **124**(1996), no.8, 2391–2399.

[130]. D. O'Regan, A fixed point theorem for weakly condensing operators, *Proc. Roy. Soc. Edinburgh Sect. A,* **126**(1996), no.2, 391–398.

[131]. D. O'Regan, A continuation method for weakly condensing operators, *Z. Anal. Anwendungen,* **15**(1996), no.3, 565–578.

[132]. D. O'Regan, Existence results for nonlinear integral equations, *J. Math. Anal. Appl.*, **192**(1995), no.5, 705–726.

[133]. D. O'Regan, Existence theory for nonlinear Volterra and Hammerstein integral equations, *Dynamical Systems and Appl.*, (edited by R.P. Agarwal), *World Scientific Series in Applicable Analysis,* River Edge, New Jersey, Vol.4, (1995), 601–615.

[134]. D. O'Regan, Theory of Singular Boundary Value Problems, *World*

Scientific Press, Singapore, 1994.

[135]. D. O'Regan, A multiplicity fixed point theorem in Fréchet spaces, *Z. Anal. Anwendungen,* **19**(2000), 853–862.

[136]. D. O'Regan, Fixed points and random fixed points in cones of Banach spaces, *Math. Comput. Modelling,* **32**(2000), 1473–1483.

[137]. D. O'Regan and M. Meehan, Existence Theory for Nonlinear Integral and Integrodifferential Equations, *Kluwer Acad. Publ.,* Dordrecht, 1998.

[138]. D. O'Regan and R. Precup, Fixed point theorems for set valued maps and existence principles for integral inclusions, *J. Math. Anal. Appl.,* **245**(2000), no.2, 594–612.

[139]. D. O'Regan and R. Precup, Theorems of Leray-Schauder Type and Applications, *Gordon and Breach, forthcoming.*

[140]. D. O'Regan and R. Precup, Existence criteria for integral equations in Banach spaces, *J. Inequalities Appl., to appear.*

[141]. B.G. Patchpatte, Applications of Leray-Schauder alternative to some Volterra integral and integrodifferential equations, *Indian J. Pure Appl. Math.,* **26**(1995),1161–1168.

[142]. W.V. Petryshyn, Existence of fixed points of positive k-set contractive maps as consequences of suitable boundary conditions, *J. London Math. Soc.,* **38**(1988), 503–512.

[143]. W.V. Petryshyn, Multiple positive solutions of multivalued condensing mappings with some applications, *J. Math. Anal. Appl,* **124**(1987), 237–253.

[144]. R. Precup, Discrete continuation method for boundary value problems on bounded sets in Banach spaces, *J. Comput. Appl. Math., to appear.*

[145]. R. Precup, Discrete continuation method for nonlinear integral equations in Banach spaces, *Pure Math. Appl., to appear.*

[146]. T. Pruszko, Topological degree methods in multivalued boundary value problems, *Nonlinear Anal.,* **5**(1981), 953–973.

[147]. W. Rudin, Functional Analysis, *McGraw Hill,* New York, 1973.

[148]. V. Šeda and Z. Kubáček, On the connectedness of the set of fixed points of a compact operator in the Fréchet space $C^m(\langle b, \infty\rangle, \mathbf{R}^n)$, *Czech.*

Math. J., **42** (117) (1992), 577–588.

[149]. C.A. Stuart, Integral equations with decreasing nonlinearities and applications, *J. Differential Equations*, **18**(1975), 202–217.

[150]. C.A. Stuart, Concave solutions of singular nonlinear differential equations, *Math. Z.*, **136**(1974), 117–135.

[151]. C.A. Stuart, Existence theorems for a class of nonlinear integral equations, *Math. Z.*, **137**(1974), 49–66.

[152]. C.A. Stuart, Positive solutions of a nonlinear integral equation, *Math. Ann.*, **192** (1971), 119–124.

[153]. C. Swartz, Measure, Integration and Function Spaces, *World Scientific*, Singapore, 1994.

[154]. S. Szufla, On the differential equation $x^{(m)} = f(t,x)$ in Banach spaces, *Funkcial. Ekvac.*, **41**(1998), 101–105.

[155]. S. Szufla, On the Kneser-Hukuhara property for integral equations in locally convex spaces, *Bull. Austral. Math. Soc.*, **36**(1987), 353–360.

[156]. S. Szufla, On the existence of L^p-solutions of Volterra integral equations in Banach spaces, *Funkcial. Ekvac.*, **27**(1984), 157–172.

[157]. S. Szufla, Set of fixed points of nonlinear mappings in function spaces, *Funkcial. Ekvac.*, **22**(1979), 121–126.

[158]. K.K. Tan and X.Z. Yuan, Random fixed points theorems and approximation in cones, *J. Math. Anal. Appl.*, **185**(1994), 378–390.

[159]. A. Torchinsky, Real Variables, *Addison Wesley*, Redwood City, 1988.

[160]. M. Vath, Fixed point theorems and fixed point index for countably condensing maps, *Topol. Methods Nonlinear Anal.*, **13**(1999), 341–363.

[161]. L. Von Wolfersdorf, Travelling wave solutions of a nonlinear diffusion equation with integral term, *Z. Anal. Anwendungen*, **9**(1990), no. 4, 303–312.

[162]. Dong Wei, Uniqueness of solutions for a class of nonlinear Volterra integral equations without continuity, *Appl. Math. Mech. (English Ed.)*, **18**(1997), no.12, 1191–1196.

[163]. R.L. Wheeden and A. Zygmund, Measure and Integral, *Monographs and Textbooks in Pure and Applied Mathematics, Marcel Dekker Inc.*, New York and Basel, 1977.

[164]. P.J.Y. Wong and R.P. Agarwal, Criteria for multiple solutions of difference and partial difference equations subject to multipoint conjugate conditions, *Nonlinear Anal., to appear.*

[165]. P.J.Y. Wong and R.P. Agarwal, Results and estimates on multiple solutions of Lidstone boundary value problems, *Acta Math. Hungarica, to appear.*

[166]. P.J.Y. Wong and R.P. Agarwal, Multiple solutions of generalized multipoint conjugate boundary value problems, *Georgian J. Math., to appear.*

[167]. K. Yosida, Functional Analysis, (5th edition), *Springer-Verlag,* Berlin, 1978.

[168]. P.P. Zabreyko, A.I. Koshelev, M.A. Krasnoselskii, S.G. Milkin, L.S. Rakoshchik and V.Y. Stetsenko, Integral Equations-a Reference Text, *Noordhoff International Publ.,* Leyden, 1975.

INDEX

A

B

C

D

L

M

N

O

P

R

S

T

U

V

W